Learning the Physics of Einstein with Georges Lemaître

Georges Lemaître

Jan Govaerts
Jean-François Stoffel *Editors*

Learning the Physics of Einstein with Georges Lemaître

Before the Big Bang Theory

Translated by Christine Leroy and Stephen Lyle

Springer

Author
Georges Lemaître
Louvain-la-Neuve, Belgium

Editors
Jan Govaerts
Mathematics and Physics
Catholic University of Louvain
Louvain-la-Neuve, Belgium

Jean-François Stoffel
Catégorie paramédicale
Haute École Louvain-en-Hainaut
Montignies-sur-Sambre, Belgium

Translated by
Christine Leroy
Louvain-la-Neuve, Belgium

Stephen N. Lyle
Alzen, France

ISBN 978-3-030-22029-7 ISBN 978-3-030-22030-3 (eBook)
https://doi.org/10.1007/978-3-030-22030-3

Translation of Georges Lemaître's French language text: *La Physique d'Einstein (1922)* by Stephen Lyle and Christine Leroy

This Springer imprint is published by the registered company Springer Nature Switzerland AG
The registered company address is: Gewerbestrasse 11, 6330 Cham, Switzerland

L'univers n'est pas trop grand pour l'homme,
il n'excède pas les possibilités de la science
ni la capacité de l'esprit humain

Monseigneur Georges Lemaître (1894-1966)

The universe is not too large for humankind,
neither does it exceed the reach of science
nor the abilities of the human mind

Monsignor Georges Lemaître (1894-1966)

Preface

On the 29th October 2018 the International Astronomical Union (IAU) acted on historical and scientific fact by renaming the law for the recession of galaxies and the expansion of the Universe as the "Hubble–Lemaître Law". By doing so the IAU duly recognised the contributions of Monsignor Georges Lemaître as the genuine founder of the theory of the Big Bang and a visionary architect of modern cosmology. It was in 1927 that Georges Lemaître predicted the law for the recession of galaxies as a direct consequence of his theory of the expansion of the Universe in a publication with the *Annales de la Société Scientifique de Bruxelles*.[1] The astronomical and observational corroboration for such a behaviour of the galaxies was convincingly established by Edwin Hubble in a famous publication of 1929.[2]

In this early 21st century the scientific community is finally acknowledging the full measure of the scientific legacy of a towering figure of 20th century European physics. Against the best advice of the greatest names of his time, the young Lemaître was convinced, solely on the basis of the physics of Albert Einstein's theory of General Relativity, not only that the Universe is expanding but even more remarkably, in 1931, that space and time must have had a beginning from a "primeval atom" with an event known as the "Big Bang" – a fireworks of sorts with a truly singular geometric quantum state – in other words, the emergence of the quantum universe and the beginning of the quest for a quantum theory of gravity, with the hypothesis of an ever expanding Universe and a positive cosmological constant.[3]

---

[1] G. Lemaître, *Un univers homogène de masse constante et de rayon croissant rendant compte de la vitesse radiale des nébuleuses extra-galactiques*, *Annales de la Société Scientifique de Bruxelles*, série A : *Sciences Mathématiques*, 1ère partie : *Comptes rendus des Séances*, t. 47, 1927, pp. 49–59.

[2] E. Hubble, *A relation between distance and radial velocity among extra-galactic nebulae*, *Proceedings of the National Academy of Sciences* 15 (3), 15 March 1929, pp. 168–173.

[3] G. Lemaître, *The beginning of the world from the point of view of quantum theory*, *Nature* 127, 9 May 1931, p. 706.

But how did the young Lemaître, essentially on his own, learn "The Physics of Einstein" for the gravitational interaction? While studying for the priesthood (Lemaître was ordained as a diocesan priest in September 1923), Georges Lemaître submitted a dissertation dated 31 May 1922 and entitled "*La Physique d'Einstein*", or "*The Physics of Einstein*". The thesis was to earn him Travel Fellowships and launched Lemaître onto a scientific path of ground breaking discoveries. The Fellowships enabled him to spend the years 1923-1924 with Arthur S. Eddington at Cambridge University (UK) and 1924-1925 at Harvard University with Harlow Shapley and MIT (USA), where Lemaître was eventually granted a PhD in 1927.

Almost a century after Lemaître's seminal publications of 1927 and 1931 which were to provide the basis of modern cosmology, this most pedagogical treatise of 1922 remains of timely interest for young minds as well as from a history of science perspective. For the first time ever, the present volume makes available an English translation of Lemaître's early thesis together with a revisited critical edition of the original manuscript in French and includes recently discovered notes and corrections that Lemaître himself had added to his original text. The original manuscript alongside these recently discovered additions are part of the invaluable resources held by the Georges Lemaître Archives at the *Université catholique de Louvain* (UCLouvain) in Louvain-la-Neuve, Belgium. The Georges Lemaître Archives also house numerous other scientific and personal effects belonging to Lemaître and have been granted the status of "a treasure and intangible heritage of the Wallonia–Brussels Federation" placed under the auspices of UCLouvain. A digitisation project is currently making these documents accessible worldwide *via* the internet through the web site `https://archives.uclouvain.be/`.

Following Georges Lemaître's passing on 20 June 1966, the family specifically bequeathed all of Lemaître's archives to the French speaking UCLouvain, in Louvain-la-Neuve. These comprise many personal effects, scientific items, numerous scientific documents including a rich correspondence with renowned scientists of his day. As the facts would have it, from 1925 till 1966 Lemaître remained a Professor solely within the French speaking section of the then single *Université de Louvain* geographically situated in the Flemish city of Leuven. In other words, Georges Lemaître, proud of his francophone extraction, remained for the entire period of his academic career Full professor of the French speaking section of the *Université de Louvain*. It was shortly after Lemaître's death that the University officially separated into two linguistically independent Institutions sharing a common heritage and proud history dating back to 1425, renamed the Flemish speaking *Katholieke Universiteit te Leuven* (KU Leuven) situated in Leuven, and the French speaking *Université catholique de Louvain* (UCLouvain) situated in Ottignies–Louvain-la-Neuve.

According to historical fact, up until the early 1930s professors at the *Université de Louvain* did not have a personal office at the University. Consequently upon his return to Belgium in 1925 and as a young academic, Lemaître pursued his research in his living quarters as a resident of the *Collège du Saint-Esprit* (*Holy Spirit College*), currently the *Heilige Geestcollege* situated at Naamsestraat 40, Leuven. Hence the aforementioned articles of 1927 and 1931 – and specifically the latter one – are signed by Georges Lemaître with the address *Rue de Namur 40* and as a member of the *Université de Louvain.* All his life Georges Lemaître remained deeply attached to his family, profoundly faithful to his spiritual vocation, and ever mindful of his roots in the "black mining country" of his youth in Charleroi, where he was born on 17 July 1894. Georges Lemaître now rests in the family vault at the Cemetery of Marcinelle near Charleroi, Belgium.

Much could and should be said about the remarkable scientist, the gentle humanist ever inquisitive of the deeper questions of life, and the vibrant spiritual thinker who liked to metaphorically represent his core motivation as the coherent adoption "of two separate paths towards Truth". The first bio-bibliography of Lemaître's career and publications was published in 1996 as part of a volume of Proceedings of a Colloquium held to mark the centenary of Lemaître's birth.

- *Mgr Georges Lemaître savant et croyant. La physique d'Einstein. Actes du colloque tenu à Louvain-la-Neuve le 4 novembre 1994.* Ed. J.-Fr. Stoffel, Centre interfacultaire d'étude en histoire des sciences. Collection Réminisciences 3, Université catholique de Louvain, 1996.

An authoritative biography by Dominique Lambert published in 2000 has since been made available in English in 2015.

- Dominique Lambert, *Un Atome d'Univers. La vie et l'œuvre de Georges Lemaître*, Bruxelles : Racine & Lessius, 2000.
- Dominique Lambert, *The Atom of the Universe. The Life and Work of Georges Lemaître*, Krakow: Copernicus Press, 2015.

At the initiative of the Belgian Physical Society (BPS) and its current President, Dr. Jozef Ongena, the European Physical Society (EPS), on 23 May 2019, bestowed on the *Heilige Geestcollege* in Leuven an EPS Historic Sites Award for being the location in Europe that witnessed the emergence of the theory of the Big Bang and the foundation of modern cosmology by Monsignor Georges Lemaître. Truly a visionary European physicist of the 20th century, proudly Belgian, an elected member and later President of the Pontifical Academy of Sciences, that humble humanist resolutely mindful of his origins from Charleroi, postulated the expansion of the Universe and the existence of the "quantum primeval atom". The recognition of the scientific legacy of Georges Lemaître by the EPS provides a timely incentive for the present volume and fittingly follows the above cited first ever publication in 1996 of the original French version of Lemaître's 1922 dissertation, "*La Physique d'Einstein*" on the occasion of the centenary of Lemaître's birth.

With the first ever English translation of Lemaître's early dissertation, "*The Physics of Einstein*", the reader is able to acquire some insight into the sheer audacity of Lemaître's approach to Albert Einstein's theory of the gravitational interaction. Contemporary scholars – scientists, philosophers and historians of science alike – are given a unique opportunity of accompanying the young Lemaître in his personal study and reconstruction of Einstein's theories of relativity. Lemaître's preliminary study of the physics of Einstein provides scholars with an open window into the premises of Lemaître's reflection on the new physics of his day. From Lemaître's personal understanding and interpretation of Albert Einstein's physics would emerge the revolutionary ideas that were to lay the very foundation of Lemaître's theory of the quantum universe and the context for his ideas of its expanding evolution.

The present volume is divided into four main parts, the first and the last being chapters on their own. Concluding with an invitation to, and some suggestions for further reading of available documents addressing Lemaître's life and rich and diverse scientific legacy, the first part provides a historical perspective of the particular scientific context in which Georges Lemaître conducted the research that led him to submit a dissertation in May 1922 entitled in French, "*La Physique d'Einstein*". This introductory chapter provides a translation of the contribution by Professor Lucien Bossy to the Colloquium in honor of the centenary of Lemaître's birth held on the 4th November 1994 at the *Université catholique de Louvain*, Louvain-la-Neuve. Professor at UCLouvain from 1964 till 1985 and member of the Belgian Royal Meteorological Institute, Lucien Bossy (1918-1996) was amongst Lemaître's early collaborators.

The following two parts, each with its specific chapters which are in correspondence through translation, present Lemaître's dissertation of 1922. The critical revisited edition of the original manuscript in French is to be found in the third part, following its English translation in the second part.

The fourth and last part and chapter of the volume provides recently uncovered material (in French) from the Georges Lemaître Archives of personal correspondence between Georges Lemaître and Maurice Alliaume, and includes additional corrections by Lemaître to his original manuscript, "*La Physique d'Einstein*".

As mentioned at the beginning of our Preface, Lemaître published his groundbreaking study on the expansion of the Universe and the recession of galaxies in 1927. We wish readers an enjoyable journey of rediscovery into the mind of the founder of the theory of the Big Bang by learning the physics of Einstein in the excellent company of Georges Lemaître himself.

Acknowledgements

The editors would like to express sincere thanks to Dr Christian Caron and Dr Angela Lahee, Executive Publishing Editors with Springer, for accepting this volume into the Springer book programme.

We are indebted to Ms Liliane Moens–Haulotte for her untiring dedication as voluntary curator ensuring the proper conservation and preservation of the Georges Lemaître Archives. Whenever the opportunity arose Liliane sought to expand the original collection through on-going contact with the Lemaître family all the while updating information and international publications surrounding the scientific contribution of Georges Lemaître. The current digitisation project of the "treasure of the Georges Lemaître Archives" by the *Service des Archives* UCLouvain will henceforth ensure the transmission of the legacy of Georges Lemaître to future generations.

Professional advice was provided by the *Service des Archives* UCLouvain. In particular sincere thanks go to the Directrice, Professor Aurore François, to Caroline Derauw, Delphine Picron, and especially Véronique Fillieux for locating the relevant information and original documents now in digitised form in the Georges Lemaître Archives, UCLouvain.

Finally a word of genuine gratitude and profoud appreciation should go to Ms Christine Leroy for her expert translation skills and for her precious revision of the English, thereby making the volume all the more readable for a wider readership.

Louvain-la-Neuve, Belgium

Jan Govaerts
Jean-François Stoffel

Contents

The Physics of Einstein by Georges Lemaître (1922)

The Historical Context[1]

The text entitled "*La Physique d'Einstein*" was submitted by Georges Lemaître as part of the competition for Travel Scholarships awarded by the Belgian Government for the year 1923. Its redaction in the form of a typed manuscript was undertaken at the Major Seminary in Mechelen at the end of 1921 and concluded on 31 May 1922.

It would have been interesting to discover the very reasons which prompted Georges Lemaître to pursue the solitary study of a new and difficult field of modern physics. It would have been enlightening to identify those professors whom he knew and who may have positively inspired Lemaître to undertake such a study. Did he have contacts with Belgian experts or, much less likely, experts from abroad? Could he have been in touch with Théophile de Donder who, with such younger collaborators as H. Vanderlinden and M. Nuyens, was at the time preparing his important treatise, "*La gravifique einsteinienne*"? But there is no evidence to that effect.

What kind of reference material did Lemaître have at his disposal? The typed manuscript remains equally silent in that respect and provides no bibliography nor any reference except for Einstein's *Grundlagen* of 1916.

Thanks to the Georges Lemaître Archives held at the *Université catholique de Louvain* some of these questions can be clarified and reveal to some extent the personality of Georges Lemaître or at least provide some understanding of the course of his scientific trajectory.

What was *grosso modo* the nature of the available literature on the theory of relativity at the time of the redaction of this dissertation?

Ever since 1905, when Einstein proposed his theory of special relativity and in 1915, his theory of general relativity, the great physicists and mathematicians of the day were bent on analysing the hidden subtleties of these theories.

[1] This presentation of the historial context for Georges Lemaître's thesis is a translation by Christine Leroy of, Lucien Bossy (1918-1996), "*La physique d'Einstein de Georges Lemaître (1922)*", in *Mgr Georges Lemaître savant et croyant. La physique d'Einstein. Actes du colloque tenu à Louvain-la-Neuve le 4 novembre 1994.* Ed. J-Fr. Stoffel, Centre interfacultaire d'étude en histoire des sciences. Collection Réminisciences 3, Université catholique de Louvain, 1996, pp. 9-19.

As a result, by 1921, besides Einstein's treatise of 1916, *Die Grundlagen der allgemeinen Relativitätstheorie*, a popular French publication of Einstein's *La théorie de la relativité restreinte et généralisée* – already in its tenth edition for the original publication in German – presented the basic concepts for these theories; Th. de Donder had just published *La gravifique einsteinienne*; H. Weyl's *Raum, Zeit und Materie*, first published in 1918, was available in its fourth edition; A.S. Eddington's *Space, Time and Gravitation* of 1920 had just been edited in French with additional mathematical material; M. von Laue's second volume of *Die Relativitätstheorie* devoted to general relativity appeared in 1921. And finally, both W. Pauli's *Relativitätstheorie* and J. Becquerel's *Le principe de relativité* are dated 1922 while A.S. Eddington's *The Mathematical Theory of Relativity* only appeared in 1923. Furthermore the first works by Einstein and de Sitter in cosmology can be dated back to 1917. In other words, theories than remain challenging to grasp were at least fairly easy to access in diverse publications.

Georges Lemaître, freshly graduated as a Doctor in mathematical and physical sciences in 1920, submitted work for a Travel Scholarship with firm hopes to study abroad, particularly at Cambridge (UK), in order to deepen his understanding of astronomy in relation to the new science of relativity.

In order to achieve such ambitions, all on his own Lemaître had to master the specific concepts and reasonings of the theories of relativity with the corresponding mathematical methods. Furthermore, he had to write up a well structured and argued synthesis of his readings and demonstrate an aptitude for original research.

One may safely presume that, except for some review articles, his main references would have been limited to Einstein's *Die Grundlagen der allgemeinen Relativitätstheorie* of 1916, Weyl's *Raum, Zeit und Materie* of 1918, and Eddington's *Space, Time and Gravitation* of 1920 in the French edition of 1921. Presumably Lemaître also had available *Das Relativitätsprinzip* by Lorentz, Einstein and Minkowski of 1920, and Schwarzschild and de Sitter's important publications, alongside those of French scientists such as Langevin.

The result of these efforts is the present dissertation – a piece of research of some 130 pages – which Lemaître submitted in the first half of 1922.

The Jury given the task of examining this work as well as that of another candidate (on a topic dealing with probabilities) included H. Janne from the *Université de Liège*, M. Alliaume from the *Université catholique de Louvain*, and Th. de Donder from the *Université Libre de Bruxelles*. At the time, the latter was considered a renowned authority on the subject of relativity. Furthermore, since 1916 Th. de Donder was in close correspondence with Einstein and a little later with Eddington.

The Jury requested that Georges Lemaître provide a note detailing the aspects of his work which he believed to be his most original contributions.

The text of this note is conserved at the Georges Lemaître Archives.[2] Lemaître responded in a note that refers to work by Pauli published after Lemaître had submited his dissertation, as well as to other specific references, while also providing

[2] See Annex 1, on p. 246.

errata[3] and one further remark[4] in which he mentions the more recent results presented in publications by Eddington and Becquerel.

Following the Jury's decision on the eligibility of the two dissertations submitted by the two candidates, Lemaître was finally asked to prepare, besides the public defence, some further comments related to proper coordinates in light of a recent article by Fokker; clarifications regarding his proposed definition of the concept of simultaneity; to "consult the recent book by Eddington"; and to expect "to discuss his on-going research since the submitted dissertation".[5] In all fairness it must be said that the requests made by the Jury constitute quite a number of additional requirements.

Georges Lemaître was awarded the Travel Scholarship. He spent a year with A.S. Eddington at the University of Cambridge (UK) and the first milestone of the promising scientific trajectory as we know it.

It has not been possible to find in the archives of the relevant Ministry in Belgium a trace related to Travel bursaries and the Jury's deliberations for 1923. The documents were apparently destroyed long ago.

Basically, Lemaître's dissertation addresses all the fundamental principles of general relativity, Einstein's own developments in his *Grundlagen*, and concludes with the first basic formulations of the equations for the gravitational field.

The notations used by Lemaître are very close to those of Einstein, except for the Riemann and Ricci tensors for which Lemaître prefers the conventions used by Weyl and Pauli. Lemaître does not use the summation convention but denotes the Christoffel symbols in the same way as Einstein.

In his Introduction, Lemaître is true to what will always remain his approach to scientific research. It is an act of faith in experimental science. Indeed he writes that experimental science should always help us "adapt to the universe in which we live",[6] that the knowledge of the universe perfects itself naturally "using the experience accumulated over many centuries and assiduously transmitted from generation to generation",[7] that disconnected empirical laws call for a synthesis and that the progress that results from such a unification consists in "the discovery of a more comprehensive simplicity".[8]

Placing Einstein's work into perspective, Lemaître outlines what he refers to as "the main lines of these attempts by human intelligence to apprehend a universe so well suited to our various activities".[9] First, he writes, it was geometric notions that laid the foundations of our knowledge of the sensitive world. Next, thanks to Newton's contributions, it was mechanics that would make use of geometry in order to dominate the world of physics and introduce the world of actions effectuated at

[3] See Annex 2, on p. 249.

[4] See G. Lemaître's letter to M. Alliaume, on p. 257.

[5] See M. Alliaume's letter of 13 March 1923 to G. Lemaître, on p. 252.

[6] See p. 11, and for the original text in French p. 125.

[7] See p. 11, and for the original text in French p. 125.

[8] See p. 12, and for the original text in French p. 126.

[9] See p. 12, and for the original text in French p. 126.

a distance. Finally, electrical actions introduce us to the third level of science in which geometry and mechanics become useful tools while actions at a distance are no longer instantaneous but are now retarded.

It is Einstein's physics that can claim to unify physical laws through an interpenetration of these three realms based on the principle of relativity according to which, as Lemaître writes, "The equations expressing a physical law must retain the same algebraic form when we make an arbitrary change of coordinates",[10] while the means to avoid any subjective bias is provided by absolute differential calculus that enables one to reach laws that are "truly objective".[11]

Hence Lemaître's Introduction, no doubt somewhat influenced by Eddington, sets the perspective for the development of the chapters that follow and highlights the objectivity which characterises the whole essay.

It would be quite difficult to detail the dissertation's original insights and only some aspects may be put forward based on the author's own note and errata in this regard.

From a general point of view, Lemaître's approach is fundamentally different from that of Einstein in the *Grundlagen*. Indeed, for Einstein the theory of general relativity is an extension of special relativity. For Lemaître on the other hand, the basic conceptual starting point is the principle of general relativity. As a result, Lemaître follows an intellectual process somewhat analogous to that of de Donder who develops the theory as a generalisation of Hamilton's principle.

This explains why, after having introduced the fundamental concepts of space-time, Lemaître considers exploring the different fields using small test bodies. On the one hand, Lemaître considers the inertial and gravitational fields and on the other hand, the electromagnetic fields.

Next, and reciprocally, he shows how, based on the principle of relativity, the nonuniform motion of a material body relates to the potentials determining the line element ds^2. A much detailed discussion is provided with the case of a uniformly accelerated motion and that of a uniform rotation. Lemaître then concludes that chapter with the identification of the Riemann tensor.

Finally, in a manner similar to Einstein, Lemaître uses the Poisson equation to introduce the matter tensor in the Einstein-de Sitter form and includes the cosmological constant. In particular, his discussion develops along the following lines.

Early on in the discussion Lemaître already establishes the necessary concepts of differential geometry to show, among other things, that the trajectory of a free point has a total proper time value that is a local maximum.

He develops what comprises, to the best of his knowledge, a complete study of the motion of a solid body with uniform acceleration in the inertial field thereby generated. He also outlines the general problem of the motion of a solid body in a Galilean field. He then goes on to solve it, and also discusses it in the case of an arbitrary rectilinear motion and thereby determines the inertial field in a form analogous to that of the Schwarzschild field. He addresses the complex problem

[10] See p. 14, and for the original text in French p. 128.

[11] See p. 14, and for the original text in French p. 128.

of a collection of masses moving through their own gravitational field, provides an approximate solution and suggests a possible approach towards their generalisation.

By making a clear distinction between direct measurements (on a given body) and indirect measurements (from an observation point), taking into account the invariance of the spacetime line element and by using his characterisation of simultaneity in analytic form, Lemaître avoids the difficulties inherent to the interpretation of the Lorentz equations and achieves a new solution for those equations related to the Sagnac experiment.

Furthermore, at the outset of his dissertation, Lemaître introduces a quadratic form that includes a potential, which to first approximation corresponds to classical mechanics. This quadratic form is instrumental throughout the dissertation in the successive phases of building up the demonstration of his synthesis.

Finally, in order to achieve the dissertation's main objective and establish Einstein's equations for the gravitational field, Lemaître does not follow the conventional path that usually devotes an entire first chapter to tensor calculus. He opts rather for the explicit verification of the tensorial properties in a succession of results in which, for the sake of calculational efficiency, he systematically uses proper coordinates (geodetic coordinates). In this way, the concept of tensor need only be introduced at the very end when the role and meaning of the Riemann tensor proves to be fundamental. Of course the results that he thereby achieves are known, albeit presently through a most personal approach.

Once the gravitational equations with the cosmological constant have been established according to the perspective of Einstein and de Sitter, Lemaître goes on to apply them in the context of astronomy (the Schwarzschild problem, Mercury's perihelion, the deflection of light, the redshift of the solar spectrum), by using whenever possible the potentials that contribute to the specific quadratic differential form already mentioned above.

Finally, in the paragraph entitled *Fixed Stars* (*Les étoiles fixes*), Lemaître briefly considers the use of gravitational equations for a set of stars in a manner analogous to molecules of a gas. Like Einstein, by assuming the local density to be uniform, in a very direct way Lemaître finds the potentials of the metric tensor and establishes the equivalence of all points in space (the cosmological condition). Lemaître then succintly describes the resulting spherically symmetric Riemann universe with its main properties.

The above provides but a brief account of the personal synthesis that constitutes *La Physique d'Einstein* in which Georges Lemaître shows how well he mastered the modes of thought and the mathematical methods of relativity, and how aptly he could apply these to specific problems. His mathematical prowess combines a solid understanding of physics with imaginative insights, and leads Lemaître to the core of the objectives he set himself. This will remain a characteristic trait of his future research.

Quite unsurprisingly Th. de Donder warmly recommended Georges Lemaître to Eddington. And not surprisingly, no sooner had Lemaître arrived in Cambridge that

Eddington referred to the young student in his letter of Christmas 1924 to de Donder in the following terms:

> *"I found M. Le Maître a very brilliant student, wonderfully quick and clear-sighted, and of great mathematical ability. He did some excellent work whilst here, which I hope he will publish soon. I hope he will do well with Shapley at Harvard. In case his name is considered for any post in Belgium I would be able to give him my strongest recommendations."*[12]

Finally it comes as no surprise that the audacious originality of the dissertation was to lead, four years later, to the innovative ideas that Georges Lemaître would go on to develop into the brilliant synthesis of astrophysics and general relativity, better known as the expanding Universe.

Bibliography

Becquerel, J., *Le principe de relativité et la théorie de la gravitation : Leçons professées en 1921 et 1922 à l'école polytechnique et au musée d'histoire naturelle*. Paris : Gauthier-Villars et Cie, 1922.

Bosquet, J., *Théophile De Donder et la gravifique einsteinienne*, in *Bulletin de la Classe des Sciences de l'Académie Royale de Belgique*, 5e série, t. 73, 1987, n° 5, pp. 209-253.

de Donder, Th., *La gravifique einsteinienne*. Paris : Gauthier-Villars et Cie, 1921, and *Annales de l'Observatoire Royal de Belgique*, 3e série, t. 1, 1921.

Eddington, A.S., *Space, time and gravitation: An outline of the general relativity theory*. Cambridge: Cambridge University Press, 1920.

—, *Espace, temps et gravitation : La théorie de la relativité générale dans ses grandes lignes. Exposé rationnel suivi d'une étude mathématique de la théorie*, translated from English by J. Rossignol; Introduction by Paul Langevin. Paris : Librairie Scientifique J. Hermann, 1921.

—, *The mathematical theory of relativity*. Cambridge: Cambridge University Press, 1923.

Einstein, A., *Die Grundlagen der allgemeinen Relativitätstheorie*, in *Annalen der Physik*, 4th series, t. 49, 1916, pp. 769-822.

—, *Kosmologische Betrachtungen sur allgemeine Relativitätstheorie*, in *Sitzungsberichte der königlichen preussischen Akademie der Wissenschaften*, t. 6, 1917, pp. 142-152.

[12] J. Bosquet, *Théophile De Donder et la gravifique einsteinienne*, p. 250.

—, *La théorie de la relativité restreinte et généralisée (mise à la portée de tout le monde)*, translated by Miss J. Rouvière from the 10th German edition; foreword by É. Borel. Paris : Gauthier-Villars et Cie, 1921. (Actualités scientifiques).

Fokker, A.D., *De geodetische precessie : Een uitvloeisel van Einstein's gravitatietheorie*, in *Verslagen der Afdeeling Natuurkunde*, DI. 29, 1920-1921, pp. 611-621.

Langevin, P., *Sur la théorie de la relativité et l'expérience de M. Sagnac*, in *Comptes rendus hebdomadaires des séances de l'Académie des Sciences de Paris*, t. 173, 7 novembre 1921, pp. 831-834

Laue, M. von, *Die Relativitätstheorie*. – Vol. II : *Die allgemeine Relativitätstheorie*. Braunschweig : Vieweg, 1921.

Lorentz, H.A., Einstein, A., Minkowksi, H., *Das Relativitätstprinzip, eine Sammlung von Abhandlungen*. – 3 erweiterte Auflage. Leipzig: Teubner, 1920 / 5 Auflage. – Berlin : Teubner, 1920.

Pauli, W., *Relativitätstheorie*, in *Encyklopädie der mathematischen Wissenschaften*, Band 5 : *Physik*, Teil 2. Leipzig : Teubner, 1922. pp. 599-775.

Schwarzschild, K., *Das Gravitationsfeld eines Massenpunktes nach der Einsteinschen Theorie*, in *Sitzungsberichte der königlichen preussischen Akademie der Wissenschaften*, t. 5, 1916, p. 189 sqq.

Sitter, W. de, *On Einstein's theory of gravitation, and its astronomical consequences*, in *Monthly Notices of the Royal Astronomical Society*, t. 76, 1916, n° 9, pp. 699-728; t. 77, 1916, n° 2, pp. 155-184; t. 78, 1917, n° 1, pp. 3-228.

—, *On the relativity of inertia*, in *Proceedings of the Koninklijke Akademie van Wetenschappen*, t. 19, 1917, pp. 1217-1225.

Weyl, H., *Temps, espace, matière : Leçons de la théorie de la relativité générale*, translated by Gustave Juvet and Robert Leroy from the 4th German edition. Paris : Librairie Scientifique Albert Blanchart, 1922. (Collection de monographies scientifiques étrangères).

—, *Raum, Zeit, Materie : Vorlesungen über allgemeine Relativitätstheorie*. Vierte, erweiterte Auflage. Berlin : Verlag von Julius Springer, 1921.

An Invitation to Further Reading

Suggestions by chronological order of publication

L'Académie Pontificale des Sciences en mémoire de son deuxième président Georges Lemaître. À l'occasion du cinquième anniversaire de sa mort, Civitate Vaticana, Pontificia Academia Scientiarum, 1972, 296 pp. (Pontificiae academiae scientiarum scripta varia, 36).

Berger A. (ed.), *The Big-Bang and Georges Lemaître. Proceedings of a symposium in honour of G. Lemaître fifthy years after his initiation of Big-Bang cosmology. Louvain-la-Neuve, Belgium, 10-13 October 1983*, Dordrecht, Boston, Lancaster, D. Reidel Publishing Company, 1984, xxii, 420 pp.

Discours prononcé lors de la cérémonie d'ouverture du symposium international organisé en l'honneur de Lemaître cinquante ans après l'initiation de sa cosmologie du Big-Bang (Louvain-la-Neuve, 10-13 octobre 1983), in *Revue des questions scientifiques*, vol. 155, 1984, n°2, pp. 139-224.

Godart O., Heller M. (eds.), *Cosmology of Lemaître*, Tucson, Pachart Publishing House, 1985, 204 pp. (History of astronomy series, 3).

De Rath V. (éd.), *Georges Lemaître, le Père du Big Bang*, en collaboration avec Léonard J.-L. et Mayence R., Bruxelles, Labor, 1994, 156 pp.

L'Univers de Lemaître, in *Ciel et Terre*, vol. 110, 1994, n°4, pp. 90-126.

Georges Lemaître et l'Académie Royale de Belgique. Œuvres choisies et notice bibliographique, Bruxelles, Académie Royale de Belgique, 1995, 219 pp. (Mémoires de la Classe des sciences. Collection in-8, 10).

Heller M., *Lemaître, Big Bang and the Quantum Universe*, Tucson, Pachart Publishing House, 1996, 108 pp.

Kragh H., *Cosmology and Controversy. The Historical Development of Two Theories of the Universe*, Princeton (N.J.) : Princeton University Press, 1996, xiii, 500 pp.

Stoffel J.-Fr. (éd.), *Mgr Georges Lemaître savant et croyant. Actes du colloque commémoratif du centième anniversaire de sa naissance (Louvain-la-Neuve, 4 novembre 1994)*, Louvain-la-Neuve, Centre interfacultaire d'étude en histoire des sciences, 1996, 371 pp. (Réminiscences, 3).

Friedman A., Lemaître G., *Essais de cosmologie, précédés de "L'invention du Big Bang"*, Textes choisis, présentés, traduits du russe ou de l'anglais et annotés par J.-P. Luminet et A. Grib, Paris, Seuil, 1997, 337 pp. (Sources du savoir).

Lambert D., *Un Atome d'Univers. La vie et l'œuvre de Georges Lemaître*, Bruxelles, Racine & Lessius, 2000, 372 pp. (Au singulier, 2).

Luminet J.-P., *L'invention du Big Bang*, Paris, Seuil, 2004, 267 pp. (Sciences, 158).

Farrell J., *The Day Without Yesterday. Lemaître, Einstein, and the Birth of Modern Cosmology*, New York, Thunder's Mouth Press, 2005, 262 pp.

Heller M., Chernin A., *Los Origenes de la cosmologia. Friedman y Lemaître*, Libros de ciencia, 2005, 133 pp. (Serie de divulgacion cientifica).

Lambert D. (dir.), *Lemaître, le père du Big Bang*, in Pour la science, 2007, pp. 26-120. (Les génies de la science, 30).

Lambert D., *L'itinéraire spirituel de Georges Lemaître. Suivi de Univers et atome* (inédit de G. Lemaître), Bruxelles, Lessius, 2007, 222 pp.

Lambert D., Reisse J., *Charles Darwin et Georges Lemaître, une improbable, mais passionnante rencontre*, Bruxelles, Académie royale de Belgique, 2008, 288 pp. (Mémoire de la Classe des Sciences, collection in-8°, 3e série, t. 30, n°2057).

Nussbaumer H., Bieri L., *Discovering the expanding universe*, Cambridge University Press, 2009, 246 pp.

Riaza E., *La Historia del Comienzo. Georges Lemaître, padre del big bang*, Encuntro, 2010, 135 pp.

Le Père du Big Bang, Georges Lemaître, un génie oublié, in Histoire du Christianisme Magazine, 2010, n°52, pp. 24-46.

Robredo J.-Fr., *La métamorphose du ciel. De Giordano Bruno à l'abbé Lemaître*, Paris, PUF, 2011, 168 pp.

Holder R. D., Mitton S. (eds), *Georges Lemaître. Life Science and Legacy*, Heidelberg, Springer, 2012, 201 pp. (Astrophysics and Space Science Library, 395).

La théorie de l'atome primitif a 80 ans, in Revue des Questions Scientifiques, vol. 183, 2012, n°4, pp. 329–590.

Lambert D., *The Atom of the Universe. The Life and Work of Georges Lemaître*, Copernicus Center Press, 2015, 484 pp.

Kragh H., *Masters of the Universe. Conversations with Cosmologists of the Past*, Oxford, Oxford University Press, 2015, 285 pp.

Lambert D., *The Atom of the Universe. The Life and Work of Georges Lemaître*, Preface by P. J. E. Peebles, translated by J. Ampleman ; edited by K. Van Bibber, 2nd revised edition, Kraków, Copernicus Center Press, 2016, 464 pp.

Stenico M., *Dall'archè al Big Bang. Georges Edouard Lemaître e la grande narrazione cosmica*, in *Quaderni di archivio Trentino*, 2017, n°46, 134 pp.

The Physics of Einstein

Introduction[1]

We can conceive of people just like ourselves living in a universe ruled by an evil spirit with the power to change the laws of nature, and taking pleasure in doing so each time some savant was on the verge of identifying them. What would physics look like to these unlucky individuals? Each generation would decry the mistakes of their predecessors. Experiments carried out in the past would give totally different results today than those described by their authors. Perhaps one day humanity would just give up trying to establish the truth, guessing itself to be the helpless victim of some evil machinations.

This fable highlights the main assumption we must make to justify the existence of experimental science: our experiments and those of our ancestors must be useful to us and we must be able to adapt to the universe in which we live, which is only possible if our senses are able to recognise a series of sensations analogous to others held in memory and whose regular succession we expect to follow as soon as we witness the first among them.

From time to time a change occurs in the natural order of things. It attracts our attention and we start looking for a cause, trying to associate it with some new fact, unnoticed until then. Step by step we set about organising our understanding of the universe, albeit rudimentary to begin with, becoming more scientific as we go along, carefully taking into account the previous stages of our learning process, using the experience accumulated over many centuries and assiduously transmitted from generation to generation.

But the constancy of the physical world would not be enough to ensure that we could adapt to the world and pursue experimental science.

Similar relationships must occur often enough in the succession of observed phenomena and they must be simple enough for us to pick them out. This is necessary, not only at the beginning of this learning process, but at every stage in the development of science. We proceed in steps: anomalies arising in the vast array of explained phenomena must be simple enough and frequent enough to be formulated according to some new law. This in turn will extend the realm of facts we understand, that we can both predict and reproduce at will.

---

[1] For the original text of this chapter in French see p. 125.

These empirical laws build up and gradually form an ensemble as complex and interrelated as the world presented to us by our senses must have once appeared to our child's eyes, and which we so easily disentangle now. We must then produce a synthesis of these scattered laws, otherwise we could no longer even know them all, nor remember them, and scientific progress would soon be stopped in its tracks.

But once again, this has to be a real possibility. There must be some overriding simplicity that prevails over the variety of phenomena and the multiplicity of laws. And it is up to the insight of the savant to identify the standpoint from which this multiplicity resolves itself into a broader and more comprehensive synthesis. In these attempts at unification, the simplicity of a few partial laws may have to be sacrificed in favour of the simplicity of a deeper general law. The simple laws of Mariotte[2] and Gay-Lussac, first steps in understanding the properties of gases, disappear in favour of the all-embracing simplicity of the kinetic theory, which gives us a better grasp of the properties of gases, at the same time as it also elucidates those of vapours and liquids.

Scientific progress is the discovery of a more comprehensive simplicity, in the sense of extending the realm of knowledge that has already been unified. Past successes encourage us to trust in the future of science: we are growing more and more aware that the universe is intelligible. The infinite variety of different phenomena can be understood in a handful of simple propositions and in a few equations which reveal their regularities and provide a way to predict their sequence when the initial conditions are known.

Here we shall describe the main lines of these attempts by human intelligence to apprehend a universe so well suited to our various activities. We shall then have a better appreciation of the role played by Einstein's work in the development of our understanding, and a clearer view of the step he made us take toward unification.

Our very first notions when we enter this world as a child are doubtless geometric. We continually come into contact with objects that appear the same to us. We get used to seeing them again and again, and to recognising them. We notice the similarity of the motions needed to reach and interact with them. Their shape must surely be one of the first things we abstract from them.

Whatever the worth of this opinion of the way we generate our empirical knowledge, it was by establishing geometry that humanity began to build up scientific knowledge. As systematised once and for all by Euclid, geometry dominated Greek science. The permanence of the shapes of solid bodies is one of the most immediate regularities that nature presents us with. And from this we may abstract the notion of distance, while the relationships between distances are simple enough to have been discovered early on.

The whole of Greek astronomy was guided by the need for geometric simplicity: the motion of a heavenly body had to be either a simple one or a combination of simple ones. But in the end they arrived at a highly complicated construction of circular and uniform relative motions which must eventually have seemed rather

[2] Translator's note: often referred to as Boyle's law in the Anglo-Saxon world.

unsatisfactory to them. From the geometrical standpoint, the high point in the explanation of astronomical phenomena was the discovery of the elliptical motion of the planets. Kepler's laws tell us that the planets move with an ideal simplicity that would certainly have pleased the geometrical minds of the Greeks.

So according to this intelligence that would laud the simplicity of the world, how could we consider as progress the replacement of these ideal trajectories by the infinitely complex curves revealed by more careful observation and which result from calculation of the perturbations brought to our attention by Newton? Was this not a step backward? Certainly, we lost the simplicity of Kepler's motion, but at the same time we achieved a deeper simplicity, and infinitely more fertile: Newton's law of gravitational attraction and his laws of motion. All bodies attract one another and their motions can be calculated as soon as we know their initial data. All we have to do is to integrate the general laws of motion.

These laws of Newtonian mechanics dominated physics just as geometry had dominated Greek astronomy: everyone was looking for actions at a distance. The law formulated by Coulomb was modelled on the universal law of gravitational attraction, expressing the mutual action of electrical charges. It was even hoped for a certain time that the whole of physics could be reduced to mechanics, although electrical phenomena soon put an end to this dream. Under the influence of Newtonian ideas, all attention turned first to electrical charges and current conductors. Faraday showed the effects of dielectrics. The field became a reality spread throughout the given medium: electrical charges and magnetised needles revealed its presence, but it existed even in their absence. Then Maxwell discovered the differential equations which account for the properties of such fields. His ideas were decisively confirmed by the discovery of Hertzian waves.

Physics thus moved forward in three main steps, with mechanics using geometry, and electricity using geometry and mechanics. The new developments changed nothing in the older parts of science, and the latter for their part were developed as independently as possible. And in fact, the three stages of science are of very different nature, even quite opposite to one another. Geometry looks like a self-contained reality that is quite independent of the vicissitudes of matter: material bodies just live in space. It looks like space exists for all time, just waiting for bodies to pass through one or other of its parts, then leave it without affecting any of its properties.

Electrical effects are transmitted at finite speed. Classical mechanics still speaks of instantaneous action at a distance. Could we hold on to the former theory of gravitational potential when we now have the retarded potentials of Lorentz? Should not mechanics undergo some consequences of the transformations used in electricity theory? Should not the more recent developments have their effects on the older branches of science? Could this not lead to greater simplicity?

This was the unification that Einstein set out to achieve. He showed that, when taken as a whole, the laws of physics assume a more profound simplicity if we accept a certain influence of electrical phenomena on mechanical phenomena and the influence of each on the properties of material rulers. The partial unifications merge into a broader one. The simplicity of Euclidean geometry and classical mechanics disappears, sacrificed in favour of the simplicity of the whole of physics, just as the simplicity of Kepler's laws was once abandoned without regret in favour of Newton's new synthesis.

Einstein's method is of the simplest.

When studying a phenomenon, along with the facts under investigation, we are compelled to include subjective elements that are necessarily involved in understanding it. Apart from the object of study, we have to take into account an observer who, at each instant of time, identifies the positions of various points and any changes that occur in them. When seeking to express the observations in an intelligible formula, we cannot distinguish in the simplicity of this law the subjective aspects of that simplicity. It may be entirely artificial, not coming from the object but from the specific observer that we have had to associate with it and the choice of reference system adopted by that observer.

Here Einstein draws attention to the *relativity* of the observer.

A phenomenon can be observed by any observer. An object can be referred to a reference system in infinitely many different ways. The simplicity of a law will be truly objective when it perdures, whatever reference system is chosen.

The equations expressing a physical law must retain the same algebraic form when we make an arbitrary change of coordinates.

This is the *principle of relativity.*

The properties it picks out are stripped as far as possible of all subjectivity. Thus the mathematical tool used to carry out this undertaking is called *absolute* differential calculus (and also tensor calculus). It determines the various possible forms of the laws of physics which accord with the principle of relativity. It then remains to see what becomes of these laws when we use the reference system for events adopted by experimenters, and to select among the various possible laws those that best account for their observations.

In this way, we obtain truly objective laws and we may legitimately expect them to give a more perfect account of observed facts and help us to discover new ones.

Chapter 1
Space and Time[1]

1.1 Riemann's General Geometry[2]

Geometry very likely arose from the experimental need to do things like measuring land areas, but it soon went beyond this purely empirical existence to appear in the highly systematised form given to it by Euclid. It claimed its independence from the experimental facts that had given rise to it. Clearly announced postulates separated it out from contingent facts and the whole theory could be deduced without further appeal to experiment. And since then, this aspect of things has become accentuated. The debate over attempts to prove the famous parallel postulate led to quite different geometries from those suggested by our intuition of space. By altering one or other of the postulates, geometries were obtained with the same logical validity as Euclid's and whose application to the physical world was still possible even though perhaps less convenient.

Can experiment decide between these different kinds of geometry? Can we carry out a geometry experiment?

There is no purely geometrical being: apart from their geometrical properties, all such beings have other physical properties, and any experiment will depend on both. If we constructed a real triangle for which the sum of the internal angles was greater than two right angles, we would still have to prove that its sides were straight. We might do this, for example, by superposing them in pairs, then repeating by rotating one of them. But how could we be sure that they had not been deformed during the operation? We could still claim that the geometry was Euclidean. A triangle for which the sum of the internal angles is greater than two right angles is a curvilinear triangle. Its sides are deformed during the operations we have used to prove that they are straight.

The only thing an experiment can check for us is a geometry and the rest of physics and we may always choose the latter in such a way as to make any geometry agree with the experiment. We may, for instance, choose the simplest geometry, the Euclidean geometry, and it will be possible to establish the physics on the basis of this geometry.

---

[1] For the original text of this chapter in French see p. 129.

[2] For the original text of this section in French see p. 129.

G. Lemaître, *Learning the Physics of Einstein with Georges Lemaître*,
https://doi.org/10.1007/978-3-030-22030-3_1

But could we be certain that the simplest physics will correspond to this simplest geometry? What is important is not the simplicity of just a part of the construction, but the simplicity of the whole. Would it not be wiser to take advantage of the arbitrary form the geometry can have in order to simplify physics as a whole, and not be afraid to complicate to some extent the well known parts like the geometry, if we can thereby facilitate exploration of newly discovered regions?

The general geometry proposed by Riemann is a perfectly natural generalisation of Euclidean geometry according to the standard procedure commonly used in every branch of mathematical physics. It is obtained by assuming that Euclid's geometry is no longer generally valid in a finite domain, but that it still prevails in an infinitesimal domain. Put another way, it is always possible to find a small enough domain in which the metric relations of Euclidean geometry are valid to within a given approximation.

Standard geometrical notions are now only applicable in the limit, and for this reason it is generally impossible to construct superposable figures of finite dimensions. Whenever we wish to maintain the possibility of superposing geometrical figures, we obtain special cases of the geometry for which the space is homogeneous. The set of such cases is often called general geometry, yet this set is a particular case only of the general geometry we shall consider here, for which the space is not necessarily homogeneous. It encompasses Euclidean geometry, the non-Euclidean geometry that is properly called Bolyai–Lobachevsky geometry, obtained by rejecting the famous parallel postulate, and finally, spherical geometry, which extends the properties of the sphere to space, and which is often called Riemannian geometry in the narrow sense. Einstein and Weyl always use the term "Riemannian geometry" in the broad sense to refer to the general geometry that we shall study here. When they wish to refer to the narrow sense, they speak of spherical space or spherical geometry.

At first glance, it is strange to think that we can construct a geometry in which the superposition of figures would be impossible. After all, we are used to defining equality by means of such a superposition, and it looks like a very basic notion. However, we know that we can measure the length of an arc of a curve even though it may nowhere coincide with the straight meter stick that represents the unit of length, and we may not even be able to match any tiny part of the curve, no matter how small, with a part of the meter stick. We can imagine making this measurement in the following way. We may apply the points of a pair of compasses to the arc to be measured, then move the instrument in such a way that one point remains where it is and the other rotates to mark another point on the arc; this is repeated a certain number of times to different positions on the arc. The ends of the arc will not generally both be reached by either point of the compass, but it will be impossible at some juncture to mark any further points on the arc. With the same opening angle, we then observe how many times we may move the pair of compasses along the unit meter. The ratio of the two numbers obtained will tend to a limit when the whole exercise is repeated for smaller and smaller compass angles, tending to zero. This limit measures the length of the arc.

It is important that the pair of compasses should not deform during these operations, or if we prefer to imagine a series of compasses each sharing a point of contact on the arc, these compasses must all be superposable. Of course, it suffices for their points to be superposable. Moreover, this superposition only need operate in the limit when the two points tend toward each other. It is enough therefore that the notions of Euclidean geometry should be applicable in the limit of an infinitely small domain.

This definition of the distance by superposition of a measuring instrument like a pair of compasses is perfectly suited to measuring distances, but it does not tell us what distance actually is. It is clear that, when the compasses are brought into a given environment, the distance between the points of this environment already existed before we introduced the instrument used to measure it. There must be a physical reality that is measured by the distance, just as a temperature is measured by a thermometer, but already exists in places where there is no thermometer. In the same way, a magnetised needle explores the magnetic field and does not produce it.

If we adopt this standpoint, we must accept that there is a physical reality extending through space and that distance is one of its manifestations; in other words, there is a metric field which is explored by the compasses, just as a magnetic field is explored with a compass needle, and we may expect this reality, which we know through its metric aspects, to play a role in physical phenomena of another order.

The notion of equality between distance elements must therefore have a physical meaning that is independent of the instruments we use to measure this distance.

In order to study this physical reality, we must begin by referencing the positions of points. We assign to each point three numbers (x_1, x_2, x_3) called its coordinates. Furthermore, to any continuous series of points there must correspond a continuous variation of their coordinates. Apart from these general conditions of one-to-one correspondence and continuity, the means used to identify points remains completely open.

Equality of distances tells us that a certain physical quantity characterised by each pair of points is equal for each of them. This quantity depends for each pair on the coordinates of each point, coordinates that must be considered infinitely close since equality is only defined in the limit as each distance tends to zero. It is thus a function of the coordinates and the differentials of the coordinates.

What should be the form of this function?

It must clearly be homogeneous in the differentials. Moreover, it must not change when we change the sign of the differentials, because it cannot depend on the order in which we consider the points. And finally, it must be possible to use it in whatever way the points themselves are referenced, whence it must have the same algebraic form when we make an arbitrary change of coordinates.

The simplest function to satisfy all these conditions is a quadratic form in the coordinate differentials:

$$\begin{aligned}\gamma_{11}\mathrm{d}x_1^2 + 2\gamma_{12}\mathrm{d}x_1\mathrm{d}x_2 &+ 2\gamma_{13}\mathrm{d}x_1\mathrm{d}x_3 \\ + \quad \gamma_{22}\mathrm{d}x_2^2 &+ 2\gamma_{23}\mathrm{d}x_2\mathrm{d}x_3 \\ &+ \quad \gamma_{33}\mathrm{d}x_3^2 .\end{aligned}$$

Pairs of equidistant points are thus those whose coordinates deliver the same value of this function.

The distance measured along the arc of a curve can be used as one of the coordinates referencing the points of this arc. The distance element $\mathrm{d}\sigma$ must therefore be an infinitesimal of the same order as the coordinate differentials.

We must therefore set

$$\mathrm{d}\sigma^2 = \sum_\mu \sum_\nu \gamma_{\mu\nu}\mathrm{d}x_\mu \mathrm{d}x_\nu \quad (\mu,\nu = 1,2,3) , \tag{1.1}$$

where

$$\gamma_{\mu\nu} = \gamma_{\nu\mu} .$$

The length of an arc of a curve is given by the line integral

$$\int \mathrm{d}\sigma = \int \sqrt{\sum_\mu \sum_\nu \gamma_{\mu\nu} \frac{\mathrm{d}x_\mu}{\mathrm{d}\lambda}\frac{\mathrm{d}x_\nu}{\mathrm{d}\lambda}}\, \mathrm{d}\lambda , \tag{1.2}$$

where the derivatives are calculated using the parametric equations of the curve whose length is being measured,

$$x_\mu = \varphi_\mu(\lambda) \quad (\mu,\nu = 1,2,3) .$$

The geometric properties of the medium are thus specified by the value at each point of the six functions $\gamma_{\mu\nu}$ of the coordinates. These are the potentials of the metric field. It is natural to assume that these functions are continuous. This means that, around any point, there will always be a small enough domain in which the potentials differ from their value at this point by less than some given amount. We shall show that, when the potentials are constant, the geometry is Euclidean. We will then be able to conclude that the relations of Euclidean geometry are valid to within a given approximation in a small enough domain. This property characterises Riemann's general geometry.

Through a linear transformation of the coordinates, a quadratic form with constant coefficients can be transformed to an algebraic sum of squares. The calculation is the same as for the reduction of a quadric surface. Indeed, we may consider $\mathrm{d}x_1$, $\mathrm{d}x_2$, $\mathrm{d}x_3$ as the Cartesian coordinates of a point: for a given value of $\mathrm{d}\sigma$, (1.1) then represents a quadric surface with reduced equation of the form

$$\mathrm{d}\sigma^2 = \mathrm{d}x^2 + \mathrm{d}y^2 + \mathrm{d}z^2 , \tag{1.3}$$

where we have set

$$\mathrm{d}x = \sum_\sigma a_\sigma \mathrm{d}x_\sigma\,, \quad \mathrm{d}y = \sum_\sigma b_\sigma \mathrm{d}x_\sigma\,, \quad \mathrm{d}z = \sum_\sigma c_\sigma \mathrm{d}x_\sigma\,,$$

for suitable constants a_σ, b_σ, and c_σ.

Equation (1.3) expresses the extension of Pythagoras' theorem to the given space. It defines the distance element of Euclidean geometry.

The various geometrical notions of straight line, angle, and so on, are now easily extended to general geometry. We only need to find a definition that is independent of any particular way of referencing the points and which agrees with the Euclidean notions when $\mathrm{d}\sigma^2$ reduces to the form (1.3).

A straight line is a curve defined by two of its points and which is the shortest path between these two points. The equation is found by using variational calculus on the equation

$$\delta \int \mathrm{d}\sigma = 0\,, \tag{1.4}$$

where the limits of integration are assumed to be constant. A straight line is determined by the coordinates of one of its points and the coordinate differentials at that point.

The notion of perpendicularity can be defined in the following way. We consider two directions

$$\delta x_1\,,\ \delta x_2\,,\ \delta x_3\,,$$

and

$$\mathrm{d}x_1\,,\ \mathrm{d}x_2\,,\ \mathrm{d}x_3\,.$$

This pair of directions specifies an angle. The two directions are said to be perpendicular if the angle they constitute is the same as the angle obtained by replacing one of the two directions by the opposite direction.

Consider the bilinear form associated with the fundamental form (1.1):

$$\sum_\mu \sum_\nu \gamma_{\mu\nu} \mathrm{d}x_\mu \delta x_\nu\,.$$

This form is invariant. Under an arbitrary change of coordinates

$$\mathrm{d}x_\mu = \sum_\sigma \frac{\partial x_\mu}{\partial x'_\sigma} \mathrm{d}x'_\sigma\,, \qquad \delta x_\nu = \sum_\tau \frac{\partial x_\nu}{\partial x'_\tau} \delta x'_\tau\,,$$

it transforms to

$$\sum_\sigma \sum_\tau \gamma'_{\sigma\tau} \mathrm{d}x'_\sigma \delta x'_\tau\,,$$

where

$$\gamma'_{\sigma\tau} = \sum_\mu \sum_\nu \frac{\partial x_\mu}{\partial x'_\sigma} \frac{\partial x_\nu}{\partial x'_\tau} \gamma_{\mu\nu}\,. \tag{1.5}$$

The coefficients transform in the same way as those of the quadratic form that would be obtained by replacing δx_ν by $\mathrm{d}x_\nu$.

For the two directions to be perpendicular, the bilinear form must not change when we change the signs of the δx_ν. This means that it must vanish. The perpendicularity condition is therefore

$$\sum_\mu \sum_\nu \gamma_{\mu\nu} \mathrm{d}x_\mu \delta x_\nu = 0 \,. \tag{1.6}$$

In the case of Euclidean geometry (1.3), we obtain the well known equation

$$\mathrm{d}x\,\delta x + \mathrm{d}y\,\delta y + \mathrm{d}z\,\delta z = 0 \,.$$

The angle between two directions is given by the formula

$$\cos\theta = \frac{\sum_\mu \sum_\nu \gamma_{\mu\nu} \mathrm{d}x_\mu \delta x_\nu}{\sqrt{\sum_\mu \sum_\nu \gamma_{\mu\nu} \mathrm{d}x_\mu \mathrm{d}x_\nu} \sqrt{\sum_\mu \sum_\nu \gamma_{\mu\nu} \delta x_\mu \delta x_\nu}} \,. \tag{1.7}$$

The expression here is a combination of invariants and reduces to the standard form in the Euclidean case.

Finally, the volume element is

$$\sqrt{\gamma}\, \mathrm{d}x_1 \mathrm{d}x_2 \mathrm{d}x_3 \,, \tag{1.8}$$

where γ is the symmetric determinant of the metric potentials. This reduces to the product of the differentials in the Euclidean case. It is an invariant. Indeed, according to (1.5) and the properties of determinants,

$$\gamma' = |\gamma'_{\sigma\tau}| = \left| \frac{\partial x_\mu}{\partial x'_\sigma} \right|^2 \gamma \,,$$

while

$$\mathrm{d}x_1 \mathrm{d}x_2 \mathrm{d}x_3 = \left| \frac{\partial x_\mu}{\partial x'_\sigma} \right| \mathrm{d}x'_1 \mathrm{d}x'_2 \mathrm{d}x'_3 \,,$$

whence we do indeed find

$$\sqrt{\gamma}\, \mathrm{d}x_1 \mathrm{d}x_2 \mathrm{d}x_3 = \sqrt{\gamma'}\, \mathrm{d}x'_1 \mathrm{d}x'_2 \mathrm{d}x'_3 \,. \tag{1.9}$$

We can get a simple idea of the consequences of adopting the general geometry to study physics by comparing the Euclidean geometry of surfaces. Let

$$x_1 = f_1(u_1, u_2) \,, \quad x_2 = f_2(u_1, u_2) \,, \quad x_3 = f_3(u_1, u_2) \,,$$

be the parametric equations of a surface. To every pair of values of the parameters u_1 and u_2, there corresponds a point on the surface, so these parameters effectively specify points of the surface and can be treated as their coordinates. The Euclidean distance element is given as a function of the x_i by (1.3).

We now differentiate the equations for the surface:

$$\mathrm{d}x_i = \frac{\partial f_i}{\partial u_1}\mathrm{d}u_1 + \frac{\partial f_i}{\partial u_2}\mathrm{d}u_2 \qquad (i = 1,2,3)\ ,$$

and substitute these into (1.3) to obtain an expression of the form

$$\mathrm{d}\sigma^2 = \gamma_{11}\mathrm{d}u_1^2 + 2\gamma_{12}\mathrm{d}u_1\mathrm{d}u_2 + \gamma_{22}\mathrm{d}u_2^2\ , \tag{1.10}$$

where γ_{11}, γ_{12}, and γ_{22} are functions of u_1 and u_2. This expression looks just like (1.1) and shows that, on an arbitrary surface, the Euclidean geometry is equivalent to Riemann's general geometry in two dimensions.

This is just what one would expect, because Riemann geometry assumes simply that Euclidean geometry is valid in the limit in an infinitely small domain, and it is clear that the geometry of figures drawn on a portion of surface whose area tends to zero will coincide in the limit with the geometry of the projections of these figures on the tangent plane to the surface.

The geometry of the sphere is well known to us. We can work out the curvature of a spherical surface just by making measurements on that surface. But there would be no way of working out whether the surface on which we sat were concave or convex. We would simply establish that it was curved. Observers who did not know that light propagates in straight lines and for whom all geometric measurements were restricted to surveying the surface of the Earth could indeed establish that the Earth is round. They could go right round it, they could identify the shortest route between three towns, and they could show that the sum of the internal angles of the triangle determined in this way was greater than the sum of two right angles, but they could not work out whether they were on a solid sphere or a hollow sphere of the same radius. The spatial properties they could discover would all relate to the curvature of the surface, or Gaussian curvature, which is equal to the reciprocal of the square of the radius in the case of a sphere.

On the sphere the curvature is constant, so figures drawn on the surface are superposable. We can study the equality of spherical triangles in the same way as we would for plane triangles. But if we were to study the Euclidean geometry of a surface of variable curvature, like an ellipsoid with unequal axes, such a method would be useless. It would be impossible to displace a figure on the ellipsoid without deforming it. But figures drawn on such a surface would nevertheless have geometric properties like lengths, angles, areas, and so on. These can be studied directly using the methods of general geometry and using the expression for $\mathrm{d}\sigma$ in terms of arbitrary coordinates.

By comparison with the geometry of arbitrary surfaces, general geometry in three dimensions is said to be the geometry of a space of variable curvature. We may look for the quantities on which the geometric properties of this space depend and which generalise the Gaussian curvature of surfaces. The displacement without deformation of a solid in such a space is as impossible as the displacement without deformation of a figure on an ellipsoid with unequal axes, and yet the geometrical properties of length, straightness, angle, volume, and so on, can still be defined and studied just as they can in a Euclidean space.

Note finally that, just as we can make a map of a sphere or an ellipsoid on a plane, so can we make a map of a Riemannian space in a Euclidean space. We only have to consider the coordinates x_1, x_2, x_3 as Cartesian coordinates in this space. The form $d\sigma^2$ in (1.1) will then determine the scale of the map in each direction and play the same role as (1.10) in the study of geographical maps.

In this situation everything happens as though the standard of length deforms when we change its direction, and the nature of the space at each point can then be described at each point using the form (1.1) of the ellipsoid of scales. Naturally, such a description will depend in part on how the map has been made. We may consider conformal maps, for which the scale ellipsoid reduces to a sphere whose radius varies from one point to another, and which conserve angles, or equivalent maps where the volume of the ellipsoid is constant and which conserve volumes (for this, as in the case of plane maps, the determinant of the $\gamma_{\mu\nu}$ must be constant). We thus see that there is no particular difficulty in representing a Riemannian space; we just make a map of it in a Euclidean space where everyday experience provides us with suitable intuitions.

All forms of general geometry can be applied to the study of the physical world. We may select the form of the geometry which best lends itself to the study of physics, just as we choose the coordinate system which is best suited to the study of a given problem of analytical geometry. Indeed, all are equally convenient from the practical standpoint, provided that Euclidean geometry is valid for the approximate measurements we make in the rather small domain occupied by our laboratories and our instruments.

1.2 Time and Mechanics[3]

What do we mean when we say that the same time elapses when the hand on our watch moves from any given number on the dial to the neighbouring number? What is a clock? What ideal definition do we have in mind when we say that one particular clock is better than another?

To answer these questions, let us consider a simple clock, the hourglass. Each time the sand flows from one bulb to the other, we expect it to take the same time. But why? Because we see no difference in the appearance of the hourglass each time we turn it upside-down and the same phenomenon occurs, so we assert that it will last the same time.

Imagine that we obtain a soft-boiled egg by leaving it in boiling water during the time taken by the sand to flow from one bulb to the other. We then repeat the experiment with a similar egg. We would indeed be surprised if we then obtained a hard-boiled egg. We would wonder whether the first egg was not perhaps bigger than

[3] For the original text of this section in French see p. 138.

the second, whether the flow of the sand had been truly steady, and goodness knows what else. But if none of the alternative explanations we could think of turned out to be adequate, we would have no way of saying whether it was better to measure the duration by the time needed to cook a soft-boiled egg or by the time required for the sand to flow from one bulb to the other. If there were a difference in the results of the measurements made by one or other procedure, we would still claim that some circumstance had changed when reproducing the phenomena we had believed identical.

But is it even possible to reproduce the same phenomenon twice? Strictly speaking, it is not. Once we have flipped the hourglass for the very first time, we will never again have an hourglass that has never been turned round. Naturally, it is not hard to take into account differences of this kind. Even among the physical differences, we will be able to take into account factors whose action could be compensated for by a suitable device. We will generally consider that a good clock should be independent of circumstances like temperature, electric field, gravity, and so on. But we cannot take everything into account. We must in the end assume that there is a physical reality that is the same between the beginning and the end of the various seconds of all our timepieces.

In order to study the physical reality measured by clocks, we must choose a way of referencing different events.

We shall associate four numbers x_1, x_2, x_3, x_4, called coordinates, with each event. The way these are attributed is of no importance, except that two different events should not be attributed the same coordinates and neighbouring events should be represented by nearby numerical values of the coordinates.

When two time intervals are equal, this tells us that a certain quantity characterised by the two pairs of events is equal for the two of them. For each pair, this quantity depends on the coordinates of the events. For infinitely short time intervals, it is a function of the coordinates and the differentials of the coordinates.

This function must be homogeneous in the differentials. Moreover, it must be applicable whatever system is adopted to reference the events, so it must retain the same algebraic form when we make an arbitrary coordinate transformation.

The time interval can be used to reference any events so it must be of the same order of magnitude as the coordinate differentials.

The simplest forms we could use are linear or quadratic forms in the coordinate differentials. We thus set either

$$\mathrm{d}s = \sum_{\sigma} a_\sigma \mathrm{d}x_\sigma \quad (\sigma = 1,2,3,4)$$

or

$$\mathrm{d}s^2 = \sum_{\mu} \sum_{\nu} g_{\mu\nu} \mathrm{d}x_\mu \mathrm{d}x_\nu \quad (\mu, \nu = 1,2,3,4) \,, \tag{1.11}$$

where

$$g_{\mu\nu} = g_{\nu\mu} \,. \tag{1.12}$$

Note that the first form is a special case of the second, when the quadratic form is a perfect square.

In any case, the time measured by a clock is equal to the integral

$$\int \mathrm{d}s\,,$$

calculated by following the clock, or along its worldline, to use the jargon, referring to the four-dimensional continuum of physical phenomena as the world.

Observation of clocks thus informs us about the functions $g_{\mu\nu}$ of the four coordinates, the "world potentials", analogous to the metric potentials of the geometry in three dimensions.

The metric potentials suffice to determine the straight lines in space (or geodesics in two dimensions). These lines are completely specified by their initial conditions, namely a point and the direction at this point.

In the world there is an analogous situation. There are worldlines that are fully specified by the coordinates of one of their points and the differentials of these coordinates. These are the trajectories of material points left to their own devices. There is a property of the environment that forces them to follow a predetermined path. This property is called inertia or gravitational field, depending on the circumstances. Could the physical realities that trace out planetary orbits not be the same as those that determine the motion of clocks? The latter are generally based on the reproduction of mechanical phenomena. Therefore, the reality they measure is indeed of the same kind as the one manifested through the dynamics of free points.

Equation (1.11) provides a general specification of any worldline given two of its points. We need only apply the calculus of variations to the equation

$$\delta \int \mathrm{d}s = 0\,,$$

taken between these two points. The trajectory of a free point is thus computed in the same way as the straight lines of general geometry or the geodesics of a surface.

However, for this procedure to be applicable, ds must not be an exact differential. If the straight line or geodesic can be defined as the shortest path between two points, it is because the length of an arc of curve depends on the shape of the curve and is not the same for all arcs having the same endpoints.

In order for the potentials $g_{\mu\nu}$ to specify the trajectories of free points, the time interval between two events must depend on the worldline along which we move from one to the other. In other words, we must abandon the familiar idea of absolute time. Identical and identically adjusted clocks will not generally indicate the same time upon two different encounters. What they show will depend on the worldline each has followed between those encounters. Indeed, the times defined along the different worldlines need only differ from one another by extremely small amounts. All that is needed is that the time should not be a strictly exact differential. It can then be greater or smaller along one trajectory than along all neighbouring trajectories with the same endpoints. There will be geodesic worldlines, trajectories of material points left to their own devices.

Let us see how these conditions can be realised. When we use Cartesian coordinates x, y, z, and a suitably defined time t, the equations of motion of a free point coming solely under the influence of the force of gravity can be expressed according to classical mechanics in the form

$$\frac{\mathrm{d}v_x}{\mathrm{d}t}+\frac{\mathrm{d}V}{\mathrm{d}x}=0\,,\qquad \frac{\mathrm{d}v_y}{\mathrm{d}t}+\frac{\mathrm{d}V}{\mathrm{d}y}=0\,,\qquad \frac{\mathrm{d}v_z}{\mathrm{d}t}+\frac{\mathrm{d}V}{\mathrm{d}z}=0\,,$$

where v_x, v_y, and v_z are the components of the velocity of the moving point and V is the potential of the gravitational field.

These equations can be written in Hamiltonian form as

$$\delta\int_{t_0}^{t_1}\left(\frac{v^2}{2}-V\right)\mathrm{d}t=0\,. \tag{1.13}$$

Let c be a very large number compared with the speeds occurring in the experiments on which these equations are based. This could be the speed of light, for instance.

Consider the quadratic form

$$\mathrm{d}s^2=-\mathrm{d}x^2-\mathrm{d}y^2-\mathrm{d}z^2+(c^2+2V)\mathrm{d}t^2\,. \tag{1.14}$$

We will then have

$$\int\mathrm{d}s=\int_{t_0}^{t_1}\sqrt{-\left(\frac{\mathrm{d}x}{\mathrm{d}t}\right)^2-\left(\frac{\mathrm{d}y}{\mathrm{d}t}\right)^2-\left(\frac{\mathrm{d}z}{\mathrm{d}t}\right)^2+c^2+2V}\,\mathrm{d}t$$
$$=c\int_{t_0}^{t_1}\sqrt{1-\frac{1}{c^2}(v^2-2V)}\,\mathrm{d}t\,.$$

Since c^2 is much greater than v^2, and hence also much greater than V, which is of the same order of magnitude, we will have approximately

$$\int\mathrm{d}s=c\int_{t_0}^{t_1}\left[1-\frac{1}{c^2}\left(\frac{v^2}{2}-V\right)\right]\mathrm{d}t\,. \tag{1.15}$$

The equation for geodesics,

$$\delta\int\mathrm{d}s=0\,,$$

will then be equivalent to the classical equation (1.13).

We could of course obtain the same result by starting with other forms than (1.14), and in particular, by taking all the terms to be positive. We shall justify this choice of signs later.

Let us now consider the consequences of (1.15). Clock time will no longer be an exact differential and there will be small differences in the time intervals measured between two events depending on the trajectories of the clocks that measure them. The main term in (1.15) is

$$c\int_{t_0}^{t_1} \mathrm{d}t\ .$$

Up to a multiplicative factor, this is the time used in classical mechanics. The time in seconds will thus be

$$\frac{1}{c}\int \mathrm{d}s = \int_{t_0}^{t_1}\left[1-\frac{1}{c^2}\left(\frac{v^2}{2}-V\right)\right]\mathrm{d}t\ .$$

A stationary clock sitting at a point where the potential is zero,

$$v=0\ , \qquad V=0\ ,$$

indicates coordinate time:

$$T_1 = \int_{t_0}^{t_1}\mathrm{d}t = t_1 - t_0\ .$$

A second clock moving on an equipotential surface

$$V=0$$

will indicate the time

$$T_2 = \int_{t_0}^{t_1}\left(1-\frac{1}{c^2}\frac{v^2}{2}\right)\mathrm{d}t = \left(1-\frac{1}{c^2}\frac{\overline{v^2}}{2}\right)T_1\ ,$$

where $\overline{v^2}$ is the mean squared speed given by

$$\overline{v^2} = \frac{1}{t_1-t_0}\int_{t_0}^{t_1} v^2\mathrm{d}t\ .$$

Finally, consider a third clock, identical to the last two, thrown vertically upwards at time t_0 with speed v_0 and falling to join the two others at time t_1. Since (1.15) is only an approximate equation, it will suffice to calculate the speed and potential along the trajectory using the classical equations

$$x = v_0 t - \frac{gt^2}{2}\ , \qquad V = gx\ .$$

We then have

$$T_1 = t_1 - t_0 = \frac{2v_0}{g}\ ,$$

and the time indicated by the third clock will be

$$T_3 = \int_{t_0}^{t_1}\left\{1+\frac{1}{c^2}\left[g\left(v_0 t - \frac{gt^2}{2}\right) - \frac{1}{2}(v_0 - gt)^2\right]\right\}\mathrm{d}t\ .$$

By carrying out the integration, we obtain

$$T_3 = \left(1 + \frac{1}{6}\frac{v_0^2}{c^2}\right) T_1 \ .$$

Between the same two events, the third clock in free motion will thus indicate a longer time than each of the other two.

It is easy to see that, if we adopt the form (1.14) for the time element, the trajectory of a free point will always be the trajectory of maximum duration. Indeed, we can always imagine a virtual trajectory, as close as we like to the actual trajectory, for which the time taken is shorter. It suffices to consider a moving body describing a spiral winding tightly around the actual trajectory. The potential V will then have the same value along the two trajectories, while the speed v will be greater on the virtual trajectory and hence the time given by (1.15) will be shorter along the latter.

However, we should not conclude that the time along the trajectory of a free point is necessarily an absolute maximum. It is only a maximum relative to all neighbouring trajectories. So, for example, the second clock discussed above could return to its point of departure without leaving the given equipotential surface by describing a great circle around the Earth, with a speed (7.9 km s^{-1}) such that the centrifugal force exactly balances the weight. It would thus describe a geodesic, the trajectory of a free point. The time taken would be greater than for all the neighbouring virtual trajectories with the same endpoints, but it would nevertheless be less than the time T_1 indicated between these same endpoints by the stationary clock. Taking

$$c = 3 \times 10^{10}\ \mathrm{cm\,s^{-1}} \ ,$$

the difference would be two millionths of a second for a journey taking 1 h 24 m.

1.3 Simultaneity and Space[4]

The above definitions of the length of a curve and the duration of a phenomenon both reduce to the actualisation of successive periodic phenomena, namely equidistant points marked on the curve and identical seconds of a clock. For the time measurement, this periodicity is realised along the worldline followed by the clock. But along what worldline is it realised in the case of a length measurement?

Up to now we have assumed that the object being measured did not deform, in which case we can take all the time we need to mark out the equidistant points and check the equality of the intervals between them. But if we want to measure the length of an object that is deforming, we must carry out all the operations constituting the measurement at the same instant of time. If we make a mistake in our appreciation of simultaneity, our measurement will be invalidated. The periodicity of the lengths must be realised on a line of simultaneous points.

[4] For the original text of this section in French see p. 144.

When we mark two points on an object, each of them will describe a worldline. Let us represent the differences between the coordinates of the two points by

$$\mathrm{d}x_1\,,\quad \mathrm{d}x_2\,,\quad \mathrm{d}x_3\,,\quad \mathrm{d}x_4\,,$$

and the variations in the coordinates along the neighbouring worldlines described by the two points by

$$\delta x_1\,,\quad \delta x_2\,,\quad \delta x_3\,,\quad \delta x_4\,.$$

The simultaneity condition can only depend on the coordinates x_σ ($\sigma = 1,2,3,4$) and the two kinds of variations $\mathrm{d}x_\sigma$ and δx_σ. This condition must be symmetrical under exchange of the two points, it cannot change when we reverse the order so it must not change when we change the sign of the $\mathrm{d}x_\sigma$. Indeed, the notion of simultaneity, just like the notion of distance, is symmetrical. It is independent of the order in which we consider the two points.

The solution to the problem of simultaneity is analogous to the solution of the geometric problem of defining perpendicular directions. We consider the bilinear form associated with the quadratic form $\mathrm{d}s^2$,

$$\sum_\mu \sum_\nu g_{\mu\nu}\mathrm{d}x_\mu \delta x_\nu\,.$$

This is an invariant function of the coordinates and of the two kinds of variations. If it is to be symmetric under exchange of the two simultaneous events, its value must not change when we change the sign of the $\mathrm{d}x_\mu$. It must therefore vanish.

The simultaneity condition

$$\sum_\mu \sum_\nu g_{\mu\nu}\mathrm{d}x_\mu \delta x_\nu = 0 \tag{1.16}$$

must therefore hold between the points of an object whose length is being measured. We can use this relation to eliminate one of the coordinate differences $\mathrm{d}x_\mu$ from the expression

$$\mathrm{d}s^2 = \sum_\mu \sum_\nu g_{\mu\nu}\mathrm{d}x_\mu \mathrm{d}x_\nu\,, \tag{1.17}$$

thereby arriving at a quadratic form

$$\mathrm{d}s^2 = \sum_{\mu=1}^{3} \sum_{\nu=1}^{3} \gamma_{\mu\nu}\mathrm{d}x_\mu \mathrm{d}x_\nu\,, \tag{1.18}$$

with three variables, where the $\gamma_{\mu\nu}$ are functions of the world potentials $g_{\mu\nu}$ and the velocity δx_ν of the object. This expression must be the same for two pairs of equidistant points. Indeed, superposing two lengths amounts to making a change of coordinates in such a way that the coordinates of the points and the various quantities characterising one of the lengths are transformed to the coordinates and the quantities corresponding to the other length. Only then can we say that they are the same, and that they differ only by the choice of coordinates used to reference them. But the bilinear form in the simultaneity equation and the expression for ds are invariant expressions, independent of the choice of coordinates. Their values must therefore be the same for two pairs of equidistant points. ds is thus the distance element. When the quadratic form is negative, we will set it equal to $-\mathrm{d}\sigma^2$, whence dσ will represent the distance element.

We thus obtain the geometry in the general Riemannian form of Section 1.1. The length of an arc will be found using (1.2). The metric potentials are functions of the world potentials $g_{\mu\nu}$ and the velocities δx_σ of the different points of the objects we are measuring.

Consider, for instance, the form (1.14)

$$\mathrm{d}s^2 = -\mathrm{d}x^2 - \mathrm{d}y^2 - \mathrm{d}z^2 + (c^2 + 2V)\mathrm{d}t^2 ,$$

obtained in the previous section. The equation of simultaneity for stationary points,

$$\delta x = \delta y = \delta z = 0 ,$$

will be

$$(c^2 + 2V)\delta t \mathrm{d}t = 0 ,$$

or

$$\mathrm{d}t = 0 .$$

The geometry of these points will thus be characterised by

$$\mathrm{d}\sigma^2 = \mathrm{d}x^2 + \mathrm{d}y^2 + \mathrm{d}z^2 ,$$

which specifies the distance element of the Euclidean geometry.

If we adjust our stationary clocks by simultaneity, namely in the present case, we arrange for them to indicate the time t, these clocks could no longer generally be considered identical and will no longer be interchangeable. Indeed, stationary identical clocks would indicate the time

$$\frac{1}{c}\int \sqrt{c^2 + 2V}\,\mathrm{d}t .$$

The definition of simultaneity on a body depends not only on the velocity δx_σ of the body, but generally also on the curve along which it is defined step by step. We shall see an example of this later on.

These various complications of our usual notion of time disappear in the limit of an infinitely small domain. In general geometry, we can always find a small enough domain for the relations of Euclidean geometry to be valid to within a given approximation. Likewise in Einstein's mechanics, in a neighbourhood of any point, we can adopt a reduced form of ds^2 analogous to the distance element of Euclidean geometry.

In a small domain, the potentials $g_{\mu\nu}$ can be replaced by their values at a point M. The quantity $\mathrm{d}s^2$ is then a quadratic form with constant coefficients. It can be reduced to an algebraic sum of squares by a linear change of coordinates.

Imaginary coordinates are generally avoided by choosing the sign combination

$$\mathrm{d}s^2 = -\mathrm{d}\xi_1^2 - \mathrm{d}\xi_2^2 - \mathrm{d}\xi_3^2 + \mathrm{d}\xi_4^2 \, .$$

This reduction can be achieved in several ways. If

$$\delta x_1 \, , \quad \delta x_2 \, , \quad \delta x_3 \, , \quad \delta x_4 \, ,$$

represent the velocity of a point of a body, we may impose the condition

$$\frac{\partial \xi_4}{\partial x_\mu} = K \sum_\nu g_{\mu\nu} \delta x_\nu \tag{1.19}$$

on the functions

$$\xi_\sigma = \xi_\sigma(x_1, x_2, x_3, x_4) \qquad (\sigma = 1,2,3,4)$$

used to carry out the reduction, where K is an arbitrary constant. The relation (1.19) need only hold exactly at M.

The simultaneity equation (1.16) will then be

$$\sum_\mu \frac{\partial \xi_4}{\partial x_\mu} \mathrm{d}x_\mu = \mathrm{d}\xi_4 = 0$$

on the body. Lengths on the body do not depend on ξ_4 and the geometry is defined by the distance element

$$\mathrm{d}\sigma^2 = \mathrm{d}\xi_1^2 + \mathrm{d}\xi_2^2 + \mathrm{d}\xi_3^2 \, ,$$

so ξ_1, ξ_2, and ξ_3 are Cartesian coordinates. In the new coordinate system, the simultaneity equation becomes

$$-\mathrm{d}\xi_1 \delta\xi_1 - \mathrm{d}\xi_2 \delta\xi_2 - \mathrm{d}\xi_3 \delta\xi_3 + \mathrm{d}\xi_4 \delta\xi_4 = 0 \, .$$

If this is to reduce to

$$\mathrm{d}\xi_4 = 0 \, ,$$

we must have

$$\delta\xi_1 = \delta\xi_2 = \delta\xi_3 = 0 \, ,$$

where the variations $\delta\xi_\sigma$ are taken along the worldlines of the points of the body. It is along these worldlines that time is calculated on the body according to

$$\int \mathrm{d}\xi_4 \, .$$

The new coordinates therefore immediately give the result of space and time measurements carried out at a point of the body. We call these the *proper coordinates* of the body at this point. We shall generally use the notation

$$ds^2 = -dx^2 - dy^2 - dz^2 + c^2 dt^2 ,$$

where c is a constant, x, y, z are Cartesian coordinates, and t the time read off stationary clocks. When this form of ds can be rigorously adopted, we shall say that we are dealing with a Galilean field.

These were the fields that Einstein first studied in his 1905 paper on the electrodynamics of moving bodies. He derived the final form of his theory (1915) using Riemann's procedure for generalising Euclidean geometry, that is, by assuming that the primitive form of his theory was valid only in the limit of an infinitely small domain. This is the usual way to carry out generalisations in physics.

1.4 Indirect Measurements of Space and Time[5]

We have seen how to measure the dimensions of a body by surveying it using identical metre sticks placed end to end on the body. The result of these measurements will depend on the motion of the body, which may deform during the motion. We only assume that the velocities of its points vary in a continuous manner from one point to another. The dimensions of a body are then perfectly determined when we know the equations of the worldlines followed by each of its points.

The results of such measurements do not depend on the coordinate system used. These are absolute measurements.

It is often impossible to make direct measurements by placing measuring instruments on the body we would like to measure. We must then deduce the lengths from indirect measurements carried out with whatever equipment is available to us.

If we wish to measure the length of a shell as it leaves the cannon, we will not have a ruler moving with it. However, we could photograph it, for example. We would then have to deduce, from the dimensions of the photo, the result we would obtain if we were able to follow the shell in its motion and measure it in the usual way.

Whatever method is used, we will be able to reduce it to this. On a body carrying our instruments, we mark at the same instant of time the points coinciding with those of the moving body. We will then obtain an accessible image of the moving body. From this we must deduce the dimensions of the moving body from those of the image.

We mark the points at the same instant of time relative to our measuring instruments. This is the only simultaneity we will be able to realise. Let ξ, η, ζ, and τ be the proper coordinates of the body carrying the instruments at the place where the measurement is carried out. Simultaneous points will be ones for which the values

[5] For the original text of this section in French see p. 149.

of τ are equal. Let x, y, z, and t be the proper coordinates of the moving object at the same point. We must have

$$\mathrm{d}s^2 = -\mathrm{d}\xi^2 - \mathrm{d}\eta^2 - \mathrm{d}\zeta^2 + c^2\mathrm{d}\tau^2 , \tag{1.20}$$

and

$$\mathrm{d}s^2 = -\mathrm{d}x^2 - \mathrm{d}y^2 - \mathrm{d}z^2 + c^2\mathrm{d}t^2 . \tag{1.21}$$

For a suitable choice of orientation of the two systems of axes, we will be able to set

$$\begin{cases} \mathrm{d}\xi = \dfrac{1}{\sqrt{1-v^2/c^2}}(\mathrm{d}x + v\,\mathrm{d}t) , \\ \mathrm{d}\eta = \mathrm{d}y , \\ \mathrm{d}\zeta = \mathrm{d}z , \\ \mathrm{d}\tau = \dfrac{1}{\sqrt{1-v^2/c^2}}\left(\dfrac{v}{c^2}\,\mathrm{d}x + \mathrm{d}t\right) . \end{cases} \tag{1.22}$$

This transformation will be compatible with (1.20) and (1.21) whatever the value of v. The speed of the object is obtained by setting

$$\mathrm{d}x = \mathrm{d}y = \mathrm{d}z = 0 .$$

We find

$$\frac{\mathrm{d}\xi}{\mathrm{d}\tau} = v , \qquad \frac{\mathrm{d}\eta}{\mathrm{d}\tau} = 0 , \qquad \frac{\mathrm{d}\zeta}{\mathrm{d}\tau} = 0 .$$

The equations in (1.22) are called the Lorentz transformations. They can also be written

$$\begin{cases} \mathrm{d}x = \dfrac{1}{\sqrt{1-v^2/c^2}}(\mathrm{d}\xi - v\,\mathrm{d}\tau) , \\ \mathrm{d}y = \mathrm{d}\eta , \\ \mathrm{d}z = \mathrm{d}\zeta , \\ \mathrm{d}t = \dfrac{1}{\sqrt{1-v^2/c^2}}\left(-\dfrac{v}{c^2}\,\mathrm{d}\xi + \mathrm{d}\tau\right) . \end{cases}$$

If we make indirect measurements, we obtain the coordinate differences

$$\mathrm{d}\xi , \mathrm{d}\eta , \mathrm{d}\zeta ,$$

between points related by the simultaneity condition

$$\mathrm{d}\tau = 0 .$$

The true lengths

$$\mathrm{d}x, \mathrm{d}y, \mathrm{d}z ,$$

are therefore

$$\begin{cases} dx = \dfrac{1}{\sqrt{1 - v^2/c^2}} d\xi \,, \\ dy = d\eta \,, \\ dz = d\zeta \,. \end{cases}$$

The dimensions of the image differ from those of the object. The indirect measurement process does not immediately give the dimensions of the object. The dimensions normal to the velocity are conserved, but measurements carried out in the direction of this velocity must be corrected. The object appears to contract in the direction of motion. Its lengths parallel to the velocity appear to be reduced by a factor

$$\sqrt{1 - v^2/c^2} \,.$$

We can also make indirect measurements of time. Let us measure the time interval dt between two events occurring at the same point of a moving body, supposing it is impossible to measure this directly using a clock moving with the object. However, we suppose we do have clocks placed on an accessible body and adjusted by simultaneity. We note the difference $d\tau$ between the times indicated by the two clocks next to which the two events occur. The problem now is to deduce, from the indirectly measured time $d\tau$, the time dt that would have been measured by a clock moving with the moving body.

Since the events occur at the same point on the moving object, we must set

$$dx = 0$$

in the Lorentz transformations, whence

$$d\tau = \frac{1}{\sqrt{1 - v^2/c^2}} dt \,.$$

We can interpret this correction to indirect measurements by saying that phenomena occurring on the moving body appear to be retarded by a factor

$$\frac{1}{\sqrt{1 - v^2/c^2}} \,.$$

We have seen how absolute quantities like the duration of a phenomenon and the length of a body can be determined by indirect measurements relative to another body. We shall now consider a quantity that is fundamentally relative, namely, the velocity.

We have already mentioned the velocity of a point several times. It is characterised by the coordinate differentials

$$dx_1 \,, \quad dx_2 \,, \quad dx_3 \,, \quad dx_4 \,.$$

It depends in an essential way on the choice of coordinates. When we use measurement devices located on a body, the velocity of a moving object relative to this body will be understood to be the differentials

$$\mathrm{d}x\,,\quad \mathrm{d}y\,,\quad \mathrm{d}z\,,\quad \mathrm{d}t\,,$$

of the proper coordinates of the body at the point where the moving object is located. Then we shall say that

$$\frac{\mathrm{d}x}{\mathrm{d}t}\,,\quad \frac{\mathrm{d}y}{\mathrm{d}t}\,,\quad \frac{\mathrm{d}z}{\mathrm{d}t}\,,$$

are the components of the velocity and

$$\sqrt{\left(\frac{\mathrm{d}x}{\mathrm{d}t}\right)^2+\left(\frac{\mathrm{d}y}{\mathrm{d}t}\right)^2+\left(\frac{\mathrm{d}z}{\mathrm{d}t}\right)^2}$$

the magnitude of the velocity, or speed.

We sometimes consider what is known as the proper velocity, which has components

$$\frac{\mathrm{d}x}{\mathrm{d}s}\,,\quad \frac{\mathrm{d}y}{\mathrm{d}s}\,,\quad \frac{\mathrm{d}z}{\mathrm{d}s}\,.$$

The Lorentz transformation can then be used to calculate the components

$$\frac{\mathrm{d}\xi}{\mathrm{d}\tau}\,,\quad \frac{\mathrm{d}\eta}{\mathrm{d}\tau}\,,\quad \frac{\mathrm{d}\zeta}{\mathrm{d}\tau}\,,$$

of the velocity of a point relative to a body with proper coordinates ξ, η, ζ, τ, when we know the velocity of the same point relative to another body with proper coordinates x, y, z, t, and moving with speed v along the ξ axis.

We immediately obtain

$$\begin{cases} \dfrac{\mathrm{d}\xi}{\mathrm{d}\tau} = \dfrac{\dfrac{\mathrm{d}x}{\mathrm{d}t}+v}{1+\dfrac{v}{c^2}\dfrac{\mathrm{d}x}{\mathrm{d}t}}\,, \\[2ex] \dfrac{\mathrm{d}\eta}{\mathrm{d}\tau} = \dfrac{\sqrt{1-v^2/c^2}}{1+\dfrac{v}{c^2}\dfrac{\mathrm{d}x}{\mathrm{d}t}}\dfrac{\mathrm{d}y}{\mathrm{d}t}\,, \\[2ex] \dfrac{\mathrm{d}\zeta}{\mathrm{d}\tau} = \dfrac{\sqrt{1-v^2/c^2}}{1+\dfrac{v}{c^2}\dfrac{\mathrm{d}x}{\mathrm{d}t}}\dfrac{\mathrm{d}z}{\mathrm{d}t}\,. \end{cases}$$

To a first approximation (when we neglect terms involving $1/c^2$), these give the usual formulas for composition of velocities. These standard formulas remain applicable when the velocities being combined are small enough. If they are large enough, on the other hand, the difference between the new equations and the old ones may become accessible to experiment.

Some phenomena result from the difference between the components of the velocity of light relative to the direction of motion of the body with respect to which we measure it.

Let us suppose that the components of the velocity being measured are of the same order of magnitude as c, while the relative speed v of the compared systems is much smaller than c. The first of the previous equations can be written approximately

$$\frac{\mathrm{d}\xi}{\mathrm{d}\tau} = \frac{\mathrm{d}x}{\mathrm{d}t} + v\left[1 - \frac{1}{c^2}\left(\frac{\mathrm{d}x}{\mathrm{d}t}\right)^2\right] ,$$

or again

$$\frac{\mathrm{d}\xi}{\mathrm{d}\tau} - \frac{\mathrm{d}x}{\mathrm{d}t} = v\left(1 - \frac{1}{n^2}\right) ,$$

where

$$n = \frac{c}{\mathrm{d}x/\mathrm{d}t} .$$

This is the formula found by Fizeau and checked experimentally to great accuracy. Fizeau created interference between light rays from the same source propagating in water flowing at speed v, one ray in the direction of flow and the other in the opposite direction. The shift in the interference fringes when the direction of flow is reversed results from a velocity difference equal to

$$v\left(1 - \frac{1}{n^2}\right) ,$$

and not v, the value that would have been expected according to the classical velocity addition law. Here n is the the refractive index of the medium, namely the ratio of the speed in vacuum and the speed in the medium.

The new velocity composition law immediately accounts for Fizeau's experiment if we take c to be the speed of light in vacuum. Following a light ray, we will thus have

$$\mathrm{d}s = 0 .$$

It follows that the magnitude of the velocity of light is constant at a point, regardless of the direction of the ray and the motion of the system relative to which it is measured.

Michelson tried in vain to detect a difference in the speeds of light rays propagating in opposite directions. He thought he would thus detect the motion of the Earth relative to whatever medium transmits light. The negative result of his experiment played a major role in establishing and spreading Einstein's theories.

The fact that the velocity of light is independent of the direction of the light ray gives a convenient way to check the simultaneity of clocks. Two rays meeting each other in opposite directions can be considered symmetric with respect to one another.

Consider two light rays propagating in opposite directions along the straight line segment joining two points A and B. The first signal leaves A at the instant when a clock placed at A indicates the time t'_A, passes by intermediate clocks indicating the time t', and reaches B when the clock there indicates the time t'_B. The signal sent from B at time t''_B passes by intermediate clocks indicating the time t'', and reaches A at time t''_A. Then the indications

$$\frac{t'_A + t''_A}{2}\,, \qquad \frac{t' + t''}{2}\,, \qquad \frac{t'_B + t''_B}{2}\,,$$

on the corresponding clocks are simultaneous. This follows simply from the fact that the speed of light is the same in both directions. The speed can vary from one point to another; it only has to be the same at each point during the time separating the passage of the two signals at that point.

This procedure can be used along an arbitrary polygonal chain, arranging mirrors to reflect the light rays at points where it changes direction. We can then compare experimentally the way simultaneity is realised between two points connected by different paths.

For example, if the polygonal chain followed by the signals is closed, the clocks at A and B will be at the same point, and their simultaneity realised directly amounts to their equivalence, namely the fact that they indicate the same time for any event happening nearby. We could, if we wished, use just one clock.

When we define simultaneity by going round the polygon, we must take the indication

$$\frac{t'_A + t''_A}{2}$$

of the clock considered as being at A to be simultaneous with the indication

$$\frac{t'_B + t''_B}{2}$$

of the same clock considered as being at B. If the two signals leave at the same time, so that

$$t'_A = t''_B\,,$$

the difference between their times of return,

$$t'_B - t''_A\,,$$

will be equal to twice the difference Δt resulting from the definition of simultaneity around the polygon.

This experiment could be carried out by splitting a light beam into two and sending them in opposite directions around the same polygon to meet up again. The difference in arrival times could be measured by examining interference patterns.

This has indeed been done by Sagnac in a centrifugal field with equation [see (1.17)]

$$\mathrm{d}s^2 = -\mathrm{d}r^2 - r^2(\mathrm{d}\theta - \omega \mathrm{d}t)^2 - \mathrm{d}z^2 + c^2\mathrm{d}t^2\,,$$

expressed in polar coordinates, where r is the distance from the axis of rotation. The simultaneity equation for a stationary body, such that

$$\delta r = \delta\theta = \delta z = 0\,,$$

is

$$r^2(\mathrm{d}\theta - \omega \mathrm{d}t)\omega\delta t + c^2 \mathrm{d}t\delta t = 0\,,$$

or

$$\omega r^2 \mathrm{d}\theta + (c^2 - \omega^2 r^2)\mathrm{d}t = 0\,.$$

The change in time t around a closed contour of simultaneous points is

$$\Delta t = -\frac{\omega}{c^2}\int_0^{2\pi}\frac{r^2\mathrm{d}\theta}{1-\omega^2 r^2/c^2} \approx -\frac{2\omega S}{c^2}\,,$$

where

$$S = \int_0^{2\pi}\frac{r^2\mathrm{d}\theta}{2}$$

is the area of the projection of the polygon on the plane normal to the axis of rotation. The time difference for the two rays must therefore be equal to

$$2\Delta t = \frac{4\omega S}{c^2}\,.$$

This has been confirmed experimentally.

Chapter 2
Force Fields[1]

Newton's law of universal gravitation and the notion of action and reaction forces exerted by one body on another suggest that this force is a reality seated in each of the masses. There does not seem to be anything that corresponds to it outside the masses. The role of the surroundings is reduced to a minimum. However, it is important to note that the force varies as the inverse square of the distance separating the bodies, and thus depends on the amount of space separating them.

This idea of action at a distance certainly caused Newton some discomfort. As he put it, it looks exactly as though the bodies attract one another. But in the general enthusiasm quite justifiably inspired by the success of the theory, his successors forgot his initial reservations, and explanations for the various physical phenomena were sought in the form of action at a distance.

The role of the surrounding environment however soon returned to the scene. For Newton, light was a rapid emission of material particles. This idea, which later found its application in the theory of cathode rays, proved perfectly inadequate to account for the laws of optics. A transmitting medium, or ether, was needed. In line with the way of thinking at that time, it was endowed with mechanical properties. It had to be more rigid than steel but lighter than air, which was rather strange, but it fitted into the general scheme of solid bodies whose properties could be interpreted in terms of actions that were a function of the distance.

The role of the surrounding medium in the transmission of forces came out more clearly in electricity. The reciprocal action of one charge on another was expressed by a law that was completely analogous to gravity's universal law of attraction, namely Coulomb's universal law. Electrical charges of the same sign repel one another in direct proportion to the product of their charges and in inverse proportion to the square of the distance separating them. But the repulsion no longer depends only on the amount of space between the charges. It also depends on the nature of the medium separating them. The force changes when a glass plate is placed between them. Attention gradually turned away from conductors carrying electric charges, to focus more and more on phenomena arising in dielectrics.

[1] For the original text of this chapter in French see p. 155.

G. Lemaître, *Learning the Physics of Einstein with Georges Lemaître*,
https://doi.org/10.1007/978-3-030-22030-3_2

As the theory developed, it soon transpired that the fields induced by currents had to propagate at a finite speed, equal to the ratio of the electrical quantities measured in the two systems of electrostatic and electromagnetic units. Currents do not act on one another instantaneously. There is a delay between cause and effect. The action must therefore be transmitted through the medium separating the bodies.

The speed of this transmission, the ratio of the units in the two systems, is precisely equal to the speed of light. There is therefore a connection between these dissimilar phenomena. Maxwell showed that the propagation of an induction wave has all the properties of light. These waves were produced by Hertz, who thus provided conclusive confirmation of Maxwell's theory. We are all aware of the unexpected practical importance of these theoretical studies in the development of wireless telegraphy.

These results made it necessary to rethink the forces of action and reaction between bodies. They do not act directly on one another, but they sit in a medium which they modify and which acts upon them. We may study the state of the medium by observing the motion that it imposes on neutral or electrically charged particles, or on magnetised needles. We can thereby explore inertial and gravitational fields, just as we can probe electric and magnetic fields.

A second problem is the study of the reaction that matter exerts on a field. A complete theory must synthesise these two points of view. It must account for the production of fields by masses or charges, and also the exploration of these fields by observation of the motions of small test bodies.

In this chapter, we shall study the first problem: the exploration of fields using test bodies and the definition of the quantities characterising the state of the medium. In the following chapters, we shall examine the various ways to produce fields, e.g., by relative motion (the centrifugal force field), or by the presence of sources in the form of mass or electrical charge. We shall conclude with the general laws governing the two kinds of phenomenon, namely the action and reaction of fields on sources.

Fields can be immediately divided into two main classes: those that act in the same way on a particle whatever its nature, electric charge, etc., and those whose action depends on the electrical state of the test body. We shall begin by considering fields in the first class, namely inertial and gravitational fields. We have already said a few words about this in Section 1.2. We shall now pick up from there and develop the theory.

2.1 Inertial and Gravitational Fields[2]

We have seen that the spatiotemporal properties of a medium can be represented by the fundamental form, involving four variables,

$$\mathrm{d}s^2 = \sum_{\mu}\sum_{\nu} g_{\mu\nu}\mathrm{d}x_\mu \mathrm{d}x_\nu \,. \tag{2.1}$$

[2] For the original text of this section in French see p. 158.

Particular worldlines, namely geodesics, specify the free motion of material particles. We have seen that the form (2.1) can be chosen in such a way that the equations of free points differ only very slightly from those observed experimentally. We shall now calculate these equations when the potentials are arbitrary functions of the coordinates.

This calculation is of course the same as the one used to determine the geodesics on a surface, or the straight lines of the general geometry, when we know the distance element for a certain coordinate system. We must compare the value of

$$\int \mathrm{d}s$$

along the actual trajectory of a free point and along virtual trajectories with the same endpoints. To calculate the line integrals, we take a variable λ which will be a suitable function of the coordinates at each point.

The line integrals then become definite integrals with the same limits,

$$\int_{\lambda_1}^{\lambda_2} F \mathrm{d}\lambda \,, \tag{2.2}$$

where

$$F = \sqrt{\sum_\mu \sum_\nu g_{\mu\nu} \frac{\mathrm{d}x_\mu}{\mathrm{d}\lambda} \frac{\mathrm{d}x_\nu}{\mathrm{d}\lambda}} \,. \tag{2.3}$$

The derivatives, which we shall denote by

$$u^\sigma = \frac{\mathrm{d}x_\sigma}{\mathrm{d}\lambda} \qquad (\sigma = 1,2,3,4) \,, \tag{2.4}$$

must be calculated along the relevant trajectory.

Let

$$x_\sigma = x_\sigma(\lambda) \qquad (\sigma = 1,2,3,4)$$

be the equations of the actual trajectory. Then equations

$$x_\sigma = x_\sigma(\lambda) + \alpha \varpi_\sigma(\lambda) \qquad (\sigma = 1,2,3,4)$$

will be taken to represent the virtual trajectories, where the functions $\varpi_\sigma(\lambda)$ are continuous and vanish at λ_1 and λ_2.

When α tends to zero, the virtual trajectories become as close as we like to the actual trajectory. For this to be a geodesic, the difference between the value of the integral (2.2) along the actual trajectory and its value along the virtual trajectory must not depend on the sign of α when α tends to zero.

This difference can be expressed in the form

$$\int \Big[F(x_\sigma, u^\sigma) - F\big(x_\sigma + \alpha\varpi_\sigma, u^\sigma + \alpha \mathrm{d}\varpi_\sigma / \mathrm{d}\lambda\big) \Big] \mathrm{d}\lambda \,,$$

in which the main term when α tends to zero is

$$\alpha \int \sum_{\sigma} \left(\frac{\partial F}{\partial x_\sigma} \varpi_\sigma + \frac{\partial F}{\partial u^\sigma} \frac{\mathrm{d}\varpi_\sigma}{\mathrm{d}\lambda} \right) \mathrm{d}\lambda \, .$$

This term must vanish. Integrating by parts, we obtain

$$\int_{\lambda_1}^{\lambda_2} \frac{\partial F}{\partial u^\sigma} \frac{\mathrm{d}\varpi_\sigma}{\mathrm{d}\lambda} \mathrm{d}\lambda = \left[\frac{\partial F}{\partial u^\sigma} \varpi_\sigma \right]_{\lambda_1}^{\lambda_2} - \int_{\lambda_1}^{\lambda_2} \varpi_\sigma \frac{\mathrm{d}}{\mathrm{d}\lambda} \left(\frac{\partial F}{\partial u^\sigma} \right) \mathrm{d}\lambda \, .$$

And since ϖ_σ vanishes at the integration limits,

$$\int \sum_{\sigma} \left[\frac{\partial F}{\partial x_\sigma} - \frac{\mathrm{d}}{\mathrm{d}\lambda} \left(\frac{\partial F}{\partial u^\sigma} \right) \right] \varpi_\sigma \mathrm{d}\lambda = 0 \, .$$

This equation must be satisfied for any continuous and continuously differentiable functions ϖ_σ.

It follows that, at any point on the actual trajectory, we must have

$$\frac{\partial F}{\partial x_\sigma} - \frac{\mathrm{d}}{\mathrm{d}\lambda} \left(\frac{\partial F}{\partial u^\sigma} \right) = 0 \qquad (\sigma = 1,2,3,4) \, . \tag{2.5}$$

If the left-hand side of one of these equations were different from zero at a point, we could easily find functions ϖ_σ for which the integral did not vanish. Indeed, in the neighbourhood of this point, we could find an interval

$$\ell < \lambda < L$$

where one of the expressions between square brackets did not change sign. It is straightforward to find a function ϖ_σ that does not change sign in this interval,

$$\varpi_\sigma = (\ell - \lambda)^2 (L - \lambda)^2 \, .$$

The function ϖ_σ can then be set to zero outside this interval. It will thus be a continuous function with continuous first derivative. Setting the other functions ϖ_σ equal to zero, the integrand would have the same sign everywhere without vanishing everywhere, and the integral could not be zero.

In order to proceed with (2.5), it is convenient to choose the function λ in such a way that, on the actual trajectory, we have

$$\mathrm{d}\lambda = \mathrm{d}s \, .$$

After calculating the partial derivatives of

$$F = \sqrt{\sum_{\mu} \sum_{\nu} g_{\mu\nu} u^\mu u^\nu} \, ,$$

we can replace the quantity under the square root sign by its value 1 on the actual trajectory. We thus have

$$\frac{\partial F}{\partial x_\sigma} = \frac{1}{2}\sum_\mu\sum_\nu \frac{\partial g_{\mu\nu}}{\partial x_\sigma} u^\mu u^\nu$$

and

$$\frac{\partial F}{\partial u^\sigma} = \frac{1}{2}\left(\sum_\nu g_{\sigma\nu}u^\nu + \sum_\mu g_{\mu\sigma}u^\mu\right).$$

Equation (2.5) for the geodesics then becomes

$$\frac{1}{2}\sum_\mu\sum_\nu \frac{\partial g_{\mu\nu}}{\partial x_\sigma}\frac{\mathrm{d}x_\mu}{\mathrm{d}s}\frac{\mathrm{d}x_\nu}{\mathrm{d}s} - \frac{1}{2}\frac{\mathrm{d}}{\mathrm{d}s}\left(\sum_\nu g_{\sigma\nu}\frac{\mathrm{d}x_\nu}{\mathrm{d}s} + \sum_\mu g_{\mu\sigma}\frac{\mathrm{d}x_\mu}{\mathrm{d}s}\right) = 0.$$

We thus carry out the differentiation, noting that

$$\frac{\mathrm{d}}{\mathrm{d}s}g_{\sigma\nu} = \sum_\mu \frac{\partial g_{\sigma\nu}}{\partial x_\mu}\frac{\mathrm{d}x_\mu}{\mathrm{d}s}$$

and

$$\frac{\mathrm{d}}{\mathrm{d}s}g_{\mu\sigma} = \sum_\nu \frac{\partial g_{\mu\sigma}}{\partial x_\nu}\frac{\mathrm{d}x_\nu}{\mathrm{d}s}.$$

It follows that

$$\frac{1}{2}\sum_\mu\sum_\nu\left(\frac{\partial g_{\mu\nu}}{\partial x_\sigma} - \frac{\partial g_{\sigma\nu}}{\partial x_\mu} - \frac{\partial g_{\mu\sigma}}{\partial x_\nu}\right)\frac{\mathrm{d}x_\mu}{\mathrm{d}s}\frac{\mathrm{d}x_\nu}{\mathrm{d}s}$$

$$-\frac{1}{2}\left(\sum_\nu g_{\sigma\nu}\frac{\mathrm{d}^2x_\nu}{\mathrm{d}s^2} + \sum_\mu g_{\mu\sigma}\frac{\mathrm{d}^2x_\mu}{\mathrm{d}s^2}\right) = 0.$$

The two terms in the last bracket are equal due to the symmetry of the $g_{\mu\nu}$ in μ and ν [see (1.12)]:

$$g_{\mu\nu} = g_{\nu\mu}.$$

We introduce the following notation due to Christoffel:

(2.6) $$\begin{bmatrix}\mu\ \nu\\ \sigma\end{bmatrix} = \frac{1}{2}\left(\frac{\partial g_{\mu\sigma}}{\partial x_\nu} + \frac{\partial g_{\nu\sigma}}{\partial x_\mu} - \frac{\partial g_{\mu\nu}}{\partial x_\sigma}\right).$$

The geodesic equations now become

(2.7) $$\sum_\mu g_{\mu\sigma}\frac{\mathrm{d}^2x_\mu}{\mathrm{d}s^2} + \sum_\mu\sum_\nu \begin{bmatrix}\mu\ \nu\\ \sigma\end{bmatrix}\frac{\mathrm{d}x_\mu}{\mathrm{d}s}\frac{\mathrm{d}x_\nu}{\mathrm{d}s} = 0.$$

It is convenient to re-express these as equations for the second derivatives. This is done by introducing the quantities $g^{\sigma\tau}$ defined by the relations

$$\sum_\sigma g_{\mu\sigma} g^{\sigma\tau} = g_\mu^\tau \qquad (\mu, \tau = 1,2,3,4) , \tag{2.8}$$

where

$$g_\mu^\tau = \begin{cases} 1 & \text{if } \mu = \tau , \\ 0 & \text{if } \mu \neq \tau . \end{cases}$$

For each value of τ, these constitute a system of four linear equations in four unknowns $g^{1\tau}$, $g^{2\tau}$, $g^{3\tau}$, $g^{4\tau}$, from which they can be calculated. We multiply the first of the geodesic equations ($\sigma = 1$) by $g^{1\tau}$, the second by $g^{2\tau}$, and the third and fourth by $g^{3\tau}$ and $g^{4\tau}$, respectively, then sum all the resulting expressions. It follows that

$$\sum_\sigma \sum_\mu g_{\mu\sigma} g^{\sigma\tau} \frac{\mathrm{d}^2 x_\mu}{\mathrm{d}s^2} + \sum_\sigma \sum_\mu \sum_\nu g^{\sigma\tau} \begin{bmatrix} \mu \ \nu \\ \sigma \end{bmatrix} \frac{\mathrm{d}x_\mu}{\mathrm{d}s} \frac{\mathrm{d}x_\nu}{\mathrm{d}s} = 0 .$$

The first term can be written

$$\sum_\mu g_\mu^\tau \frac{\mathrm{d}^2 x_\mu}{\mathrm{d}s^2} = \frac{\mathrm{d}^2 x_\tau}{\mathrm{d}s^2} ,$$

while the second simplifies by introducing the Christoffel symbol of the second kind defined by

$$\begin{Bmatrix} \mu \ \nu \\ \tau \end{Bmatrix} = \sum_\sigma g^{\sigma\tau} \begin{bmatrix} \mu \ \nu \\ \sigma \end{bmatrix} . \tag{2.9}$$

Finally, the geodesic equations take the form

$$\frac{\mathrm{d}^2 x_\tau}{\mathrm{d}s^2} + \sum_\mu \sum_\nu \begin{Bmatrix} \mu \ \nu \\ \tau \end{Bmatrix} \frac{\mathrm{d}x_\mu}{\mathrm{d}s} \frac{\mathrm{d}x_\nu}{\mathrm{d}s} = 0 \qquad (\tau = 1,2,3,4) . \tag{2.10}$$

These are the equations of motion for the new mechanics.

Note 1. Associated Potentials

The quantities $g^{\sigma\tau}$ we have introduced into the calculation are said to be associated with the potentials $g_{\sigma\tau}$. It is straightforward to express them in terms of these potentials. The equations used to define them can be decomposed into four systems ($\tau = 1,2,3,4$) of four equations in four unknowns:

$$g_{\mu 1} g^{1\tau} + g_{\mu 2} g^{2\tau} + g_{\mu 3} g^{3\tau} + g_{\mu 4} g^{4\tau} = g_\mu^\tau \qquad (\mu = 1,2,3,4) .$$

The determinant of each of these systems is the determinant

$$g = |g_{\mu\nu}|$$

of the matrix of potentials.

As an example, let us calculate g^{23} in the system ($\tau = 3$):

$$g^{23} = \frac{1}{g}\begin{vmatrix} g_{11} & 0 & g_{31} & g_{41} \\ g_{12} & 0 & g_{32} & g_{42} \\ g_{13} & 1 & g_{33} & g_{43} \\ g_{14} & 0 & g_{34} & g_{44} \end{vmatrix},$$

whence g^{23} is the quotient by g of the minor of the potential g_{23} with the same indices in the determinant g. In general, $g^{\mu\nu}$ is the quotient by g of the minor of $g_{\mu\nu}$ in the determinant g.

This property can be used to calculate the derivative of the determinant of the potentials. The derivative of a determinant is obtained by differentiating each element, multiplying it by its minor, and taking the sum of all such terms. We thus have in this case

$$\frac{\partial g}{\partial x_\mu} = g\sum_\sigma\sum_\tau g^{\sigma\tau}\frac{\partial g_{\sigma\tau}}{\partial x_\mu}. \tag{2.11}$$

Note 2. Symmetry of Symbols

Looking at the calculation of the $g^{\mu\nu}$, it follows that the symmetry

$$g_{\mu\nu} = g_{\nu\mu}$$

of the potentials implies the symmetry of the associated potentials,

$$g^{\mu\nu} = g^{\nu\mu}.$$

The definition of the Christoffel symbols shows that they are symmetric in the upper indices,

$$\begin{bmatrix} \mu\,\nu \\ \sigma \end{bmatrix} = \begin{bmatrix} \nu\,\mu \\ \sigma \end{bmatrix}$$

and

$$\begin{Bmatrix} \mu\,\nu \\ \sigma \end{Bmatrix} = \begin{Bmatrix} \nu\,\mu \\ \sigma \end{Bmatrix}.$$

The symmetry of these symbols can often be used to simplify sums.

For example, in the expression

$$\sum_\tau \begin{Bmatrix} \mu\,\tau \\ \tau \end{Bmatrix} = \frac{1}{2}\sum_\sigma\sum_\tau g^{\sigma\tau}\left(\frac{\partial g_{\sigma\tau}}{\partial x_\mu} + \frac{\partial g_{\mu\sigma}}{\partial x_\tau} - \frac{\partial g_{\mu\tau}}{\partial x_\sigma}\right),$$

the last two terms cancel under exchange of the summation indices σ and τ. They can be dropped thanks to the symmetry of $g^{\sigma\tau}$. From (2.11), we will then have

$$\sum_{\tau} \left\{ \begin{matrix} \mu\ \tau \\ \tau \end{matrix} \right\} = \frac{1}{2g} \frac{\partial g}{\partial x_\mu} . \tag{2.12}$$

In most applications, $\mathrm{d}s^2$ has the form

$$\mathrm{d}s^2 = g_{11}\mathrm{d}x_1^2 + g_{22}\mathrm{d}x_2^2 + g_{33}\mathrm{d}x_3^2 + g_{44}\mathrm{d}x_4^2 ,$$

so that the determinant of the $g_{\mu\nu}$ reduces to its leading diagonal and all the $g^{\mu\nu}$ are zero except for

$$g^{\sigma\sigma} = \frac{1}{g_{\sigma\sigma}} \qquad (\sigma = 1, 2, 3, 4) .$$

The Christoffel symbols are all zero except for

$$\left[\begin{matrix} \sigma\ \sigma \\ \tau \end{matrix} \right] = -\frac{1}{2} \frac{\partial g_{\sigma\sigma}}{\partial x_\tau} \qquad (\sigma \neq \tau) ,$$

$$\left[\begin{matrix} \sigma\ \tau \\ \sigma \end{matrix} \right] = \frac{1}{2} \frac{\partial g_{\sigma\sigma}}{\partial x_\tau} = \left[\begin{matrix} \tau\ \sigma \\ \sigma \end{matrix} \right] ,$$

for the symbols of the first kind, and

$$\begin{aligned} \left\{ \begin{matrix} \sigma\ \sigma \\ \tau \end{matrix} \right\} &= -\frac{1}{2} \frac{1}{g_{\tau\tau}} \frac{\partial g_{\sigma\sigma}}{\partial x_\tau} \qquad (\sigma \neq \tau) , \\ \left\{ \begin{matrix} \sigma\ \tau \\ \sigma \end{matrix} \right\} &= \frac{1}{2} \frac{1}{g_{\sigma\sigma}} \frac{\partial g_{\sigma\sigma}}{\partial x_\tau} = \left\{ \begin{matrix} \tau\ \sigma \\ \sigma \end{matrix} \right\} , \end{aligned} \tag{2.13}$$

for those of the second kind.

In the special case considered in the first chapter, with

$$\mathrm{d}s^2 = -\mathrm{d}x_1^2 - \mathrm{d}x_2^2 - \mathrm{d}x_3^2 + g_{44}\mathrm{d}x_4^2 ,$$

where we then had $g_{44} = c^2 + 2V$, we will obtain

$$\begin{aligned} \left\{ \begin{matrix} 4\ 4 \\ i \end{matrix} \right\} &= \frac{1}{2} \frac{\partial g_{44}}{\partial x_i} \qquad (i = 1, 2, 3) , \\ \left\{ \begin{matrix} 4\ \sigma \\ 4 \end{matrix} \right\} &= \frac{1}{2} \frac{1}{g_{44}} \frac{\partial g_{44}}{\partial x_\sigma} = \left\{ \begin{matrix} \sigma\ 4 \\ 4 \end{matrix} \right\} \qquad (\sigma = 1, 2, 3, 4) , \end{aligned} \tag{2.14}$$

while all other symbols of the second kind then vanish.

We found the equations of motion by taking

$$\int \mathrm{d}s$$

as independent variable along the trajectory, namely the time indicated by a clock moving with the moving body. But it is often convenient to take as the main variable one of the coordinates x_4 which generally represents the indication of clocks positioned along the trajectory. For $i = 1,2,3,4$, we thus consider

$$\frac{\mathrm{d}^2 x_i}{\mathrm{d}s^2} = \frac{\mathrm{d}}{\mathrm{d}s}\left(\frac{\mathrm{d}x_i}{\mathrm{d}x_4}\frac{\mathrm{d}x_4}{\mathrm{d}s}\right) = \frac{\mathrm{d}^2 x_i}{\mathrm{d}x_4^2}\left(\frac{\mathrm{d}x_4}{\mathrm{d}s}\right)^2 + \frac{\mathrm{d}x_i}{\mathrm{d}x_4}\frac{\mathrm{d}^2 x_4}{\mathrm{d}s^2} ,$$

whence

$$\frac{\mathrm{d}^2 x_i}{\mathrm{d}x_4^2} = \frac{\mathrm{d}^2 x_i}{\mathrm{d}s^2}\left(\frac{\mathrm{d}s}{\mathrm{d}x_4}\right)^2 - \frac{\mathrm{d}^2 x_4}{\mathrm{d}s^2}\frac{\mathrm{d}x_i}{\mathrm{d}x_4}\left(\frac{\mathrm{d}s}{\mathrm{d}x_4}\right)^2 .$$

Substituting the expressions for $\mathrm{d}^2x_i/\mathrm{d}s^2$ and $\mathrm{d}^2x_4/\mathrm{d}s^2$ obtained from the equations of motion (2.10), it follows that

$$\text{(2.15)}\qquad \frac{\mathrm{d}^2 x_i}{\mathrm{d}x_4^2} + \sum_\mu\sum_\nu \left(\begin{Bmatrix}\mu\ \nu\\ i\end{Bmatrix} - \begin{Bmatrix}\mu\ \nu\\ 4\end{Bmatrix}\frac{\mathrm{d}x_i}{\mathrm{d}x_4}\right)\frac{\mathrm{d}x_\mu}{\mathrm{d}x_4}\frac{\mathrm{d}x_\nu}{\mathrm{d}x_4} = 0 \qquad (i = 1,2,3) .$$

In particular, in the case where

$$\mathrm{d}s^2 = -\mathrm{d}x^2 - \mathrm{d}y^2 - \mathrm{d}z^2 + (c^2 + 2V)\mathrm{d}t^2 ,$$

we will have

$$\text{(2.16)}\qquad \frac{\mathrm{d}^2 x}{\mathrm{d}t^2} = -\frac{\partial V}{\partial x} + \frac{1}{c^2 + 2V}\frac{\mathrm{d}x}{\mathrm{d}t}\left[2\left(\frac{\partial V}{\partial x}\frac{\mathrm{d}x}{\mathrm{d}t} + \frac{\partial V}{\partial y}\frac{\mathrm{d}y}{\mathrm{d}t} + \frac{\partial V}{\partial z}\frac{\mathrm{d}z}{\mathrm{d}t}\right) + \frac{\partial V}{\mathrm{d}t}\right] ,$$

and analogous equations in y and z, where we have used the expressions for the Christoffel symbols calculated above.

We have already seen that this form of $\mathrm{d}s^2$ leads to equations which coincide to a first approximation with the equations of classical mechanics when the speeds are small compared with c. When these speeds are not negligible, there is an extra term in the classical expression for the force and it is a function of the velocity.

When all the derivatives of the potentials vanish, the same is true for the Christoffel symbols, and the equations of motion reduce to

$$\frac{\mathrm{d}^2 x_\sigma}{\mathrm{d}s^2} = 0 ,$$

or

$$\frac{\mathrm{d}^2 x_i}{\mathrm{d}x_4^2} = 0 .$$

We shall refer to this as uniform straight line motion.

Conversely, for the motion at a point to be uniform and in a straight line whatever the velocity of the moving body, the derivatives of all the potentials must vanish at this point. Indeed, according to the equations of motion, all the Christoffel symbols of the second kind must vanish. Solving the equations for the Christoffel symbols which define them in terms of the derivatives of the potentials, we have

$$\frac{\partial g_{\mu\nu}}{\partial x_\tau} = \sum_\sigma \left(g_{\mu\sigma} \left\{ \begin{matrix} \nu\ \tau \\ \sigma \end{matrix} \right\} + g_{\nu\sigma} \left\{ \begin{matrix} \mu\ \tau \\ \sigma \end{matrix} \right\} \right) . \tag{2.17}$$

This relation is easily established as follows. According to the definition (2.9) of the Christoffel symbols and the definition (2.8) of the $g^{\mu\nu}$,

$$\sum_\sigma g_{\mu\sigma} \left\{ \begin{matrix} \nu\ \tau \\ \sigma \end{matrix} \right\} = \sum_\sigma \sum_\rho g_{\mu\sigma} g^{\sigma\rho} \begin{bmatrix} \nu\ \tau \\ \rho \end{bmatrix} = \sum_\rho g_\mu^\rho \begin{bmatrix} \nu\ \tau \\ \rho \end{bmatrix} = \begin{bmatrix} \nu\ \tau \\ \mu \end{bmatrix} , \tag{2.18}$$

and it remains to check that

$$\frac{\partial g_{\mu\nu}}{\partial x_\tau} = \begin{bmatrix} \nu\ \tau \\ \mu \end{bmatrix} + \begin{bmatrix} \mu\ \tau \\ \nu \end{bmatrix} . \tag{2.19}$$

But this follows immediately from the definition (2.6) of the Christoffel symbols of the first kind.

We may choose coordinates ξ_1, ξ_2, ξ_3, ξ_4 such that any free point particle passing through a point P has uniform straight line motion. The derivatives of the potentials will be zero at this point. Let

$$\xi_\tau = \xi_\tau(x_1, x_2, x_3, x_4) \qquad (\tau = 1, 2, 3, 4)$$

be the equations defining the coordinate system whose existence we would like to demonstrate, where x_1, x_2, x_3, x_4 are arbitrary coordinates. We expand these functions in Taylor series:

$$\begin{aligned} \xi_\tau = \xi_\tau\big|_0 &+ \sum_\sigma \frac{\partial \xi_\tau}{\partial x_\sigma}\bigg|_0 (x_\sigma - x_\sigma^0) \\ &+ \frac{1}{2} \sum_\mu \sum_\nu \frac{\partial^2 \xi_\tau}{\partial x_\mu \partial x_\nu}\bigg|_0 (x_\mu - x_\mu^0)(x_\nu - x_\nu^0) + \cdots , \end{aligned}$$

where the functions are evaluated at the point P with coordinates $(x_1^0, x_2^0, x_3^0, x_4^0)$. We must show that we can arrange for

$$\frac{\partial \xi_\tau}{\partial x_\sigma}\bigg|_0 \quad \text{and} \quad \frac{\partial^2 \xi_\tau}{\partial x_\mu \partial x_\nu}\bigg|_0$$

to be such that

$$\frac{\mathrm{d}^2 \xi_\tau}{\mathrm{d}s^2} = 0$$

for any free point particle passing through the point P.

Carrying out the differentiation, we find

$$\frac{\mathrm{d}^2\xi_\tau}{\mathrm{d}s^2} = \sum_\sigma \left.\frac{\partial \xi_\tau}{\partial x_\sigma}\right|_0 \frac{\mathrm{d}^2 x_\sigma}{\mathrm{d}s^2}$$
$$+\frac{1}{2}\sum_\mu\sum_\nu \left.\frac{\partial^2\xi_\tau}{\partial x_\mu \partial x_\nu}\right|_0 \left[(x_\mu - x_\mu^0)\frac{\mathrm{d}^2 x_\nu}{\mathrm{d}s^2} + 2\frac{\mathrm{d}x_\mu}{\mathrm{d}s}\frac{\mathrm{d}x_\nu}{\mathrm{d}s} + (x_\nu - x_\nu^0)\frac{\mathrm{d}^2 x_\mu}{\mathrm{d}s^2}\right] + \cdots .$$

At the point P, all terms not written in the above expansion vanish and, replacing $\mathrm{d}^2x_\sigma/\mathrm{d}s^2$ by its value according to the equation of motion (2.10) of a free point particle, we obtain

$$\frac{\mathrm{d}^2\xi_\tau}{\mathrm{d}s^2} = \sum_\mu\sum_\nu \left(\frac{\partial^2\xi_\tau}{\partial x_\mu \partial x_\nu} - \sum_\sigma \frac{\partial \xi_\tau}{\partial x_\sigma}\left\{ {\mu\ \nu \atop \sigma} \right\}\right)\frac{\mathrm{d}x_\mu}{\mathrm{d}s}\frac{\mathrm{d}x_\nu}{\mathrm{d}s} .$$

This expression must vanish for all $\mathrm{d}x_\mu/\mathrm{d}s$ and $\mathrm{d}x_\nu/\mathrm{d}s$.

We conclude that, at the point P,

(2.20) $$\frac{\partial^2\xi_\tau}{\partial x_\mu \partial x_\nu} = \sum_\sigma \frac{\partial \xi_\tau}{\partial x_\sigma}\left\{ {\mu\ \nu \atop \sigma} \right\} ,$$

for all values of μ, ν, and τ. We can thus attribute arbitrary values for the first derivatives of the ξ_τ and calculate the second derivatives using the relation above.

We have seen that, for each point of a body, there is a coordinate system that makes use of measurements performed at this point on the body. At the given point, the potentials are then given by the form

$$\mathrm{d}s^2 = -\mathrm{d}\xi_1^2 - \mathrm{d}\xi_2^2 - \mathrm{d}\xi_3^2 + c^2\mathrm{d}\xi_4^2 .$$

We referred to this system as the proper coordinates of the body at this point and we obtained them by a linear coordinate transformation, given the first derivatives $\partial\xi_\tau/\partial x_\sigma$ of the functions $\xi_\tau(x_\sigma)$ specifying the coordinate change (see (1.19) of Section 1.3). We now see that we can also choose the second derivatives of the functions $\xi_\tau(x_\sigma)$ according to (2.20), so that, in the new coordinate system ξ_τ, the first derivatives of the potentials vanish at P.

The motion of a free point will then be uniform and in a straight line at the point. These coordinates are called Galilean coordinates. Special relativity studies the situation where they can be used not only at a point P, but throughout a finite domain. We conclude from the above discussion that, at any given point, there is always a Galilean system and that it is always possible to find a sufficiently small neighbourhood of this point where the equations of special relativity are valid to within a given approximation.

2.2 Electric Fields[3]

The electromagnetic properties of a medium are determined by its action on charged particles, current-carrying conductors, and magnets. The action of a field on electrically charged particles divides into two distinct effects which can be attributed to the electric field and the magnetic field. Regarding these effects, the first imparts parallel accelerations independent of their velocities, while the second imparts accelerations perpendicular to their velocities and lying in a same plane.

The motion of a given particle can be represented by the following equations:

$$\left\{\begin{aligned} m\frac{\mathrm{d}^2x}{\mathrm{d}t^2} &= X\,, \\ m\frac{\mathrm{d}^2y}{\mathrm{d}t^2} &= Y\,, \\ m\frac{\mathrm{d}^2z}{\mathrm{d}t^2} &= Z\,, \end{aligned}\right. \tag{2.21}$$

where

$$\left\{\begin{aligned} X &= e(e_x + h_z v_y - h_y v_z)\,, \\ Y &= e(e_y + h_x v_z - h_z v_x)\,, \\ Z &= e(e_z + h_y v_x - h_x v_y)\,, \end{aligned}\right. \tag{2.22}$$

with m, e, v_x, v_y, v_z the material mass, electric charge, and velocity components of the particle in rectangular coordinates. The quantities e_x, e_y, and e_z are the components of a vector characterising the electric field. The corresponding force is parallel to this vector and independent of the direction and magnitude of the particle velocity. The quantities h_x, h_y, and h_z are the components of a vector representing the magnetic field. The corresponding force is perpendicular to this vector and also to the particle velocity, and it is proportional to the area of the parallelogram specified by these two vectors.

The wires conducting an electric current derive their properties from electrically charged particles, electrons, able to react to the actions of the medium.

Actions on the electrons normal to the cross-section of the wire tend to displace them along the wire and this constitutes the electric current. This motion is soon dissipated when the electrons collide with molecules in the wire, and is transformed into thermal agitation. The current can only be maintained if the electrons are subject to a continuous action. This is the electromotive force, which causes the current.

Actions on the electrons normal to the direction of the wire cannot produce an electric current. They are transmitted to the mass of the conductor and are manifested in the form of a mechanical stress exerted by the surrounding medium on the conductor.

---

[3] For the original text of this section in French see p. 168.

Hence, (2.22) can be applied to a conductor to give the mechanical action on it, provided we take e to be its electric charge and

$$ev_x\,,\; ev_y\,,\; ev_z\,,$$

the components of the current. The actions on the electrons in the direction of the wire maintain the current. They constitute the electromotive force of the current, proportional to e_x, e_y, e_z. This force is due to the difference in potential along the wire and to induction effects.

The law of induction is generally expressed as follows. The total electromotive force on a closed circuit is equal to the change in the flux of the magnetic field through a surface bounded by the circuit. This law is not very satisfactory. To begin with, it is clear that the total electromotive force can only be the sum of the electromotive forces at each point of the conductor; it would be useful to know these. But above all, the action on the electrons in the wire cannot depend on the changes in the magnetic field outside the wire. This law must be considered as a logical consequence of another law involving only quantities localised at each point of the wire.

We know that the flux of a vector field through a surface can be expressed as a line integral around the boundary of this surface. By Stokes' theorem,

$$\iint_S (h_x\,\mathrm{d}y\,\mathrm{d}z + h_y\,\mathrm{d}z\,\mathrm{d}x + h_z\,\mathrm{d}x\,\mathrm{d}y) = \oint_C (\varphi_x\,\mathrm{d}x + \varphi_y\,\mathrm{d}y + \varphi_z\,\mathrm{d}z)\;,$$

where

$$\text{(2.23)}\qquad \begin{cases} h_x = \dfrac{\partial \varphi_y}{\partial z} - \dfrac{\partial \varphi_z}{\partial y}\,, \\[2ex] h_y = \dfrac{\partial \varphi_z}{\partial x} - \dfrac{\partial \varphi_x}{\partial z}\,, \\[2ex] h_z = \dfrac{\partial \varphi_x}{\partial y} - \dfrac{\partial \varphi_y}{\partial x}\,. \end{cases}$$

Up to a sign, the vector φ_x, φ_y, φ_z is the vector potential. If we denote the scalar potential by φ, the electromotive force due to the potential difference and the induction can be written

$$\text{(2.24)}\qquad \begin{cases} e_x = \dfrac{\partial \varphi_x}{\partial t} - \dfrac{\partial \varphi}{\partial x}\,, \\[2ex] e_y = \dfrac{\partial \varphi_y}{\partial t} - \dfrac{\partial \varphi}{\partial y}\,, \\[2ex] e_z = \dfrac{\partial \varphi_z}{\partial t} - \dfrac{\partial \varphi}{\partial z}\,. \end{cases}$$

These equations express everything we can learn from the exploration of the fields by the motion of particles and conductors. The motion of magnetised needles can teach us nothing new, because the properties of magnets can be interpreted in terms of particle currents or closed electron trajectories.

The formulas obtained here are the same as the standard formulas of electromagnetism when we use a system of units such that the electrostatic units and electromagnetic units are the same. We know that, to achieve this, we only need to choose the unit of time, for example, such that the ratio of the units in the two systems is taken as unity. This ratio has the dimensions of speed and is equal to the speed of light in vacuum. It thus suffices to take this speed as the unit of speed.

If we wish to use C.G.S. units, we must introduce numerical factors and specify whether we are using EMU or ESU.

We have obtained the equations of electromagnetism for a particular coordinate system. We shall now look for equations with an algebraic form that does not depend on the choice of coordinates and which reduces to the equations just found when we use those particular coordinates.

These equations can be written in a condensed form by adopting the following notation. We denote the potentials by

$$\varphi_1, \varphi_2, \varphi_3, \varphi_4,$$

rather than

$$\varphi_x, \varphi_y, \varphi_z, \varphi,$$

and we set

$$\begin{aligned} F_{14} &= e_x, & F_{23} &= h_x, \\ F_{24} &= e_y, & F_{31} &= h_y, \\ F_{34} &= e_z, & F_{12} &= h_z. \end{aligned} \tag{2.25}$$

Equations (2.23) and (2.24) can then be written

$$F_{\alpha\beta} = \frac{\partial \varphi_\alpha}{\partial x_\beta} - \frac{\partial \varphi_\beta}{\partial x_\alpha} \qquad (\alpha, \beta = 1,2,3,4). \tag{2.26}$$

There are only 6 distinct equations because

$$F_{\alpha\beta} = -F_{\beta\alpha},$$

and $F_{\alpha\beta}$ is zero when the two indices are equal.

The relations expressed in (2.22) can be transformed in an analogous way. Replacing

$$ev_x, ev_y, ev_z, e,$$

by

$$\mathscr{J}^1, \mathscr{J}^2, \mathscr{J}^3, \mathscr{J}^4,$$

respectively, and denoting the components

$$X, Y, Z,$$

of the force by

$$-f_1\,,\ -f_2\,,\ -f_3\,,$$

they take the form[4]

$$f_\sigma + \sum_\sigma \mathscr{J}^\alpha F_{\alpha\sigma} = 0 \qquad (\sigma = 1,2,3)\,. \tag{2.27}$$

When the fields are explored by uncharged particles, the trajectories are characterised by the variational equations

$$\delta \int \mathrm{d}s = 0\,,$$

where $\mathrm{d}s^2$ is a quadratic form in the coordinate differentials. When the field is explored by a charged particle with mass m and charge e, this variation will no longer vanish. The simplest hypothesis one can make is that this variation will be equal to the variation of a linear form in these differentials,

$$-\varphi_1 \mathrm{d}x_1 - \varphi_2 \mathrm{d}x_2 - \varphi_3 \mathrm{d}x_3 - \varphi_4 \mathrm{d}x_4\,.$$

We will then have

$$\delta \int \left(m\,\mathrm{d}s + e \sum_\alpha \varphi_\alpha \mathrm{d}x_\alpha \right) = 0\,. \tag{2.28}$$

This law has the same algebraic form under an arbitrary coordinate transformation. It thus applies whatever coordinate system is used to reference the various events.

In (2.7), we calculated the contribution coming from the variation of the first term to be

$$f_\sigma = m \left\{ \sum_\mu g_{\mu\sigma} \frac{\mathrm{d}^2 x_\mu}{\mathrm{d}s^2} + \sum_\mu \sum_\nu \begin{bmatrix} \mu\ \nu \\ \sigma \end{bmatrix} \frac{\mathrm{d}x_\mu}{\mathrm{d}s} \frac{\mathrm{d}x_\nu}{\mathrm{d}s} \right\}\,. \tag{2.29}$$

This must be equal and of opposite sign to the one we obtain by applying the formula (2.5) from the calculus of variations to the function

$$F = \sum_\alpha \varphi_\alpha u^\alpha\,, \quad \text{where } u^\alpha = \frac{\mathrm{d}x_\alpha}{\mathrm{d}s}\,,$$

which yields

$$\frac{\partial F}{\partial x_\sigma} - \frac{\mathrm{d}}{\mathrm{d}s}\left(\frac{\partial F}{\partial u^\sigma} \right) = 0\,.$$

[4] Correction by G. Lemaître: "in formula (2.27), rather than $F_{\alpha\sigma}$, read $F_{\sigma\alpha}$. The proof which follows must be modified, by changing the sign in formula (2.28)."

Carrying out the differentiation,

$$\frac{\partial F}{\partial x_\sigma}=\sum_\alpha \frac{\partial \varphi_\alpha}{\partial x_\sigma}u^\alpha\,, \qquad \frac{\partial F}{\partial u^\sigma}=\varphi_\sigma\,,$$

$$\frac{\mathrm{d}}{\mathrm{d}s}\left(\frac{\partial F}{\partial u^\sigma}\right)=\sum_\alpha \frac{\partial \varphi_\sigma}{\partial x_\alpha}\frac{\mathrm{d}x_\alpha}{\mathrm{d}s}=\sum_\alpha \frac{\partial \varphi_\sigma}{\partial x_\alpha}u^\alpha\,.$$

It then follows that

$$f_\sigma+\sum_\alpha e u^\alpha\left(\frac{\partial \varphi_\alpha}{\partial x_\sigma}-\frac{\partial \varphi_\sigma}{\partial x_\alpha}\right)=0\,.$$

Setting

(2.30) $$\mathscr{J}^\alpha=e\frac{\mathrm{d}x_\alpha}{\mathrm{d}s}$$

and

(2.31) $$F_{\alpha\sigma}=\frac{\partial \varphi_\alpha}{\partial x_\sigma}-\frac{\partial \varphi_\sigma}{\partial x_\alpha}\,,$$

we obtain

(2.32) $$f_\sigma+\sum_\alpha \mathscr{J}^\alpha F_{\alpha\sigma}=0\,,$$

which is the same as (2.27).

Equations (2.29), (2.30), (2.31), and (2.32) express the action of the fields on electrically charged material particles in a way that agrees with the principle of relativity.

With the coordinates used at the beginning of this discussion, and multiplying each side of (2.32) ($\sigma=1,2,3$) by $\mathrm{d}s/\mathrm{d}t$, this equation reduces to (2.22), whereas (2.29) differs from (2.21), except as an approximation.

The equations in (2.21) must be replaced by

$$-m\frac{\mathrm{d}^2x}{\mathrm{d}s^2}+X\frac{\mathrm{d}s}{\mathrm{d}t}=0\,,$$

and analogous equations.

We may also write

(2.33)
$$m\frac{\mathrm{d}}{\mathrm{d}t}\frac{v_x}{\sqrt{1-v^2}}=X\,,$$
$$m\frac{\mathrm{d}}{\mathrm{d}t}\frac{v_y}{\sqrt{1-v^2}}=Y\,,$$
$$m\frac{\mathrm{d}}{\mathrm{d}t}\frac{v_z}{\sqrt{1-v^2}}=Z\,.$$

These are the dynamical equations of the theory of special relativity. They can be interpreted in the following way. Comparing them with the classical equations, we may say that we replace the constant m by a variable factor depending on the velocity. We then speak of variable masses. On the other hand, we can transform the equations in such a way that they assume the same form as (2.21) by replacing X, Y, and Z by a function of these quantities and the velocities. We then speak of variable forces depending on the velocity. The latter point of view looks more logical since X, Y, and Z are already functions of the velocity and we are then simply modifying the form of this function.

Variable Masses

Suppose to begin with that the force is perpendicular to the velocity, whence the magnitude of the velocity will not change along the trajectory. Then the equations in (2.33) become

$$\frac{m}{(1-v^2)^{1/2}}\frac{\mathrm{d}^2x}{\mathrm{d}t^2}=X\,,$$

with two analogous equations.

Suppose on the other hand that there is a force in the direction of the velocity and take the x axis in this common direction. Then

$$m\frac{\mathrm{d}}{\mathrm{d}t}\frac{v_x}{(1-v_x^2)^{1/2}}=\frac{m}{(1-v_x^2)^{3/2}}\frac{\mathrm{d}^2x}{\mathrm{d}t^2}=X\,.$$

We can thus keep the fundamental equations of mechanics if we replace the constant mass by a variable mass depending on the relative directions of the force and the velocity.

We will then have a transverse mass

$$\frac{m}{(1-v^2/c^2)^{1/2}}\,.$$

Here we have reinstated the factor c which must be included for arbitrary time units. The longitudinal mass will be

$$\frac{m}{(1-v^2/c^2)^{3/2}}\,.$$

Variable Forces

The equations in (2.33) can be interpreted just as simply without introducing variable masses. We have

$$\frac{\mathrm{d}}{\mathrm{d}t}\frac{v_x}{\sqrt{1-v^2}}=\frac{1}{(1-v^2)^{1/2}}\frac{\mathrm{d}v_x}{\mathrm{d}t}+\frac{v_x}{(1-v^2)^{3/2}}\frac{1}{2}\frac{\mathrm{d}v^2}{\mathrm{d}t}\,.$$

The first equation of (2.33) then becomes

$$X = \frac{m}{(1-v^2)^{1/2}}\frac{\mathrm{d}v_x}{\mathrm{d}t} + \frac{m}{(1-v^2)^{3/2}}\left(v_x\frac{\mathrm{d}v_x}{\mathrm{d}t} + v_y\frac{\mathrm{d}v_y}{\mathrm{d}t} + v_z\frac{\mathrm{d}v_z}{\mathrm{d}t}\right) .$$

Multiplying this equation and the analogous equations in y and z by v_x, v_y, and v_z, then adding term by term, it follows that

$$Xv_x + Yv_y + Zv_z = \left[\frac{m}{(1-v^2)^{1/2}} + \frac{mv^2}{(1-v^2)^{3/2}}\right]\left(v_x\frac{\mathrm{d}v_x}{\mathrm{d}t} + v_y\frac{\mathrm{d}v_y}{\mathrm{d}t} + v_z\frac{\mathrm{d}v_z}{\mathrm{d}t}\right) ,$$

or

$$Xv_x + Yv_y + Zv_z = \frac{m}{(1-v^2)^{3/2}}\left(v_x\frac{\mathrm{d}v_x}{\mathrm{d}t} + v_y\frac{\mathrm{d}v_y}{\mathrm{d}t} + v_z\frac{\mathrm{d}v_z}{\mathrm{d}t}\right) .$$

The equations of motion can then be written

$$m\frac{\mathrm{d}^2x}{\mathrm{d}t^2} = \left[X - v_x(Xv_x + Yv_y + Zv_z)\right]\sqrt{1-v^2} ,$$

or, in arbitrary units,

$$m\frac{\mathrm{d}^2x}{\mathrm{d}t^2} = \left[X - \frac{v_x}{c^2}(Xv_x + Yv_y + Zv_z)\right]\sqrt{1-\frac{v^2}{c^2}} .$$

We may thus claim that the fundamental equation of mechanics is conserved but that the expression for the force has changed. The notion of variable mass can be useful in certain applications, but we should not forget the arbitrary nature of these considerations. We can always multiply both sides of the equations of motion by an arbitrary variable factor and consider the product of this factor and the mass as a variable mass.

Chapter 3
Field Production by Relative Motion[1]

3.1 Uniform Linear Acceleration[2]

We have seen that a coordinate system can be chosen in such a way that, in the neighbourhood of a point, the geometry will be Euclidean and the motion of free points will be uniform and in straight lines. When these conditions are satisfied in a certain domain, the field can be represented in the form

$$ds^2 = -d\xi^2 - d\eta^2 - d\zeta^2 + c^2 d\tau^2 ,$$

and the geometry of stationary bodies will be Euclidean there. Then ξ, η, and ζ are Cartesian coordinates relative to a rectangular axis system and τ is the time indicated by stationary clocks, these being interchangeable and such that identical indications are simultaneous. And finally, the speed of light is constant at every point. We call such a field a Galilean field.

Consider a body moving in the Galilean field and suppose an observer is exploring the field with instruments moving with this body. He measures lengths and times, and observes the motions of free particles, or light rays. What will the field properties look like to this observer?

The solution in the case of classical mechanics is simple enough. The observer moving in this way refers his measurements to three rectangular axes in motion. Everything looks as though these axes are at rest, but relative to the moving system, free points in uniform straight line motion in the fixed system follow trajectories with accelerations determined by inertial forces: the rectilinear force, the centrifugal force, and the composite centrifugal force (or Coriolis force). The theory of relative motion calculates these forces when the motion of the moving system is given. If this motion is a uniformly accelerated rectilinear translation, bodies will be subject to a single rectilinear acceleration equal to the constant acceleration but in the opposite direction. Everything looks as though the system is stationary and there is a constant gravitational field. This acceleration will be the same for all moving bodies. Light rays will also be subject to this acceleration.

[1] For the original text of this chapter in French see p. 176.

[2] For the original text of this section in French see p. 176.

G. Lemaître, *Learning the Physics of Einstein with Georges Lemaître*,
https://doi.org/10.1007/978-3-030-22030-3_3

In the new mechanics, the theory of relative motion is less obvious. We know that a moving body appears to contract in the direction of motion. If a body is set in motion, it must appear to contract more and more as its speed increases. For this to happen, the various points in the body must catch up with one another. On the other hand, if the body is suddenly brought to a standstill, we could not stop all material points in it simultaneously. We must allow it to lose its apparent contraction. If we stop its endpoints at the same time, we will deform it. Measuring its length just after stopping it, we will no longer find the same value as before. We will observe a real contraction, precisely equal to the apparent contraction which distorted the indirect measurements of the fixed observer before it stopped.

A solid body in motion is therefore a body which appears to be contracted by a factor

$$\sqrt{1-\frac{v^2}{c^2}}$$

in the direction of the velocity. A body whose various material points moved in the fixed system according to the classical equations of motion of a solid would not appear to deform and would thus deform in reality. An observer who measured its dimensions directly would observe variations in them.

The equations of motion of a solid can only be conserved when the speed is constant at the same point (ξ,η,ζ), which happens for a uniform rectilinear translation, a uniform rotation, and any combination of such motions (spiral motion).

Apart from these cases, is solid motion even possible? In particular, is it possible to achieve the motion of a solid in such a way as to create a constant gravitational field artificially? Can we find the equivalent of uniformly accelerated relative motion in this new mechanics?

To do this, we must write down, in the fixed system ξ, η, ζ, τ, the equations of motion of a body whose various points follow parallel trajectories, e.g., along the ξ axis, and which undergoes a contraction proportional to

$$\sqrt{1-\frac{v^2}{c^2}}$$

in this direction. We plot axes x, y, z in this body which coincide at time

$$\tau = 0$$

with the axes ξ, η, ζ of the fixed system. The coordinates x, y, z for each point of the moving body measured by direct measurement relative to this system of axes are therefore constant. The equations of motion of an arbitrary point (x,y,z) are of the form

$$\begin{cases} \xi = f(\tau,x)\,, \\ \eta = y\,, \\ \zeta = z\,, \end{cases} \tag{3.1}$$

since measurements normal to the velocity do not undergo any contraction.

The velocity of the point is

$$v = \frac{\partial \xi}{\partial \tau} .$$

The apparent contraction is

$$\frac{\partial \xi}{\partial x} = \sqrt{1 - \frac{v^2}{c^2}} .$$

We must therefore have

$$\left(\frac{\partial \xi}{\partial x}\right)^2 + \frac{1}{c^2}\left(\frac{\partial \xi}{\partial \tau}\right)^2 = 1 . \tag{3.2}$$

This is a first order partial differential equation. We shall integrate it using the method of characteristics. We set

$$p = \frac{\partial \xi}{\partial x} , \qquad q = \frac{\partial \xi}{\partial \tau} .$$

Then we have

$$p^2 + \frac{1}{c^2} q^2 = 1 .$$

The equations for the characteristics are then

$$\frac{\mathrm{d}x}{p} = \frac{\mathrm{d}\tau}{q/c^2} = \frac{-\mathrm{d}p}{0} = \frac{-\mathrm{d}q}{0} = \frac{\mathrm{d}\xi}{p^2 + q^2/c^2} .$$

These can be integrated immediately and we conclude that p and q must be constants. The characteristics are the loci of points where the speed q is constant.

The integral is a locus of characteristics. It can be determined if we are given the motion of a point M_0 on the moving body:

$$\begin{cases} x_0 = 0 , \\ \xi_0 = \varphi(\tau_0) , \end{cases}$$

where φ is an arbitrary function of τ_0. We will then have

$$q_0 = \frac{\mathrm{d}\varphi}{\mathrm{d}\tau_0} , \qquad p_0 = \sqrt{1 - \frac{1}{c^2}\left(\frac{\mathrm{d}\varphi}{\mathrm{d}\tau_0}\right)^2} .$$

A characteristic passing through M_0 will have equations

$$\frac{x}{p_0} = \frac{\tau - \tau_0}{q_0/c^2} = \frac{\xi - \xi_0}{1} , \qquad p = p_0 , \qquad q = q_0 ,$$

or

$$\xi = \varphi(\tau_0) + \frac{x}{\sqrt{1 - \frac{1}{c^2}\left(\frac{\mathrm{d}\varphi}{\mathrm{d}\tau_0}\right)^2}}\,,$$

(3.3)

$$\tau = \tau_0 + \frac{1}{c^2}\frac{x}{\sqrt{1 - \frac{1}{c^2}\left(\frac{\mathrm{d}\varphi}{\mathrm{d}\tau_0}\right)^2}}\frac{\mathrm{d}\varphi}{\mathrm{d}\tau_0}\,.$$

We thus obtain a parametric representation of the integral. We would obtain the function $\xi = f(\tau, x)$ by eliminating τ_0 between these two equations.

Let us now see how the field properties will look to an observer using measuring instruments moving with the solid body. We know that these properties will depend on the invariant $\mathrm{d}s^2$, which is given in terms of the coordinates of the fixed system by

$$\mathrm{d}s^2 = -\mathrm{d}\xi^2 - \mathrm{d}\eta^2 - \mathrm{d}\zeta^2 + c^2\mathrm{d}\tau^2\,.$$

We express the differentials in terms of the coordinates of the moving system:

$$\mathrm{d}\xi = \frac{\mathrm{d}x}{\sqrt{1 - \frac{1}{c^2}\left(\frac{\mathrm{d}\varphi}{\mathrm{d}\tau_0}\right)^2}} + \left\{1 + \frac{1}{c^2}\frac{x}{\left[1 - \frac{1}{c^2}\left(\frac{\mathrm{d}\varphi}{\mathrm{d}\tau_0}\right)^2\right]^{3/2}}\frac{\mathrm{d}^2\varphi}{\mathrm{d}\tau_0^2}\right\}\frac{\mathrm{d}\varphi}{\mathrm{d}\tau_0}\mathrm{d}\tau_0\,,$$

$$\mathrm{d}\eta = \mathrm{d}y\,,$$

$$\mathrm{d}\zeta = \mathrm{d}z\,,$$

$$\mathrm{d}\tau = \frac{1}{c^2}\frac{\mathrm{d}\varphi}{\mathrm{d}\tau_0}\frac{\mathrm{d}x}{\sqrt{1 - \frac{1}{c^2}\left(\frac{\mathrm{d}\varphi}{\mathrm{d}\tau_0}\right)^2}} + \left\{1 + \frac{1}{c^2}\frac{x}{\left[1 - \frac{1}{c^2}\left(\frac{\mathrm{d}\varphi}{\mathrm{d}\tau_0}\right)^2\right]^{3/2}}\frac{\mathrm{d}^2\varphi}{\mathrm{d}\tau_0^2}\right\}\mathrm{d}\tau_0\,.$$

This implies that $\mathrm{d}s^2$ can be written

$$\begin{aligned}\mathrm{d}s^2 &= -\mathrm{d}x^2 - \mathrm{d}y^2 - \mathrm{d}z^2 \\ &\quad + c^2\left[1 - \frac{1}{c^2}\left(\frac{\mathrm{d}\varphi}{\mathrm{d}\tau_0}\right)^2\right]\left\{1 + \frac{1}{c^2}\frac{x}{\left[1 - \frac{1}{c^2}\left(\frac{\mathrm{d}\varphi}{\mathrm{d}\tau_0}\right)^2\right]^{3/2}}\frac{\mathrm{d}^2\varphi}{\mathrm{d}\tau_0^2}\right\}^2\mathrm{d}\tau_0^2\,.\end{aligned}$$

This expression completely specifies the properties of the field.

What form should φ take if the acceleration of a body left to its own devices should not depend on the time measured by a clock moving with the body? In (2.14) we obtained the values of the Christoffel symbols when g_{44} is the only potential that is not constant. It follows that the acceleration is

$$\frac{\mathrm{d}^2x}{\mathrm{d}s^2} = -\left\{\begin{matrix}4\,4\\ i\end{matrix}\right\} = -\frac{1}{2}\frac{\partial g_{44}}{\partial x} .$$

The time t indicated by a clock moving with the body at the given point satisfies

$$\mathrm{d}t = \sqrt{1-\frac{1}{c^2}\left(\frac{\mathrm{d}\varphi}{\mathrm{d}\tau_0}\right)^2}\,\mathrm{d}\tau_0 .$$

If instead of τ_0 we take the time t defined by

$$t = \int_0^{\tau_0}\sqrt{1-\frac{1}{c^2}\left(\frac{\mathrm{d}\varphi}{\mathrm{d}\tau_0}\right)^2}\,\mathrm{d}\tau_0 , \tag{3.4}$$

the invariant then becomes

$$\mathrm{d}s^2 = -\mathrm{d}x^2-\mathrm{d}y^2-\mathrm{d}z^2+c^2\left\{1+\frac{1}{c^2}\frac{x}{\left[1-\frac{1}{c^2}\left(\frac{\mathrm{d}\varphi}{\mathrm{d}\tau_0}\right)^2\right]^{3/2}}\frac{\mathrm{d}^2\varphi}{\mathrm{d}\tau_0^2}\right\}^2\mathrm{d}t^2 . \tag{3.5}$$

If the acceleration is to be independent of t at the point

$$x=0 ,$$

the coefficient of x must be equal to a constant:

$$\frac{\mathrm{d}^2\varphi/\mathrm{d}\tau_0^2}{\left[1-\frac{1}{c^2}\left(\frac{\mathrm{d}\varphi}{\mathrm{d}\tau_0}\right)^2\right]^{3/2}} = g . \tag{3.6}$$

The invariant then becomes

$$\mathrm{d}s^2 = -\mathrm{d}x^2-\mathrm{d}y^2-\mathrm{d}z^2+c^2\left(1+\frac{gx}{c^2}\right)^2\mathrm{d}t^2 . \tag{3.7}$$

The equation of motion of a free point is

$$\frac{1}{c^2}\frac{\mathrm{d}^2x}{\mathrm{d}s^2} = -g\left(1+\frac{gx}{c^2}\right) .$$

The observer moving with the moving body will have no way of detecting any variation in the field. He will attribute the acceleration of moving bodies to the presence of a constant gravitational field.

Making the change of variable

$$\left(1+\frac{gx}{c^2}\right)^2 = 1+\frac{2gX}{c^2}\,, \tag{3.8}$$

the invariant $\mathrm{d}s^2$ can be set in the form

$$\mathrm{d}s^2 = -\frac{\mathrm{d}X^2}{1+2gX/c^2} - \mathrm{d}y^2 - \mathrm{d}z^2 + c^2\left(1+\frac{2gX}{c^2}\right)\mathrm{d}t^2\,. \tag{3.9}$$

Let us see what becomes of the general equations of rectilinear motion of a solid in the special case we have just investigated. We shall thereby obtain the equations corresponding to the uniformly accelerated motion of the former mechanics. The function $\varphi(\tau_0)$ then satisfies the relation (3.6). This can be integrated immediately. Choosing the constant of integration such that $\mathrm{d}\varphi/\mathrm{d}\tau_0$ vanishes with τ_0, we obtain

$$\frac{\mathrm{d}\varphi/\mathrm{d}\tau_0}{\left[1-\frac{1}{c^2}\left(\frac{\mathrm{d}\varphi}{\mathrm{d}\tau_0}\right)^2\right]^{1/2}} = g\tau_0\,.$$

Solving this for $\mathrm{d}\varphi/\mathrm{d}\tau_0$, we find

$$\frac{\mathrm{d}\varphi}{\mathrm{d}\tau_0} = \frac{g\tau_0}{\sqrt{1+\left(\frac{g\tau_0}{c}\right)^2}}\,.$$

Now integrating,

$$\varphi = \int_0^{\tau_0} \frac{g\tau_0\,\mathrm{d}\tau_0}{\sqrt{1+\left(\frac{g\tau_0}{c}\right)^2}} = \frac{c^2}{g}\left[\sqrt{1+\left(\frac{g\tau_0}{c}\right)^2}-1\right]\,.$$

Substituting the expressions for φ and $\mathrm{d}\varphi/\mathrm{d}\tau_0$ into (3.3), we obtain the equations of motion in the parametric form

$$\begin{cases} \xi = \sqrt{1+\left(\frac{g\tau_0}{c}\right)^2}\left(\frac{c^2}{g}+x\right)-\frac{c^2}{g}\,, \\ \tau = \tau_0\left(1+\frac{gx}{c^2}\right)\,. \end{cases} \tag{3.10}$$

It is now straightforward to eliminate τ_0 from these two equations, which yields

$$\xi = \frac{c^2}{g}\left[-1+\sqrt{\left(1+\frac{gx}{c^2}\right)^2+\left(\frac{g\tau}{c}\right)^2}\right]. \tag{3.11}$$

Let us now find the equations for changing from the coordinates x and t of the moving system to the coordinates ξ and τ of the fixed system. The time t is defined by (3.4), which can be written

$$t = \int_0^{\tau_0} \frac{\mathrm{d}\tau_0}{\sqrt{1+\left(\frac{g\tau_0}{c}\right)^2}} = \frac{c}{g}\,\mathrm{arcsinh}\,\frac{g\tau_0}{c},$$

where we have used the expression for $\mathrm{d}\varphi/\mathrm{d}\tau_0$. We deduce that

$$\tau_0 = \frac{c}{g}\sinh\frac{gt}{c}.$$

Expressing τ_0 in terms of t, the equations specifying the transformation become

$$\begin{cases} \xi = \dfrac{c^2}{g}\left[-1+\left(1+\dfrac{gx}{c^2}\right)\cosh\dfrac{gt}{c}\right], \\ \tau = \dfrac{c}{g}\left(1+\dfrac{gx}{c^2}\right)\sinh\dfrac{gt}{c}. \end{cases} \tag{3.12}$$

Solving these for x and t, we find

$$\begin{cases} x = \dfrac{c^2}{g}\left[-1+\sqrt{\left(1+\dfrac{g\xi}{c^2}\right)^2-\left(\dfrac{g\tau}{c}\right)^2}\right], \\ t = \dfrac{1}{2}\dfrac{c}{g}\ln\dfrac{1+\dfrac{g\xi}{c^2}+\dfrac{g\tau}{c}}{1+\dfrac{g\xi}{c^2}-\dfrac{g\tau}{c}}. \end{cases}$$

If we used the coordinate X defined in (3.8) instead of x, we would immediately obtain analogous equations to the previous ones.

3.2 Uniform Rotation[3]

The Galilean field can be represented in semi-polar coordinates r, θ, z such that

$$x = r\cos\theta\,, \qquad y = r\sin\theta\,,$$

with the result

$$\mathrm{d}s^2 = -\mathrm{d}r^2 - r^2\mathrm{d}\theta^2 - \mathrm{d}z^2 + c^2\mathrm{d}t^2\,.$$

If we consider a system rotating about the z axis of the Galilean system with a constant angular speed ω, then in the new system $\mathrm{d}s^2$ will become

$$\mathrm{d}s^2 = -\mathrm{d}r^2 - r^2(\mathrm{d}\theta - \omega\mathrm{d}t)^2 - \mathrm{d}z^2 + c^2\mathrm{d}t^2\,, \tag{3.13}$$

where θ is now measured relative to these axes following this motion. We shall investigate the geometric properties of bodies that are stationary in this system. We set

$$\delta r = \delta\theta = \delta z = 0\,,$$

in the simultaneity equation (1.16). We obtain

$$-r^2(\mathrm{d}\theta - \omega\mathrm{d}t)(-\omega\delta t) + c\mathrm{d}t\delta t = 0\,,$$

or

$$\omega r^2(\mathrm{d}\theta - \omega\mathrm{d}t) + c^2\mathrm{d}t = 0\,. \tag{3.14}$$

We have already discussed the main features of simultaneity around a closed contour in this case, and we described how Sagnac's experiment demonstrated these.

We now eliminate $\mathrm{d}t$ from the fundamental equation (3.13) and the simultaneity equation, thereby obtaining the distance element. Changing the signs, we find

$$\mathrm{d}\sigma^2 = \mathrm{d}r^2 + \frac{r^2\mathrm{d}\theta^2}{1 - \omega^2 r^2/c^2} + \mathrm{d}z^2\,. \tag{3.15}$$

We see that the length of a radius, with

$$\mathrm{d}\theta = 0 = \mathrm{d}z\,,$$

will be given by

$$\int_0^r \mathrm{d}r = r\,,$$

while the length of the circumference, with

$$\mathrm{d}r = 0 = \mathrm{d}z\,,$$

[3] For the original text of this section in French see p. 183.

will be

$$\int_0^{2\pi} \frac{r\mathrm{d}\theta}{\sqrt{1-\omega^2 r^2/c^2}} = \frac{2\pi r}{\sqrt{1-\omega^2 r^2/c^2}} .$$

The ratio of the circumference to the diameter will thus be greater than π.

We would obtain this result immediately by applying the Lorentz correction to the indirect measurements read on stationary instruments in the Galilean field.

An observer photographing a disk rotating about the axis from a point on the axis would obtain an image of the disk on which the ratio of the circumference to the diameter would be equal to π, since the geometry of stationary bodies in the Galilean field is Euclidean.

The Lorentz correction is zero for the diameter because its motion is everywhere normal to its direction. Only the circumference will be greater than it appears, in the ratio

$$\frac{1}{\sqrt{1-v^2/c^2}} .$$

Observers carrying out highly accurate geometric measurements with identical measuring sticks will find that the geometry on the disk is not Euclidean.

Naturally, they could always reject this conclusion if they prefer to say that their rulers deform due to the presence of the centrifugal force. They could make various hypotheses about these deformations. The simplest is to assume that the metre gets shorter in the ratio

$$\sqrt{1-\frac{\omega^2}{c^2}r^2} ,$$

when it lies in a direction normal to the radius in order to measure the circumference. We make the change of variable

$$\frac{r^2}{1-\frac{\omega^2}{c^2}r^2} = \rho^2 , \tag{3.16}$$

whence

$$\mathrm{d}r = \frac{\mathrm{d}\rho}{\left(1+\frac{\omega^2}{c^2}\rho^2\right)^{3/2}} ,$$

where ρ is thus the length of the circumference divided by 2π. Then the distance element takes the form

$$\mathrm{d}\sigma^2 = \frac{\mathrm{d}\rho^2}{\left(1+\frac{\omega^2}{c^2}\rho^2\right)^{3}} + \rho^2\mathrm{d}\theta^2 + \mathrm{d}z^2 . \tag{3.17}$$

We may therefore say that the geometry remains Euclidean but that rulers lying along the direction of the centrifugal force are longer by a factor of

$$\left(1+\frac{\omega^2}{c^2}\rho^2\right)^{3/2} = \frac{1}{\left(1-\frac{\omega^2}{c^2}r^2\right)^{3/2}} .$$

We see that the geometry is Euclidean on a circular cylinder whose axis is the axis of rotation.

Finally, we may ask what dilation or contraction, independent of the direction, must be attributed to rulers in order to be able to say that the geometry on a rotating disk is planar Euclidean geometry. To achieve this, the distance element on the disk must be written in the form

$$\mathrm{d}r^2 + \frac{r^2\mathrm{d}\theta^2}{1-\frac{\omega^2}{c^2}r^2} = \frac{\mathrm{d}y^2 + y^2\mathrm{d}\theta^2}{x^2} ,$$

where x is the supposed dilation of the rulers and y the length of the radius corrected for this dilation.

We must integrate the two equations

$$\mathrm{d}r = \frac{\mathrm{d}y}{x}$$

and

$$\frac{r}{\sqrt{1-\frac{\omega^2}{c^2}r^2}} = \frac{y}{x} .$$

Dividing the first by the second to obtain

$$\frac{\mathrm{d}r}{r}\sqrt{1-\frac{\omega^2}{c^2}r^2} = \frac{\mathrm{d}y}{y} ,$$

then integrating, we arrive at

$$y = \frac{2r}{1+\sqrt{1-\frac{\omega^2}{c^2}r^2}} \exp\left(\sqrt{1-\frac{\omega^2}{c^2}r^2} - 1\right)$$

and

$$x = \frac{2\sqrt{1-\frac{\omega^2}{c^2}r^2}}{1+\sqrt{1-\frac{\omega^2}{c^2}r^2}} \exp\left(\sqrt{1-\frac{\omega^2}{c^2}r^2} - 1\right) .$$

Note that x tends to zero when r tends to c/ω. When $\omega r/c$ is small enough, we have

$$x = 1 - \frac{3}{4}\frac{\omega^2}{c^2}r^2 - \frac{19}{132}\frac{\omega^6}{c^6}r^6 + \cdots .$$

Let us ask what kind of shape a surface of revolution must have for the geometry to be Euclidean when it is rotating with angular speed ω about its axis. According to (3.17), the profile of the surface must satisfy

$$\frac{\mathrm{d}\rho^2}{\left(1+\frac{\omega^2}{c^2}\rho^2\right)^3}+\mathrm{d}z^2=\mathrm{d}\rho^2\,.$$

On the surface, we will then have

$$\mathrm{d}\sigma^2=\mathrm{d}\rho^2+\rho^2\mathrm{d}\theta^2\,,$$

which shows that the geometry there is Euclidean.

If we use the variable r related to ρ by (3.16),

$$\mathrm{d}r^2+\mathrm{d}z^2=\frac{\mathrm{d}r^2}{\left(1-\frac{\omega^2}{c^2}r^2\right)^3}\,,$$

the profile of the generatrix of the required surface of revolution will be given by

$$z=\int\sqrt{\frac{1}{\left(1-\frac{\omega^2}{c^2}r^2\right)^3}-1}\,\mathrm{d}r\,.$$

We note that z tends to infinity when r tends to c/ω.

For small values of r, we may expand this in the series

$$z=\frac{\sqrt{3}}{2}\frac{\omega}{c}r^2\left(1+\frac{1}{2}\frac{\omega^2}{c^2}r^2+\cdots\right)\,.$$

This surface therefore osculates a paraboloid at the origin and is asymptotic to the cylinder $r=c/\omega$. On a surface rotating with angular speed ω, the geometry is the same as on a Euclidean plane. Suppose we stopped the rotation. As the motion comes to a standstill, an observer surveying the surface would note the same deformations of the surface as those that would be noted at rest by observing the deformations of a plane surface being mapped onto the surface of revolution.

The motion of moving bodies relative to the rotating system would not have any remarkable features. The standard theory of relative motion is applicable since the equations for the transformation between the coordinates of the two systems can be maintained.

Note finally that, if the angular velocity were changed, the moving body would necessarily have to deform, since the ratio

$$\frac{\pi}{\sqrt{1-\frac{\omega^2}{c^2}r^2}}$$

of the circumference to the diameter would have to change. The circumference or the diameter would therefore have to change length, or both would have to deform.

3.3 General Equation for Inertial Fields[4]

When the potentials are constant, the motion of a free point is uniform and in a straight line. We are then dealing with a Galilean field. If we make a change of coordinates, the potentials will no longer be constant. The second derivatives of the coordinates will no longer necessarily vanish along the trajectory of a free point and we will obtain what is known as an inertial field.

The inertial field is therefore characterised by the fact that, by a change of coordinates

$$\xi_\sigma = \xi_\sigma(x_1, x_2, x_3, x_4) \qquad (\sigma = 1,2,3,4)\,, \tag{3.18}$$

the fundamental form ds^2 can be transformed into a quadratic form with constant coefficients,

$$ds^2 = \sum_\sigma \sum_\tau \gamma_{\sigma\tau} d\xi_\sigma d\xi_\tau\,. \tag{3.19}$$

We have seen that, in this case, the derivatives of the transformation equations will satisfy the relations (2.20),

$$\frac{\partial^2 \xi_\tau}{\partial x_\alpha \partial x_\beta} = \sum_\sigma \frac{\partial \xi_\tau}{\partial x_\sigma} \left\{ \begin{matrix} \alpha\,\beta \\ \sigma \end{matrix} \right\} \qquad (\alpha, \beta, \tau = 1,2,3,4)\,, \tag{3.20}$$

where the Christoffel symbol of the second kind is found from the potentials $g_{\mu\nu}$ of the coordinate system x_1, x_2, x_3, x_4.

We shall now eliminate the functions ξ_σ from these equations. Taking the derivative of (3.20) with respect to x_γ, we find

$$\frac{\partial^3 \xi_\tau}{\partial x_\alpha \partial x_\beta \partial x_\gamma} = \sum_\sigma \frac{\partial^2 \xi_\tau}{\partial x_\sigma \partial x_\gamma} \left\{ \begin{matrix} \alpha\,\beta \\ \sigma \end{matrix} \right\} + \sum_\sigma \frac{\partial \xi_\tau}{\partial x_\sigma} \frac{\partial}{\partial x_\gamma} \left\{ \begin{matrix} \alpha\,\beta \\ \sigma \end{matrix} \right\}.$$

[4] For the original text of this section in French see p. 188.

Transforming the second derivative of the right-hand side by applying (3.20) and replacing the dummy index σ by ρ, we obtain

$$\frac{\partial^2 \xi_\tau}{\partial x_\rho \partial x_\gamma} = \sum_\sigma \frac{\partial \xi_\tau}{\partial x_\sigma} \begin{Bmatrix} \rho\,\gamma \\ \sigma \end{Bmatrix} .$$

It follows that

$$\frac{\partial^3 \xi_\tau}{\partial x_\alpha \partial x_\beta \partial x_\gamma} = \sum_\sigma \frac{\partial \xi_\tau}{\partial x_\sigma} \left[\sum_\rho \begin{Bmatrix} \rho\,\gamma \\ \sigma \end{Bmatrix} \begin{Bmatrix} \alpha\,\beta \\ \rho \end{Bmatrix} + \frac{\partial}{\partial x_\gamma} \begin{Bmatrix} \alpha\,\beta \\ \sigma \end{Bmatrix} \right] .$$

This equation must be satisfied for all values

$$\alpha, \beta, \gamma, \tau \in \{1, 2, 3, 4\} .$$

It must therefore still be satisfied if we transpose the indices α and γ. The left-hand side remains unchanged. Subtracting one expression from the other, we obtain

$$\sum_\sigma \frac{\partial \xi_\tau}{\partial x_\sigma} R^\sigma_{\alpha\beta\gamma} = 0 , \tag{3.21}$$

where we have set[5]

$$R^\sigma_{\alpha\beta\gamma} = -\frac{\partial}{\partial x_\alpha} \begin{Bmatrix} \beta\,\gamma \\ \sigma \end{Bmatrix} + \frac{\partial}{\partial x_\gamma} \begin{Bmatrix} \alpha\,\beta \\ \sigma \end{Bmatrix}$$

$$-\sum_\rho \begin{Bmatrix} \rho\,\alpha \\ \sigma \end{Bmatrix} \begin{Bmatrix} \beta\,\gamma \\ \rho \end{Bmatrix} + \sum_\rho \begin{Bmatrix} \rho\,\gamma \\ \sigma \end{Bmatrix} \begin{Bmatrix} \alpha\,\beta \\ \rho \end{Bmatrix} . \tag{3.22}$$

Equation (3.21) for $\tau = 1, 2, 3, 4$ can be considered as a homogeneous system of four linear equations in four unknowns

$$R^\sigma_{\alpha\beta\gamma} \qquad (\sigma = 1, 2, 3, 4) .$$

The determinant of this system is the functional determinant of the transformation equations (3.18). We shall of course assume that it differs from zero. The equations in (3.21) are therefore equivalent to

$$R^\sigma_{\alpha\beta\gamma} = 0 \qquad (\alpha, \beta, \gamma, \sigma = 1, 2, 3, 4) . \tag{3.23}$$

These are the general equations for inertial fields. Geometrically speaking, these are the equations of a Euclidean space. In two variables, they express the condition for a surface to be mappable onto a Euclidean plane.

---

[5] Correction by G. Lemaître: "The Riemann tensor is defined by an expression with a sign opposite to the usual definition. The present definition is maintained throughout the dissertation, except in Sects. 4.3 and 4.4. In the latter case there results an error in the calculation of ρ! When correcting, we obtain (see Eq.(4.56)) $\rho' = (1 + \frac{2V}{c^2})\rho + \frac{4}{\kappa c^2} g^2$."

When the fundamental form can be reduced by a change of coordinates to a form with constant coefficients, the expressions in (3.22) no longer vanish. They have important properties on which Einstein based his study of gravitational fields in the most general case. To establish these properties, we shall use a special coordinate system for which the calculations are greatly simplified.

We have seen that, at any given point M, we may always choose the value of the second derivatives of the transformation functions (3.18) in such a way that, at this point, the derivatives of the potentials vanish in the system ξ_σ. To achieve this, we simply give them the value specified in (3.20) at that point M. We may also arrange for the first derivatives

$$\frac{\partial \xi_\tau}{\partial x_\sigma}$$

to be such that the potentials $\gamma_{\sigma\tau}$ reduce at M to the values

$$\gamma_{\sigma\tau} = g_\sigma^\tau = \begin{cases} 1 & \text{if } \sigma = \tau \,, \\ 0 & \text{if } \sigma \neq \tau \,. \end{cases}$$

Some coordinates will be imaginary. This presents no difficulty for the calculations we shall carry out here and it is a straightforward matter to go back to real variables.

By substituting the values of the differentials obtained from (3.18) into the expression (3.19) for $\mathrm{d}s^2$, we should obtain

$$\mathrm{d}s^2 = \sum_\alpha \sum_\beta g_{\alpha\beta} \mathrm{d}x_\alpha \mathrm{d}x_\beta \,,$$

and it follows that

(3.24)
$$g_{\alpha\beta} = \sum_\sigma \sum_\tau \frac{\partial \xi_\sigma}{\partial x_\alpha} \frac{\partial \xi_\tau}{\partial x_\beta} \gamma_{\sigma\tau} \,.$$

The Christoffel symbols which feature in (3.22) are defined by

(3.25)
$$\left\{ \begin{matrix} \alpha\ \beta \\ \sigma \end{matrix} \right\} = \frac{1}{2} \sum_\tau g^{\sigma\tau} \left(\frac{\partial g_{\alpha\tau}}{\partial x_\beta} + \frac{\partial g_{\beta\tau}}{\partial x_\alpha} - \frac{\partial g_{\alpha\beta}}{\partial x_\tau} \right) .$$

Apart from the potentials $g_{\sigma\tau}$, these expressions contain the quantities $g^{\sigma\tau}$ associated with these potentials. These were defined by the equations

(3.26)
$$\sum_\rho g_{\alpha\rho} g^{\beta\rho} = g_\alpha^\beta \,.$$

If we also set

(3.27)
$$\sum_\rho \gamma_{\alpha\rho} \gamma^{\beta\rho} = g_\alpha^\beta \,,$$

then the $g^{\alpha\beta}$ will be given as a function of the $\gamma^{\sigma\tau}$ by the equation

(3.28)
$$g^{\alpha\beta} = \sum_\sigma \sum_\tau \frac{\partial x_\alpha}{\partial \xi_\sigma} \frac{\partial x_\beta}{\partial \xi_\tau} \gamma^{\sigma\tau} .$$

We now just have to show that this expression satisfies the definition (3.26). Indeed, using (3.24) and (3.28), we find

$$\sum_\rho g_{\alpha\rho} g^{\beta\rho} = \sum_\rho \sum_\sigma \sum_\tau \sum_{\sigma_1} \sum_{\tau_1} \frac{\partial \xi_\sigma}{\partial x_\alpha} \frac{\partial \xi_\tau}{\partial x_\rho} \frac{\partial x_\rho}{\partial \xi_{\tau_1}} \frac{\partial x_\beta}{\partial \xi_{\sigma_1}} \gamma_{\sigma\tau} \gamma^{\sigma_1\tau_1} .$$

By carrying out the sum over ρ, we obtain

$$\sum_\rho \frac{\partial \xi_\tau}{\partial x_\rho} \frac{\partial x_\rho}{\partial \xi_{\tau_1}} = \frac{\mathrm{d}\xi_\tau}{\mathrm{d}\xi_{\tau_1}} = g_{\tau_1}^{\tau} .$$

Then carrying out the sums over τ and τ_1 and using (3.27), we find

$$\sum_\tau \sum_{\tau_1} g_{\tau_1}^{\tau} \gamma_{\sigma\tau} \gamma^{\sigma_1\tau_1} = \sum_\tau \gamma_{\sigma\tau} \gamma^{\sigma_1\tau} = g_\sigma^{\sigma_1} .$$

This leaves

$$\sum_\sigma \sum_{\sigma_1} \frac{\partial \xi_\sigma}{\partial x_\alpha} \frac{\partial x_\beta}{\partial \xi_{\sigma_1}} g_\sigma^{\sigma_1} = \sum_\sigma \frac{\partial \xi_\sigma}{\partial x_\alpha} \frac{\partial x_\beta}{\partial \xi_\sigma} = \frac{\mathrm{d}x_\beta}{\mathrm{d}x_\alpha} = g_\alpha^\beta ,$$

as required. From (3.27) it follows that, at M, the equations

(3.29)
$$\gamma_{\alpha\beta} = g_\alpha^\beta , \qquad \frac{\partial \gamma_{\alpha\beta}}{\partial \xi_\gamma} = 0 ,$$

imply

(3.30)
$$\gamma^{\alpha\beta} = g_\alpha^\beta , \qquad \frac{\partial \gamma^{\alpha\beta}}{\partial \xi_\gamma} = 0 .$$

Let us ask what happens to the expression (3.22) at M when the Christoffel symbols are expressed in terms of the functions ξ and the γ. The terms in this expression contain the potentials $\gamma_{\alpha\beta}$ and $\gamma^{\alpha\beta}$, together with the first and second derivatives of these quantities.

The terms that contain no derivatives are those that are obtained when the derivatives vanish, namely they are obtained when we assume that the $\gamma_{\alpha\beta}$ are constants, and we have checked that these terms vanish in (3.23). The terms that contain the first derivatives vanish at M thanks to (3.29) and (3.30). It remains only to calculate those containing second derivatives.

These occur only in the two first terms of (3.22),

$$-\frac{\partial}{\partial x_\alpha} \begin{Bmatrix} \beta\,\gamma \\ \sigma \end{Bmatrix} + \frac{\partial}{\partial x_\gamma} \begin{Bmatrix} \alpha\,\beta \\ \sigma \end{Bmatrix} .$$

Let us calculate the second, but taking into account only terms containing second derivatives. We can then obtain the first by permuting the indices α and γ. We will have

$$\frac{\partial}{\partial x_\gamma}\left\{\begin{matrix}\alpha\,\beta\\ \sigma\end{matrix}\right\} = \frac{1}{2}\sum_\tau g^{\sigma\tau}\left(\frac{\partial^2 g_{\alpha\tau}}{\partial x_\beta \partial x_\gamma} + \frac{\partial^2 g_{\beta\tau}}{\partial x_\alpha \partial x_\gamma} - \frac{\partial^2 g_{\alpha\beta}}{\partial x_\tau \partial x_\gamma}\right) + \cdots .$$

According to (3.24), the second derivatives are equal to

$$\frac{\partial^2 g_{\alpha\tau}}{\partial x_\beta \partial x_\gamma} = \sum_\lambda \sum_\rho \frac{\partial \xi_\lambda}{\partial x_\alpha}\frac{\partial \xi_\rho}{\partial x_\tau}\frac{\partial^2 \gamma_{\lambda\rho}}{\partial x_\beta \partial x_\gamma} + \cdots ,$$

while

$$\frac{\partial^2 \gamma_{\lambda\rho}}{\partial x_\beta \partial x_\gamma} = \sum_\mu \sum_\nu \frac{\partial^2 \gamma_{\lambda\rho}}{\partial \xi_\mu \partial \xi_\nu}\frac{\partial \xi_\mu}{\partial x_\beta}\frac{\partial \xi_\nu}{\partial x_\gamma} + \cdots ,$$

where terms not written here do not contain second derivatives. It thus follows that

$$\frac{\partial^2 g_{\alpha\tau}}{\partial x_\beta \partial x_\gamma} = \sum_\lambda \sum_\mu \sum_\nu \sum_\rho \frac{\partial \xi_\lambda}{\partial x_\alpha}\frac{\partial \xi_\mu}{\partial x_\beta}\frac{\partial \xi_\nu}{\partial x_\gamma}\frac{\partial \xi_\rho}{\partial x_\tau}\frac{\partial^2 \gamma_{\lambda\rho}}{\partial \xi_\mu \partial \xi_\nu} . \tag{3.31}$$

The other derivatives are obtained by permuting the indices α, β, γ, and τ, which amounts to permuting the corresponding indices λ, μ, ν, and ρ.

Furthermore, from (3.28) and (3.30),

$$g^{\sigma\tau} = \sum_\eta \frac{\partial x_\sigma}{\partial \xi_\eta}\frac{\partial x_\tau}{\partial \xi_\eta} .$$

It thus follows that

$$\frac{\partial}{\partial x_\gamma}\left\{\begin{matrix}\alpha\,\beta\\ \sigma\end{matrix}\right\} = \frac{1}{2}\sum_\lambda\sum_\mu\sum_\nu\sum_\rho\sum_\eta\sum_\tau \frac{\partial \xi_\lambda}{\partial x_\alpha}\frac{\partial \xi_\mu}{\partial x_\beta}\frac{\partial \xi_\nu}{\partial x_\gamma}\frac{\partial \xi_\rho}{\partial x_\tau}\frac{\partial x_\tau}{\partial \xi_\eta}\frac{\partial x_\sigma}{\partial \xi_\eta}$$
$$\times\left(\frac{\partial^2 \gamma_{\lambda\rho}}{\partial \xi_\mu \partial \xi_\nu} + \frac{\partial^2 \gamma_{\mu\rho}}{\partial \xi_\lambda \partial \xi_\nu} - \frac{\partial^2 \gamma_{\lambda\mu}}{\partial \xi_\rho \partial \xi_\nu}\right) + \cdots .$$

Carrying out the sum over τ leads to

$$\sum_\tau \frac{\partial \xi_\rho}{\partial x_\tau}\frac{\partial x_\tau}{\partial \xi_\eta} = g_\rho^\eta ,$$

while the sum over η then leads to

$$\sum_\eta g_\rho^\eta \frac{\partial x_\sigma}{\partial \xi_\eta} = \frac{\partial x_\sigma}{\partial \xi_\rho} .$$

It thus follows that

$$\frac{\partial}{\partial x_\gamma}\left\{\begin{matrix}\alpha\,\beta\\ \sigma\end{matrix}\right\} = \frac{1}{2}\sum_\lambda\sum_\mu\sum_\nu\sum_\rho \frac{\partial \xi_\lambda}{\partial x_\alpha}\frac{\partial \xi_\mu}{\partial x_\beta}\frac{\partial \xi_\nu}{\partial x_\gamma}\frac{\partial x_\sigma}{\partial \xi_\rho}$$
$$\times\left(\frac{\partial^2\gamma_{\lambda\rho}}{\partial\xi_\mu\partial\xi_\nu}+\frac{\partial^2\gamma_{\mu\rho}}{\partial\xi_\lambda\partial\xi_\nu}-\frac{\partial^2\gamma_{\lambda\mu}}{\partial\xi_\rho\partial\xi_\nu}\right)+\cdots.$$

It remains to permute the indices α and γ and subtract the resulting expression. All terms not written explicitly will cancel out. Instead of permuting the indices α and γ, we can permute the corresponding summation indices λ and μ. It thus follows that

(3.32) $$R^\sigma_{\alpha\beta\gamma} = \sum_\lambda\sum_\mu\sum_\nu\sum_\rho \frac{\partial \xi_\lambda}{\partial x_\alpha}\frac{\partial \xi_\mu}{\partial x_\beta}\frac{\partial \xi_\nu}{\partial x_\gamma}\frac{\partial x_\sigma}{\partial \xi_\rho}P^\rho_{\lambda\mu\nu},$$

where

(3.33) $$P^\rho_{\lambda\mu\nu} = \frac{1}{2}\left(\frac{\partial^2\gamma_{\lambda\rho}}{\partial\xi_\mu\partial\xi_\nu}-\frac{\partial^2\gamma_{\mu\lambda}}{\partial\xi_\nu\partial\xi_\rho}+\frac{\partial^2\gamma_{\nu\mu}}{\partial\xi_\rho\partial\xi_\lambda}-\frac{\partial^2\gamma_{\rho\nu}}{\partial\xi_\lambda\partial\xi_\mu}\right).$$

We thus obtain a new expression for $R^\sigma_{\alpha\beta\gamma}$.

Let us just recall the meanings of the various functions appearing here: ξ_1, ξ_2, ξ_3, and ξ_4 are the coordinate functions specifying a coordinate system (3.18) with the special properties (3.29) and (3.30) at the point M where we apply the formula (3.32). The derivatives appearing in (3.33) are the values at M of the derivatives of the potentials in the system $(\xi_1,\xi_2,\xi_3,\xi_4)$. These equations can be applied at any point M, but we should not forget that the ξ will be different functions at each point. We could not, for example, differentiate the expressions in (3.32).

Suppose now that, instead of (x_1,x_2,x_3,x_4), we wish to use new coordinates (x'_1,x'_2,x'_3,x'_4) related to the first coordinates by the equations

$$x_\sigma = x_\sigma(x'_1,x'_2,x'_3,x'_4) \qquad (\sigma=1,2,3,4),$$
$$x'_\sigma = x'_\sigma(x_1,x_2,x_3,x_4) \qquad (\sigma=1,2,3,4).$$

Replacing the coordinates x_σ by their expressions in terms of the new coordinates in the functions ξ_σ, we obtain functions of the x'_σ that we shall also denote by ξ_σ. Using a prime to indicate expressions calculated in the new system, we will have

$$R'^\sigma_{\alpha\beta\gamma} = \sum_\lambda\sum_\mu\sum_\nu\sum_\rho \frac{\partial \xi_\lambda}{\partial x'_\alpha}\frac{\partial \xi_\mu}{\partial x'_\beta}\frac{\partial \xi_\nu}{\partial x'_\gamma}\frac{\partial x'_\sigma}{\partial \xi_\rho}P^\rho_{\lambda\mu\nu}.$$

By the chain rule,

$$\frac{\partial\xi_\lambda}{\partial x'_\alpha} = \sum_{\lambda_1}\frac{\partial\xi_\lambda}{\partial x_{\lambda_1}}\frac{\partial x_{\lambda_1}}{\partial x'_\alpha}, \qquad \frac{\partial\xi_\mu}{\partial x'_\beta} = \sum_{\mu_1}\frac{\partial\xi_\mu}{\partial x_{\mu_1}}\frac{\partial x_{\mu_1}}{\partial x'_\beta},$$

$$\frac{\partial \xi_\nu}{\partial x'_\gamma} = \sum_{\nu_1} \frac{\partial \xi_\nu}{\partial x_{\nu_1}} \frac{\partial x_{\nu_1}}{\partial x'_\gamma} , \qquad \frac{\partial x'_\sigma}{\partial \xi_\rho} = \sum_{\rho_1} \frac{\partial x'_\sigma}{\partial x_{\rho_1}} \frac{\partial x_{\rho_1}}{\partial \xi_\rho} .$$

According to (3.32),

$$R^{\rho_1}_{\lambda_1\mu_1\nu_1} = \sum_\lambda \sum_\mu \sum_\nu \sum_\rho \frac{\partial \xi_\lambda}{\partial x_{\lambda_1}} \frac{\partial \xi_\mu}{\partial x_{\mu_1}} \frac{\partial \xi_\nu}{\partial x_{\nu_1}} \frac{\partial x_{\rho_1}}{\partial \xi_\rho} P^\rho_{\lambda\mu\nu} ,$$

whence finally,

$$R'^\sigma_{\alpha\beta\gamma} = \sum_{\lambda_1} \sum_{\mu_1} \sum_{\nu_1} \sum_{\rho_1} \frac{\partial x_{\lambda_1}}{\partial x'_\alpha} \frac{\partial x_{\mu_1}}{\partial x'_\beta} \frac{\partial x_{\nu_1}}{\partial x'_\gamma} \frac{\partial x'_\sigma}{\partial x_{\rho_1}} R^{\rho_1}_{\lambda_1\mu_1\nu_1} . \tag{3.34}$$

This tells us how the quantities $R^\sigma_{\alpha\beta\gamma}$ transform when we change coordinates. In particular, when we use the system ξ_σ at a point M, the equations (3.32) reduce at M to

$$R^\sigma_{\alpha\beta\gamma} = P^\sigma_{\alpha\beta\gamma} .$$

Equation (3.33) is the reduced expression for $R^\sigma_{\alpha\beta\gamma}$. This is what $R^\sigma_{\alpha\beta\gamma}$ reduces to at a point M when we use a special coordinate system. This rigorous expression at M may be considered a good approximation in a small enough domain around M, because the conditions (3.29) on which it is based, and which are exactly satisfied at M, will be satisfied to within a given approximation in a small enough domain.

A set of quantities which transform according to (3.34) through a change of coordinates, or by equations of the general form

$$T'^{\beta_1\beta_2\dots}_{\alpha_1\alpha_2\dots} = \sum_{\sigma_1} \sum_{\sigma_2} \cdots \sum_{\tau_1} \sum_{\tau_2} \cdots \frac{\partial x_{\sigma_1}}{\partial x'_{\alpha_1}} \frac{\partial x_{\sigma_2}}{\partial x'_{\alpha_2}} \cdots \frac{\partial x'_{\beta_1}}{\partial x_{\tau_1}} \frac{\partial x'_{\beta_2}}{\partial x_{\tau_2}} \cdots T^{\tau_1\tau_2\dots}_{\sigma_1\sigma_2\dots} , \tag{3.35}$$

is called a *tensor*. Each $T^{\beta_1\beta_2\dots}_{\alpha_1\alpha_2\dots}$ is called a component of the tensor. The above argument regarding the Riemann tensor $R^\sigma_{\alpha\beta\gamma}$ extends easily to an arbitrary tensor. We thus deduce the following theorem:

Theorem

If we have a coordinate system $\xi_1, \xi_2, \xi_3, \xi_4$ and numbers $\Theta^{\tau_1\tau_2\dots}_{\sigma_1\sigma_2\dots}$ at an arbitrary point M such that the quantities $T^{\beta_1\beta_2\dots}_{\alpha_1\alpha_2\dots}$ are defined in an arbitrary system x_1, x_2, x_3, x_4 by the equations

$$T^{\beta_1\beta_2\dots}_{\alpha_1\alpha_2\dots} = \sum_{\sigma_1} \sum_{\sigma_2} \cdots \sum_{\tau_1} \sum_{\tau_2} \cdots \frac{\partial \xi_{\sigma_1}}{\partial x_{\alpha_1}} \frac{\partial \xi_{\sigma_2}}{\partial x_{\alpha_2}} \cdots \frac{\partial x_{\beta_1}}{\partial \xi_{\tau_1}} \frac{\partial x_{\beta_2}}{\partial \xi_{\tau_2}} \cdots \Theta^{\tau_1\tau_2\dots}_{\sigma_1\sigma_2\dots} , \tag{3.36}$$

then $T^{\beta_1\beta_2\dots}_{\alpha_1\alpha_2\dots}$ is a tensor, namely it transforms according to (3.35) under a change of coordinates.

Then $\Theta_{\sigma_1\sigma_2\ldots}^{\tau_1\tau_2\ldots}$ are the components of this tensor in the system $\xi_1, \xi_2, \xi_3, \xi_4$. Naturally, the tensor need not have both kinds of index, raised and lowered. We have encountered tensors having only lowered indices, namely the potentials $g_{\mu\nu}$. Such tensors are referred to as *covariant*. The fact that the potentials are tensors follows directly from (3.24). An example of a *contravariant* tensor, namely with raised indices, would be the proper velocity

$$u^\sigma \equiv \frac{\mathrm{d}x_\sigma}{\mathrm{d}s} = \sum_\tau \frac{\partial x_\sigma}{\partial \xi_\tau}\frac{\mathrm{d}\xi_\tau}{\mathrm{d}s} ,$$

but also the associated potentials $g^{\mu\nu}$ whose tensorial nature results from (3.28).

It follows immediately from the definition (3.35) of a tensor that the components of two tensors of the same kind can be combined by adding components with the same values of the indices. The resulting quantities are again components of a tensor of the same kind as the summands:

$$A_{\alpha_1\alpha_2\ldots}^{\beta_1\beta_2\ldots} + B_{\alpha_1\alpha_2\ldots}^{\beta_1\beta_2\ldots} = C_{\alpha_1\alpha_2\ldots}^{\beta_1\beta_2\ldots} .$$

Multiplying in pairs the components of two tensors of arbitrary kinds, we obtain the components of another tensor which we can represent by placing all the indices of the two sets of tensor components on a single letter:

$$A_{\alpha_1\alpha_2\ldots}^{\beta_1\beta_2\ldots} B_{\gamma_1\gamma_2\ldots}^{\delta_1\delta_2\ldots} = C_{\alpha_1\alpha_2\ldots\gamma_1\gamma_2\ldots}^{\beta_1\beta_2\ldots\delta_1\delta_2\ldots} .$$

Finally, we can equate and sum over all their four values one upper index and one lower index, the other indices remaining free to assume any values, thereby constructing the *contracted tensor*. Hence, by contracting $T_{\alpha_1\alpha_2\ldots}^{\beta_1\beta_2\ldots}$ with respect to α_1 and β_1, we obtain

$$\sum_\sigma T_{\sigma\alpha_2\ldots}^{\sigma\beta_2\ldots} = T_{\alpha_2\ldots}^{\beta_2\ldots} .$$

The tensorial nature of the new tensor is easily demonstrated. When we change the coordinates, we have from (3.35)

$$\begin{aligned} T'^{\beta_2\ldots}_{\alpha_2\ldots} &= \sum_\sigma T'^{\sigma\beta_2\ldots}_{\sigma\alpha_2\ldots} \\ &= \sum_\sigma\sum_{\sigma_1}\sum_{\sigma_2}\cdots\sum_{\tau_1}\sum_{\tau_2}\cdots\frac{\partial x_{\sigma_1}}{\partial x'_\sigma}\frac{\partial x_{\sigma_2}}{\partial x'_{\alpha_2}}\cdots\frac{\partial x'_\sigma}{\partial x_{\tau_1}}\frac{\partial x'_{\beta_2}}{\partial x_{\tau_2}}\cdots T_{\sigma_1\sigma_2\ldots}^{\tau_1\tau_2\ldots} . \end{aligned}$$

We can carry out the sum over σ,

$$\sum_\sigma \frac{\partial x_{\sigma_1}}{\partial x'_\sigma}\frac{\partial x'_\sigma}{\partial x_{\tau_1}} = \frac{\mathrm{d}x_{\sigma_1}}{\mathrm{d}x_{\tau_1}} = g_{\tau_1}^{\sigma_1} ,$$

then the sums over σ_1 and τ_1,

$$\sum_{\sigma_1}\sum_{\tau_1} g_{\tau_1}^{\sigma_1} T_{\sigma_1\sigma_2\ldots}^{\tau_1\tau_2\ldots} = \sum_{\sigma_1} T_{\sigma_1\sigma_2\ldots}^{\sigma_1\tau_2\ldots} = T_{\sigma_2\ldots}^{\tau_2\ldots} .$$

This leaves us with

$$T'^{\beta_2\ldots}_{\alpha_2\ldots} = \sum_{\sigma_2}\cdots\sum_{\tau_2}\cdots\frac{\partial x_{\sigma_2}}{\partial x'_{\alpha_2}}\cdots\frac{\partial x'_{\beta_2}}{\partial x_{\tau_2}}\cdots T^{\tau_2\ldots}_{\sigma_2\ldots} ,$$

which is precisely the definition (3.35) of the tensorial nature of the contracted tensor.

The last two operations, namely multiplication and contraction, are often carried out at the same time, and this is then called the *inner product*. The tensor g^{β}_{α} is thus the inner product of $g_{\alpha\beta}$ and $g^{\alpha\beta}$. Indeed, we have

$$g^{\beta}_{\alpha} = \sum_{\sigma} g_{\alpha\sigma} g^{\sigma\beta} ,$$

since this was how we defined the $g^{\alpha\beta}$.

The tensors obtained by forming the inner product of a tensor with one or other of the fundamental tensors $g_{\alpha\beta}$ and $g^{\alpha\beta}$ will be denoted by the same letter. They can be distinguished by the number and position of their indices. These are referred to as *associated tensors*. By contracting the Riemann tensor, we obtain the *contracted Riemann tensor*:

(3.37) $$R_{\alpha\beta} = \sum_{\sigma} R^{\sigma}_{\alpha\beta\sigma} .$$

The associated tensors will be

(3.38) $$R^{\beta}_{\alpha} = \sum_{\sigma} g^{\beta\sigma} R_{\alpha\sigma}$$

and

(3.39) $$R^{\alpha\beta} = \sum_{\sigma} g^{\alpha\sigma} R^{\beta}_{\sigma} .$$

By taking the inner product of these tensors with $g_{\alpha\beta}$, we do not obtain new tensors. In fact we have

(3.40) $$\sum_{\sigma} g_{\alpha\sigma} R^{\beta\sigma} = \sum_{\sigma}\sum_{\tau} g_{\alpha\sigma} g^{\sigma\tau} R^{\beta}_{\tau} = R^{\beta}_{\alpha} ,$$

because

$$\sum_{\sigma} g_{\alpha\sigma} g^{\sigma\tau} = g^{\tau}_{\alpha} .$$

Likewise,

(3.41) $$\sum_{\sigma} g_{\beta\sigma} R^{\sigma}_{\alpha} = R_{\alpha\beta} .$$

Contracting R_{α}^{β}, we obtain the invariant

$$R = \sum_{\sigma} R_{\sigma}^{\sigma} . \tag{3.42}$$

Indeed, in this case, the definition of the tensorial nature reduces to

$$R' = R .$$

We can use this tensor to form new ones

$$R g_{\alpha\beta} , \qquad R g^{\alpha\beta} . \tag{3.43}$$

They cannot be denoted by the same letter as $R_{\alpha\beta}$ and $R^{\alpha\beta}$.

This tensor calculus can thus be used to construct expressions which transform simply under coordinate transformations. The new components are linear and homogeneous functions of the old ones. It follows that, if all the components of a tensor vanish in one coordinate system, they will vanish in all other coordinate systems. This means that, if a law of physics is expressed as the vanishing of all the components of a tensor, it will be valid in any coordinate system. In conformity with the principle of relativity, we will have obtained a law whose algebraic expression is independent of the referencing system used to locate events in space and time.

The method introduced by Einstein is thus as follows: to find tensorial laws which, when we use the reference frame adopted by certain observers, coincide with the empirical laws describing their observations, to within the level of approximation appropriate to their measurements.

According to (3.32), the reduced expression for the contracted Riemann tensor is

$$\begin{aligned} P_{\lambda\mu} = \sum_{\rho} P_{\lambda\mu\rho}^{\rho} = & -\frac{1}{2} \sum_{\rho} \frac{\partial^2 \gamma_{\lambda\mu}}{\partial \xi_{\rho}^2} \\ & + \frac{1}{2} \frac{\partial}{\partial \xi_{\mu}} \sum_{\rho} \left(\frac{\partial \gamma_{\lambda\rho}}{\partial \xi_{\rho}} - \frac{1}{2} \frac{\partial \gamma_{\rho\rho}}{\partial \xi_{\lambda}} \right) \\ & + \frac{1}{2} \frac{\partial}{\partial \xi_{\lambda}} \sum_{\rho} \left(\frac{\partial \gamma_{\mu\rho}}{\partial \xi_{\rho}} - \frac{1}{2} \frac{\partial \gamma_{\rho\rho}}{\partial \xi_{\mu}} \right) . \end{aligned}$$

It follows that the contracted tensor has the symmetry

$$P_{\lambda\mu} = P_{\mu\lambda} ,$$

and by (3.37) and (3.39),

$$R_{\alpha\beta} = R_{\beta\alpha} .$$

The coordinate system ξ_{σ} was determined by imposing the conditions (3.29) at a point M. These conditions determine at M the values of the first and second derivatives $\partial\xi_\sigma/\partial x_\alpha$ and $\partial^2\xi_\sigma/\partial x_\alpha \partial x_\beta$ of the transformation functions ξ_σ given by (3.18).

Could we not find values at M of the third derivatives of these functions that would simplify the reduced expression for $R_{\alpha\beta}$? If we could choose these derivatives such that, at M,

$$\frac{\partial}{\partial \xi_\mu}\sum_\rho \left(\frac{\partial \gamma_{\lambda\rho}}{\partial \xi_\rho} - \frac{1}{2}\frac{\partial \gamma_{\rho\rho}}{\partial \xi_\lambda}\right) = 0 \qquad (\lambda,\mu = 1,2,3,4)\,, \tag{3.44}$$

then the reduced expression for $R_{\alpha\beta}$ would reduce to its first term,

$$P_{\lambda\mu} = -\frac{1}{2}\sum_\rho \frac{\partial^2 \gamma_{\lambda\mu}}{\partial \xi_\rho^2}\,. \tag{3.45}$$

Let $A(\alpha\beta)$ be the value at M of

$$\frac{\partial}{\partial x_\beta}\sum_\rho \left(\frac{\partial g_{\alpha\beta}}{\partial x_\rho} - \frac{1}{2}\frac{\partial g_{\rho\rho}}{\partial x_\alpha}\right) = A(\alpha\beta)\,.$$

We now see what becomes of this expression when we express it in terms of the $\gamma_{\mu\nu}$ and the ξ_σ and impose the conditions (3.29) and (3.30), together with the relations (3.44).

The terms containing the second derivatives of the $\gamma_{\mu\nu}$ vanish thanks to (3.44). This follows immediately from the calculation of the second derivatives carried out in (3.31). The terms containing the first derivatives vanish since these derivatives are zero according to (3.29) and (3.30).

It remains to calculate the terms that do not contain derivatives. To do this, we must calculate $A(\alpha\beta)$, assuming the $\gamma_{\sigma\tau}$ to be constant. Differentiating $g_{\alpha\beta}$ as given in (3.24), and using (3.29) to set $\gamma_{\sigma\tau} = g_\sigma^\tau$, we have

$$\frac{\partial g_{\alpha\beta}}{\partial x_\gamma} = \sum_\sigma \left(\frac{\partial \xi_\sigma}{\partial x_\alpha}\frac{\partial^2 \xi_\sigma}{\partial x_\beta \partial x_\gamma} + \frac{\partial^2 \xi_\sigma}{\partial x_\alpha \partial x_\gamma}\frac{\partial \xi_\sigma}{\partial x_\beta}\right)\,.$$

Inserting this relation into the expression for $A(\alpha\beta)$,

$$A(\alpha\beta) = \frac{\partial}{\partial x_\beta}\sum_\sigma\sum_\rho \left(\frac{\partial \xi_\sigma}{\partial x_\alpha}\frac{\partial^2 \xi_\sigma}{\partial x_\rho^2} + \frac{\partial^2 \xi_\sigma}{\partial x_\alpha \partial x_\rho}\frac{\partial \xi_\sigma}{\partial x_\rho} - \frac{1}{2}\frac{\partial \xi_\sigma}{\partial x_\rho}\frac{\partial^2 \xi_\sigma}{\partial x_\alpha \partial x_\rho} - \frac{1}{2}\frac{\partial^2 \xi_\sigma}{\partial x_\alpha \partial x_\rho}\frac{\partial \xi_\sigma}{\partial x_\rho}\right)\,,$$

then simplifying, this leads to

$$A(\alpha\beta) = \frac{\partial}{\partial x_\beta}\sum_\sigma\sum_\rho \frac{\partial \xi_\sigma}{\partial x_\alpha}\frac{\partial^2 \xi_\sigma}{\partial x_\rho^2}\,.$$

We now carry out the differentiation to obtain

$$A(\alpha\beta) = \sum_\sigma \left(\frac{\partial \xi_\sigma}{\partial x_\alpha} \frac{\partial}{\partial x_\beta} \sum_\rho \frac{\partial^2 \xi_\sigma}{\partial x_\rho^2} + \frac{\partial^2 \xi_\sigma}{\partial x_\alpha \partial x_\beta} \sum_\rho \frac{\partial^2 \xi_\sigma}{\partial x_\rho^2} \right) .$$

Can we determine the third derivatives in such a way as to satisfy this relation for $\alpha, \beta = 1,2,3,4$, whatever the values of $A(\alpha\beta)$ and of the derivatives $\partial \xi_\sigma / \partial x_\alpha$ and $\partial^2 \xi_\sigma / \partial x_\alpha \partial x_\beta$?

For $\alpha = 1,2,3,4$, the relation can be considered as a linear system in

$$\frac{\partial}{\partial x_\beta} \sum_\rho \frac{\partial^2 \xi_\sigma}{\partial x_\rho^2} .$$

The determinant of this system differs from zero. It is the functional determinant of the transformation equations (3.18). The system can therefore be solved. We obtain equations of the form

(3.46) $$\frac{\partial}{\partial x_\beta} \sum_\rho \frac{\partial^2 \xi_\sigma}{\partial x_\rho^2} = a(\beta\sigma) ,$$

where $a(\beta\sigma)$ are functions of the $A(\alpha\beta)$, $\partial \xi_\sigma / \partial x_\alpha$, and $\partial^2 \xi_\sigma / \partial x_\alpha \partial x_\beta$.

It will always be possible to satisfy (3.44) if, for any $a(\beta\sigma)$, we can determine the values at M of the third derivatives in such a way as to satisfy (3.46). This is done by setting

$$\frac{\partial^3 \xi_\sigma}{\partial x_\alpha \partial x_\beta \partial x_\gamma} = \frac{a(\alpha\sigma) a(\beta\sigma) a(\gamma\sigma)}{\sum_\rho a(\rho\sigma)^2} .$$

It is therefore always possible to determine the system ξ in such a way that the reduced expression for the contracted Riemann tensor is given by (3.45):

(3.47) $$P_{\lambda\mu} = -\frac{1}{2} \left(\frac{\partial^2 \gamma_{\lambda\mu}}{\partial \xi_1^2} + \frac{\partial^2 \gamma_{\lambda\mu}}{\partial \xi_2^2} + \frac{\partial^2 \gamma_{\lambda\mu}}{\partial \xi_3^2} + \frac{\partial^2 \gamma_{\lambda\mu}}{\partial \xi_4^2} \right) .$$

We have used a system ξ for which the fundamental form reduces at M to

$$\mathrm{d}s^2 = \mathrm{d}\xi_1^2 + \mathrm{d}\xi_2^2 + \mathrm{d}\xi_3^2 + \mathrm{d}\xi_4^2 .$$

The coordinates ξ_σ will be imaginary. We may write them

$$\xi_1 = x\sqrt{-1} , \qquad \xi_2 = y\sqrt{-1} , \qquad \xi_3 = z\sqrt{-1} , \qquad \xi_4 = t .$$

We then obtain for ds^2

$$\mathrm{d}s^2 = -\mathrm{d}x^2 - \mathrm{d}y^2 - \mathrm{d}z^2 + \mathrm{d}t^2 .$$

The $\gamma_{\lambda\mu}$ transform to $g_{\lambda\mu}$ and the $P_{\lambda\mu}$ to $R_{\lambda\mu}$. This transformation only introduces one or more constant factors of $\sqrt{-1}$. These factors cancel one another on either side of (3.47).

In the system (x, y, z, t), the reduced form of the contracted tensor will thus be

$$R_{\lambda\mu} = \frac{1}{2}\left(\frac{\partial^2 g_{\lambda\mu}}{\partial x^2} + \frac{\partial^2 g_{\lambda\mu}}{\partial y^2} + \frac{\partial^2 g_{\lambda\mu}}{\partial z^2} - \frac{\partial^2 g_{\lambda\mu}}{\partial t^2}\right). \tag{3.48}$$

Chapter 4
Gravitation[1]

4.1 Newtonian Potential and Retarded Potential[2]

Astronomical observations can be accurately encapsulated by Newton's universal law of gravitation. To obtain this law, we use a particular way of referencing events in space and time: Cartesian coordinates on three orthogonal axes, either fixed or moving at constant speed in a straight line relative to the general stellar background, while time is measured by the Earth's rotation relative to that background.

The acceleration of a free body can be attributed to the action of the various celestial bodies. The acceleration is dependent on a potential V, namely the components of the acceleration can be written

$$\begin{cases} \dfrac{\mathrm{d}^2x}{\mathrm{d}t^2} = -\dfrac{\partial V}{\partial x}\,, \\[2ex] \dfrac{\mathrm{d}^2y}{\mathrm{d}t^2} = -\dfrac{\partial V}{\partial y}\,, \\[2ex] \dfrac{\mathrm{d}^2z}{\mathrm{d}t^2} = -\dfrac{\partial V}{\partial z}\,. \end{cases} \tag{4.1}$$

The potential V is then given by the equation

$$V = -K\sum \frac{m}{r}\,, \tag{4.2}$$

where m is the mass of the given celestial body and r its distance from the point where the potential is to be calculated, while K is the gravitational constant. In C.G.S. units,

$$K = 6.7 \times 10^{-8}\,, \tag{4.3}$$

[1] For the original text of this chapter in French see p. 202.

[2] For the original text of this section in French see p. 202.

G. Lemaître, *Learning the Physics of Einstein with Georges Lemaître*,
https://doi.org/10.1007/978-3-030-22030-3_4

with physical dimensions $L^3T^{-2}M^{-1}$. Equation (4.2) expresses the action of masses on the field, while (4.1) expresses the reaction of the field on any free point particles that happen to be present there.

Equations (4.2) and (4.3) can be written in another form. Let ρ be the density of matter at a point with coordinates ξ, η, ζ, whence the mass contained in an element of volume $\mathrm{d}\xi\,\mathrm{d}\eta\,\mathrm{d}\zeta$ situated at this point will be $\rho\mathrm{d}\xi\,\mathrm{d}\eta\,\mathrm{d}\zeta$. The sum in (4.2) can be written in the form of a triple integral extended over the whole of space. The potential at a point x, y, z at time t will be

$$V(x,y,z,t) = -K\iiint \frac{\rho(\xi,\eta,\zeta,t)}{r}\mathrm{d}\xi\,\mathrm{d}\eta\,\mathrm{d}\zeta\,, \tag{4.4}$$

where r is defined by the relation

$$r^2 = (x-\xi)^2 + (y-\eta)^2 + (z-\zeta)^2\,.$$

Conversely, we can express the density ρ as a function of the potential. We then obtain the well known Poisson equation

$$\Delta V \equiv \frac{\partial^2 V}{\partial x^2} + \frac{\partial^2 V}{\partial y^2} + \frac{\partial^2 V}{\partial z^2} = 4\pi K\rho\,. \tag{4.5}$$

At a point where there is no matter, this relation reduces to

$$\Delta V = 0\,. \tag{4.6}$$

This is known as the Laplace equation and ΔV is called the Laplacian of V.

The Newtonian potential is just one solution of the Poisson equation. It is obtained by imposing boundary conditions on V: when we move toward infinity in any direction, V must tend to a limit that is independent of this direction. The acceleration of a point infinitely far away from any mass is then zero. Naturally, this condition can only be satisfied when we use a particular system of axes, or any system in uniform straight line motion relative to this system.

Newtonian gravity assumes immediate action between attracting masses. If their action were to propagate from point to point through space at a certain speed c, the influence of masses located at a distance r from a point (x,y,z) would only be felt there after a time r/c. Taking into account this delay, we would have to write the equation for the potential in the form

$$V(x,y,z,t) = -K\iiint \frac{\rho(\xi,\eta,\zeta,t-r/c)}{r}\mathrm{d}\xi\,\mathrm{d}\eta\,\mathrm{d}\zeta\,. \tag{4.7}$$

This is the equation for the retarded potential.

If we set

$$\Box V \equiv \frac{\partial^2 V}{\partial x^2} + \frac{\partial^2 V}{\partial y^2} + \frac{\partial^2 V}{\partial z^2} - \frac{1}{c^2}\frac{\partial^2 V}{\partial t^2}\,, \tag{4.8}$$

Poisson's equation becomes

$$\square V = 4\pi K\rho \,. \tag{4.9}$$

The quantity $\square V$ is the d'Alembertian of V. Laplace's equation now becomes

$$\square V = 0 \,. \tag{4.10}$$

These equations give just as good an account for observations as those of the primitive theory when the propagation speed c is high enough, e.g., of the order of the speed of light.

When we choose the unit of time in such a way that the speed of propagation is unity, the d'Alembertian can be written

$$\square V \equiv \frac{\partial^2 V}{\partial x^2} + \frac{\partial^2 V}{\partial y^2} + \frac{\partial^2 V}{\partial z^2} - \frac{\partial^2 V}{\partial t^2} \,. \tag{4.11}$$

The Gauss constant K must then be divided by the square c^2 of the speed of propagation in the C.G.S. system.

Using the operator symbols defined above, the reduced expressions for the contracted tensor obtained at the end of Section 3.3 can be written

$$\square g_{\mu\nu} = 2R_{\mu\nu} \,. \tag{4.12}$$

These are valid in a neighbourhood of a point M where the first derivatives of the $g_{\mu\nu}$ are zero and where $\mathrm{d}s^2$ reduces to

$$\mathrm{d}s^2 = -\mathrm{d}x^2 - \mathrm{d}y^2 - \mathrm{d}z^2 + \mathrm{d}t^2 \,.$$

Therefore, just as we deduce (4.7) from (4.9), we will be able to deduce that

$$g_{\mu\nu}(x,y,z,t) = -\frac{1}{2\pi}\iiint \frac{R_{\mu\nu}(\xi,\eta,\zeta,t-r)}{r}\,\mathrm{d}\xi\,\mathrm{d}\eta\,\mathrm{d}\zeta \,. \tag{4.13}$$

In Einstein's theory, the potentials $g_{\mu\nu}$ play the role of the Newtonian potential V, just as we noted in Section 1.1. We see now that $R_{\mu\nu}$ must represent the material masses. We must therefore set $R_{\mu\nu}$ equal to a tensor of the same kind depending on the distribution of matter. Apart from $R_{\mu\nu}$, the tensors $g_{\mu\nu}$ and $Rg_{\mu\nu}$ [see (3.43)] may also appear in the relation we seek here. If $T_{\mu\nu}$ is a tensor representing the distribution of matter, we may attempt to formulate the laws of gravity by a general equation of the form

$$T_{\mu\nu} = c_1 R_{\mu\nu} + c_2 R g_{\mu\nu} + c_3 g_{\mu\nu} \,, \tag{4.14}$$

where c_1, c_2, and c_3 are constants remaining to be determined. In the following we shall ask what values should be attributed to these constants for (4.14) to satisfy the conditions imposed by the law of gravity.

Note that, instead of a relation between covariant tensors, we may also consider the equivalent relation between the associated tensors,

$$T^{\mu\nu} = c_1 R^{\mu\nu} + c_2 R g^{\mu\nu} + c_3 g^{\mu\nu} \,. \tag{4.15}$$

4.2 Material Energy Tensor[3]

The state of the distribution of matter is characterised by its density and velocity. But can these physical quantities be represented by tensors? We know that the proper velocity

$$u^\sigma = \frac{\mathrm{d}x_\sigma}{\mathrm{d}s} \tag{4.16}$$

is a tensor. For its part, the density ρ changes when we change coordinate system.

If we change only the space coordinates, we may argue that the mass contained in a certain volume must be independent of the way in which we refer to the various points within this volume. As the density is the ratio of the mass to the volume it occupies, the product of the density and the volume element must be an invariant.

We shall argue in the same way in the general case. The mass contained in a region of the universe must be independent of the way in which we refer to the various points in that region. The density will thus be the mass contained in a unit volume of the universe, which means that

$$\rho \, \mathrm{d}x_1 \, \mathrm{d}x_2 \, \mathrm{d}x_3 \, \mathrm{d}x_4 \tag{4.17}$$

will be an invariant.

When we use the proper coordinates of the matter (see Section 1.3), one of the coordinates will be the invariant $\mathrm{d}s$ measured by following matter in whatever motion it may have. The product of ρ with the other three coordinate differentials will thus be another invariant. The two definitions of the density that we have just discussed are therefore equivalent.

When we studied the general geometry, in (1.9) we saw how the volume element transforms, noting that

$$\sqrt{\gamma} \, \mathrm{d}x_1 \, \mathrm{d}x_2 \, \mathrm{d}x_3$$

is an invariant, where γ is the determinant of the potentials. This property generalises immediately to four coordinates. Taking into account the fact that the determinant g of the $g_{\mu\nu}$ is negative, we deduce that

$$\sqrt{-g} \, \mathrm{d}x_1 \, \mathrm{d}x_2 \, \mathrm{d}x_3 \, \mathrm{d}x_4 \tag{4.18}$$

---

[3] For the original text of this section in French see p. 205.

is an invariant. The quotient of the two invariants (4.17) and (4.18),

$$\frac{\rho}{\sqrt{-g}},$$

is therefore another invariant.

We may now combine this invariant with the contravariant proper velocity (4.16) to form a contravariant tensor of second order by multiplication:

$$T^{\mu\nu} = \frac{\rho}{\sqrt{-g}} \frac{\mathrm{d}x_\mu}{\mathrm{d}s} \frac{\mathrm{d}x_\nu}{\mathrm{d}s}. \tag{4.19}$$

We call this quantity the material energy tensor. This is the tensor we shall introduce into (4.15). The associated covariant tensor is easily found to be

$$T_{\mu\nu} = \sum_\sigma \sum_\tau g_{\mu\sigma} g_{\nu\tau} T^{\sigma\tau}. \tag{4.20}$$

The invariant obtained by contraction is

$$T = \sum_\mu \sum_\nu g_{\mu\nu} T^{\mu\nu} = \frac{\rho}{\sqrt{-g}} \sum_\mu \sum_\nu g_{\mu\nu} \frac{\mathrm{d}x_\mu}{\mathrm{d}s} \frac{\mathrm{d}x_\nu}{\mathrm{d}s}.$$

From the definition of $\mathrm{d}s^2$, this invariant simplifies to

$$T = \frac{\rho}{\sqrt{-g}}. \tag{4.21}$$

We shall refer to any product of a tensor with $\sqrt{-g}$ as a tensor density. We shall write tensor densities in the same manner as the corresponding tensors, but using calligraphic letters. We will thus have

$$\mathcal{T}^{\mu\nu} = \rho \frac{\mathrm{d}x_\mu}{\mathrm{d}s} \frac{\mathrm{d}x_\nu}{\mathrm{d}s}. \tag{4.22}$$

The equations of motion for a free point particle can be expressed in terms of the material energy tensor. The solution to this problem will be a first step in our quest for the general gravitational equations, because the latter must describe the motions of free particles under the influence of the field as well as the production of the field by matter.

We have seen that we may always choose a coordinate system in such a way that, at a given point M, the motions of free points are uniform and rectilinear. To do this, the derivatives of the potentials must vanish at the given point. We thus seek first to use the matter tensor to express the conditions under which free points will have uniform rectilinear motion. This will give us an equation that is valid in a particular coordinate system. The next step will be to identify a tensor equation which reduces to this equation when the derivatives of the potentials are zero. This new equation will be the equation of motion of free points in the general case.

Consider a small material object in uniform motion. We may assume that all its points move with the same velocity and that it has constant density. Let us consider the expression

$$f^{\mu} = \sum_{\sigma} \frac{\partial \mathscr{T}^{\mu\sigma}}{\partial x_{\sigma}}$$

$$= \rho \sum_{\sigma} \frac{\partial(\mathrm{d}x_{\mu}/\mathrm{d}s)}{\partial x_{\sigma}} \frac{\mathrm{d}x_{\sigma}}{\mathrm{d}s} + \frac{\mathrm{d}x_{\mu}}{\mathrm{d}s} \sum_{\sigma} \frac{\partial}{\partial x_{\sigma}} \left(\rho \frac{\mathrm{d}x_{\sigma}}{\mathrm{d}s} \right) . \tag{4.23}$$

Note that

$$\sum_{\sigma} \frac{\partial(\mathrm{d}x_{\mu}/\mathrm{d}s)}{\partial x_{\sigma}} \frac{\mathrm{d}x_{\sigma}}{\mathrm{d}s} = \frac{\mathrm{d}^2 x_{\mu}}{\mathrm{d}s^2} .$$

For the various points of the moving object, we thus find

$$f^{\mu} = 0 ,$$

since the acceleration is zero and the density and velocity are constant.

Conversely, if we set

$$f^{\mu} = 0 ,$$

we may deduce that

$$0 = \sum_{\mu} \sum_{\nu} g_{\mu\nu} f^{\mu} \frac{\mathrm{d}x_{\nu}}{\mathrm{d}s}$$

$$= \rho \sum_{\mu} \sum_{\nu} g_{\mu\nu} \frac{\mathrm{d}^2 x_{\mu}}{\mathrm{d}s^2} \frac{\mathrm{d}x_{\nu}}{\mathrm{d}s} + \sum_{\sigma} \frac{\partial}{\partial x_{\sigma}} \left(\rho \frac{\mathrm{d}x_{\sigma}}{\mathrm{d}s} \right) ,$$

because

$$\sum_{\mu} \sum_{\nu} g_{\mu\nu} \frac{\mathrm{d}x_{\mu}}{\mathrm{d}s} \frac{\mathrm{d}x_{\nu}}{\mathrm{d}s} = 1 .$$

Differentiating the last expression and using the fact that the derivatives of the $g_{\mu\nu}$ are zero at the point M, we have

$$\sum_{\mu} \sum_{\nu} g_{\mu\nu} \frac{\mathrm{d}^2 x_{\mu}}{\mathrm{d}s^2} \frac{\mathrm{d}x_{\nu}}{\mathrm{d}s} = 0 .$$

The equations

$$f^{\mu} = 0 \qquad (\mu = 1,2,3,4)$$

thus imply

$$\sum_{\sigma} \frac{\partial}{\partial x_{\sigma}} \left(\rho \frac{\mathrm{d}x_{\sigma}}{\mathrm{d}s} \right) = 0 , \tag{4.24}$$

and

$$\frac{d^2x_\mu}{ds^2}=0 \qquad (\mu=1,2,3,4)\,.$$

The latter equations express the fact that the motion of a point is uniform and rectilinear. This is indeed the equation for free particles, at the point M, given the coordinate system we have used.

Equation (4.24) is the continuity equation. It expresses the conservation of mass. To see this, we use the proper coordinates x,y,z,t of the moving body (see Section 1.3). Then ds will be equal to cdt. Using the left-hand side of the equation, we may form a volume integral over the region occupied by the moving body. This integral will be zero and we will have

$$\frac{1}{c}\iiint\left[\frac{\partial}{\partial x}\left(\rho\frac{dx}{dt}\right)+\frac{\partial}{\partial y}\left(\rho\frac{dy}{dt}\right)+\frac{\partial}{\partial z}\left(\rho\frac{dz}{dt}\right)\right]dx\,dy\,dz+\frac{1}{c}\frac{d}{dt}\iiint\rho\,dx\,dy\,dz=0\,.$$

We transform the first integral using Green's theorem. If dS is the surface element on the body surface and l, m, n are the direction cosines of the outward normal to this surface, we obtain

$$\iint\rho\left(l\frac{dx}{dt}+m\frac{dy}{dt}+n\frac{dz}{dt}\right)dS\,.$$

This integral represents the amount of matter leaving the volume. But this is zero since we are integrating over the surface of the moving body.

We are thus left with

$$\frac{d}{dt}\iiint\rho\,dx\,dy\,dz=0\,,$$

which expresses the fact that the total mass of the body remains constant. We thus obtain the tensor equations which express the conservation of matter and the motion of free particles by setting a tensor f^μ equal to zero which reduces to

$$\sum_\sigma\frac{\partial\mathscr{T}^{\mu\sigma}}{\partial x_\sigma}$$

when the derivatives of the potentials are zero.

In order to form this tensor, we use the theorem already applied in Section 3.3 to establish the tensorial nature of various tensors. We shall once again adopt the coordinate system ξ_1, ξ_2, ξ_3, ξ_4 used in that section.

The material energy tensor transforms as follows:

$$T^{\alpha\beta}=\sum_\mu\sum_\nu\frac{\partial x_\alpha}{\partial\xi_\mu}\frac{\partial x_\beta}{\partial\xi_\nu}\Theta^{\mu\nu}\,, \tag{4.25}$$

where $\Theta^{\mu\nu}$ are the components of this tensor in the system ξ_σ. Conversely, we will have

(4.26) $$\Theta^{\alpha\beta} = \sum_\mu \sum_\nu \frac{\partial \xi_\alpha}{\partial x_\mu} \frac{\partial \xi_\beta}{\partial x_\nu} T^{\mu\nu} .$$

Solving these equations for the $T^{\mu\nu}$ ($\mu, \nu = 1,2,3,4$), we should recover the previous equations. We shall use this fact in a moment.

Differentiating $\Theta^{\alpha\beta}$ with respect to x_γ, we have

$$\begin{aligned}\frac{\partial \Theta^{\alpha\beta}}{\partial x_\gamma} &= \sum_\rho \frac{\partial \xi_\rho}{\partial x_\gamma} \frac{\partial \Theta^{\alpha\beta}}{\partial \xi_\rho} \\ &= \sum_\mu \sum_\nu \frac{\partial^2 \xi_\alpha}{\partial x_\mu \partial x_\gamma} \frac{\partial \xi_\beta}{\partial x_\nu} T^{\mu\nu} + \sum_\mu \sum_\nu \frac{\partial \xi_\alpha}{\partial x_\mu} \frac{\partial^2 \xi_\beta}{\partial x_\nu \partial x_\gamma} T^{\mu\nu} + \sum_\mu \sum_\nu \frac{\partial \xi_\alpha}{\partial x_\mu} \frac{\partial \xi_\beta}{\partial x_\nu} \frac{\partial T^{\mu\nu}}{\partial x_\gamma} .\end{aligned}$$

At the point M, we can calculate the second derivatives using (3.20),

$$\frac{\partial^2 \xi_\alpha}{\partial x_\mu \partial x_\gamma} = \sum_\tau \frac{\partial \xi_\alpha}{\partial x_\tau} \left\{ \begin{matrix} \gamma\mu \\ \tau \end{matrix} \right\} .$$

The first of the three terms of the previous equation thus becomes

$$\sum_\mu \sum_\nu \sum_\tau \frac{\partial \xi_\alpha}{\partial x_\tau} \frac{\partial \xi_\beta}{\partial x_\nu} \left\{ \begin{matrix} \gamma\mu \\ \tau \end{matrix} \right\} T^{\mu\nu} ,$$

or, by exchanging the summation indices μ and τ,

$$\sum_\mu \sum_\nu \sum_\tau \frac{\partial \xi_\alpha}{\partial x_\mu} \frac{\partial \xi_\beta}{\partial x_\nu} \left\{ \begin{matrix} \gamma\tau \\ \mu \end{matrix} \right\} T^{\tau\nu} .$$

The second term can likewise be written as

$$\sum_\mu \sum_\nu \sum_\tau \frac{\partial \xi_\alpha}{\partial x_\mu} \frac{\partial \xi_\beta}{\partial x_\nu} \left\{ \begin{matrix} \gamma\tau \\ \nu \end{matrix} \right\} T^{\mu\tau} .$$

We will thus have

$$\sum_\rho \frac{\partial \xi_\rho}{\partial x_\gamma} \frac{\partial \Theta^{\alpha\beta}}{\partial \xi_\rho} = \sum_\mu \sum_\nu \frac{\partial \xi_\alpha}{\partial x_\mu} \frac{\partial \xi_\beta}{\partial x_\nu} \left(\frac{\partial T^{\mu\nu}}{\partial x_\gamma} + \sum_\tau \left\{ \begin{matrix} \gamma\tau \\ \mu \end{matrix} \right\} T^{\tau\nu} + \sum_\tau \left\{ \begin{matrix} \gamma\tau \\ \nu \end{matrix} \right\} T^{\mu\tau} \right) .$$

We solve this equation for the quantity in brackets on the right-hand side. This is the same calculation as the one we performed to go from (4.26) to (4.25). The result is

$$\frac{\partial T^{\mu\nu}}{\partial x_\gamma} + \sum_\tau \left\{ \begin{matrix} \gamma\tau \\ \mu \end{matrix} \right\} T^{\tau\nu} + \sum_\tau \left\{ \begin{matrix} \gamma\tau \\ \nu \end{matrix} \right\} T^{\mu\tau} = \sum_\mu \sum_\nu \sum_\rho \frac{\partial x_\alpha}{\partial \xi_\mu} \frac{\partial x_\beta}{\partial \xi_\nu} \frac{\partial \xi_\rho}{\partial x_\gamma} \frac{\partial \Theta^{\alpha\beta}}{\partial \xi_\rho} .$$

Applying the theorem in Section 3.3, we see that the left-hand side of this equation is a tensor.

The corresponding contracted tensor,

$$f^{\mu} = \sum_{\sigma} \frac{\partial T^{\mu\sigma}}{\partial x_{\sigma}} + \sum_{\sigma}\sum_{\tau} \left\{ {\sigma\tau \atop \mu} \right\} T^{\tau\sigma} + \sum_{\sigma}\sum_{\tau} \left\{ {\sigma\tau \atop \sigma} \right\} T^{\mu\tau} ,$$

can be simplified by noting from (2.12) that

$$\sum_{\sigma} \left\{ {\sigma\tau \atop \sigma} \right\} = \frac{1}{2}\frac{1}{g}\frac{\partial g}{\partial x_{\tau}} = \frac{1}{\sqrt{-g}}\frac{\partial\sqrt{-g}}{\partial x_{\tau}} .$$

It thus follows that

$$f^{\mu} = \frac{1}{\sqrt{-g}} \sum_{\sigma} \frac{\partial}{\partial x_{\sigma}} (T^{\mu\sigma}\sqrt{-g}) + \sum_{\sigma}\sum_{\tau} \left\{ {\sigma\tau \atop \mu} \right\} T^{\tau\sigma} .$$

The corresponding tensor density will be

$$f^{\mu} = \sum_{\sigma} \frac{\partial \mathcal{T}^{\mu\sigma}}{\partial x_{\sigma}} + \sum_{\sigma}\sum_{\tau} \left\{ {\sigma\tau \atop \mu} \right\} \mathcal{T}^{\tau\sigma} . \tag{4.27}$$

The right-hand side reduces to its first term when the derivatives of the potentials are zero, because the Christoffel symbols vanish in this case.

We may conclude that the tensor equations

$$f^{\mu} = 0 \qquad (\mu = 1,2,3,4) \tag{4.28}$$

express the equations of motion of free point particles and the principle of conservation of mass in terms of the material energy tensor.

Note. In the proof of the tensorial nature of f^{μ}, we have not used the symmetry of $\mathcal{T}^{\mu\nu}$.

This proof would also work for a tensor density for which

$$\mathcal{T}^{\sigma\tau} = -\mathcal{T}^{\tau\sigma} .$$

In this case, the tensor f^{μ} reduces to its first term

$$f^{\mu} = \sum_{\sigma} \frac{\partial \mathcal{T}^{\mu\sigma}}{\partial x_{\sigma}} . \tag{4.29}$$

4.3 General Equations of Mechanics and Gravity[4]

At the end of Section 4.1, we saw that the gravitational field equations could be brought to the general form

$$T_{\mu\nu} = c_1 R_{\mu\nu} + c_2 R g_{\mu\nu} + c_3 g_{\mu\nu} \,, \tag{4.30}$$

or in the equivalent contravariant form (4.15). The task now will be to determine the arbitrary constants c_1, c_2, c_3 in such a way that the equations

$$f^\alpha = 0 \qquad (\alpha = 1,2,3,4) \tag{4.31}$$

which express the dynamical equations of free points in terms of $\mathscr{T}^{\mu\nu}$ are consequences of (4.30).

Since (4.30) and (4.31) are tensor equations, it will suffice to prove this at an arbitrary point M, using a special coordinate system in which the proof is simplified. We shall use the system adopted in Section 3.3, which we denoted there by ξ_1, ξ_2, ξ_3, ξ_4. Here it will be more convenient to use the notation x_1, x_2, x_3, x_4. With the current notation, the conditions (3.29) and (3.30) become

$$g_{\alpha\beta} = g_\alpha^\beta \,, \qquad \frac{\partial g_{\alpha\beta}}{\partial x_\gamma} = 0 \,, \tag{4.32}$$

and

$$g^{\alpha\beta} = g_\alpha^\beta \,, \qquad \frac{\partial g^{\alpha\beta}}{\partial x_\gamma} = 0 \,. \tag{4.33}$$

We have shown that it is always possible to impose these at a point M. We deduce that, at M,

$$\sqrt{-g} = 1 \,,$$

and that the Christoffel symbols vanish there.

The tensor f^α reduces to

$$f^\alpha = \sum_\tau \frac{\partial T^{\alpha\tau}}{\partial x_\tau} \,,$$

or again, using the covariant tensor associated with $T^{\alpha\beta}$ in (4.20),

$$f^\alpha = \sum_\tau \frac{\partial T_{\alpha\tau}}{\partial x_\tau} \,,$$

thanks to (4.32). Replacing $T_{\alpha\tau}$ by the expression (4.30), we must obtain

$$c_1 \sum_\tau \frac{\partial R_{\alpha\tau}}{\partial x_\tau} + c_2 \sum_\tau \frac{\partial (R g_{\alpha\tau})}{\partial x_\tau} + c_3 \sum_\tau \frac{\partial g_{\alpha\tau}}{\partial x_\tau} = 0 \,. \tag{4.34}$$

[4] For the original text of this section in French see p. 212.

The last term is zero by (4.32). The second simplifies as follows:

$$\sum_\tau \frac{\partial (R g_{\alpha\tau})}{\partial x_\tau} = \sum_\tau g_{\alpha\tau} \frac{\partial R}{\partial x_\tau} = \frac{\partial R}{\partial x_\alpha} .$$

Then, from the definition of the invariant

$$R = \sum_\sigma \sum_\tau g^{\sigma\tau} R_{\sigma\tau} ,$$

we have

$$\frac{\partial R}{\partial x_\alpha} = \sum_\sigma \sum_\tau \frac{\partial}{\partial x_\alpha} (g^{\sigma\tau} R_{\sigma\tau}) = \sum_\tau \frac{\partial R_{\tau\tau}}{\partial x_\alpha} .$$

The constants c_1 and c_2 must therefore be chosen such that

$$c_1 \sum_\tau \frac{\partial R_{\alpha\tau}}{\partial x_\tau} + c_2 \sum_\tau \frac{\partial R_{\tau\tau}}{\partial x_\alpha} = 0 . \tag{4.35}$$

Recalling the definition (3.37) of the contracted Riemann tensor and the expression (3.22), we have

$$\begin{aligned} R_{\lambda\mu} &= \sum_\sigma \left(-\frac{\partial}{\partial x_\sigma} \begin{Bmatrix} \lambda\,\mu \\ \sigma \end{Bmatrix} + \frac{\partial}{\partial x_\mu} \begin{Bmatrix} \lambda\,\sigma \\ \sigma \end{Bmatrix} \right) \\ &\quad + \sum_\sigma \sum_\tau \left(\begin{Bmatrix} \lambda\,\sigma \\ \tau \end{Bmatrix} \begin{Bmatrix} \mu\,\tau \\ \sigma \end{Bmatrix} - \begin{Bmatrix} \lambda\,\mu \\ \sigma \end{Bmatrix} \begin{Bmatrix} \sigma\,\tau \\ \tau \end{Bmatrix} \right) , \end{aligned} \tag{4.36}$$

and the definition of the Christoffel symbol is

$$\begin{Bmatrix} \lambda\,\mu \\ \nu \end{Bmatrix} = \frac{1}{2} \sum_\sigma g^{\sigma\nu} \left(\frac{\partial g_{\lambda\sigma}}{\partial x_\mu} + \frac{\partial g_{\mu\sigma}}{\partial x_\lambda} - \frac{\partial g_{\lambda\mu}}{\partial x_\sigma} \right) . \tag{4.37}$$

We now calculate at the point M the expression for

$$\frac{\partial R_{\lambda\mu}}{\partial x_\nu}$$

when we use the particular coordinates defined above. Terms arising from derivatives of products of Christoffel symbols (second line of (4.36)) vanish with these symbols. The others come from second derivatives of the Christoffel symbols. From (4.37), these derivatives reduce using (4.32) and (4.33) to

$$\frac{\partial^2}{\partial x_\alpha \partial x_\beta} \begin{Bmatrix} \lambda\,\mu \\ \nu \end{Bmatrix} = \frac{1}{2} \left(\frac{\partial^3 g_{\lambda\nu}}{\partial x_\mu \partial x_\alpha \partial x_\beta} + \frac{\partial^3 g_{\mu\nu}}{\partial x_\lambda \partial x_\alpha \partial x_\beta} - \frac{\partial^3 g_{\lambda\mu}}{\partial x_\nu \partial x_\alpha \partial x_\beta} \right) .$$

In each derivative there are five indices α, β, λ, μ, and ν. We can characterise each derivative by the two indices on the potential and use the short notation

$$\frac{1}{2}\frac{\partial^3 g_{\lambda\nu}}{\partial x_\mu \partial x_\alpha \partial x_\beta} = (\lambda\nu)\,.$$

Then we have

$$\frac{\partial^2}{\partial x_\alpha \partial x_\beta}\begin{Bmatrix}\lambda\,\mu\\ \nu\end{Bmatrix} = (\lambda\nu)+(\mu\nu)-(\lambda\mu)\,,$$

the five indices being α, β, λ, μ, and ν, and with this notation the relation

$$\frac{\partial R_{\lambda\mu}}{\partial x_\nu} = \sum_\sigma \left(-\frac{\partial^2}{\partial x_\nu \partial x_\sigma}\begin{Bmatrix}\lambda\,\mu\\ \sigma\end{Bmatrix} + \frac{\partial^2}{\partial x_\nu \partial x_\mu}\begin{Bmatrix}\lambda\,\sigma\\ \sigma\end{Bmatrix}\right)$$

can be written

$$\begin{aligned}\frac{\partial R_{\lambda\mu}}{\partial x_\nu} &= -(\lambda\sigma)-(\mu\sigma)+(\lambda\mu)\\ &\quad +(\lambda\sigma)+(\sigma\sigma)-(\lambda\sigma)\,,\end{aligned}$$

or

$$\frac{\partial R_{\lambda\mu}}{\partial x_\nu} = -(\lambda\sigma)-(\mu\sigma)+(\lambda\mu)+(\sigma\sigma)\,. \tag{4.38}$$

The five indices are

$$\lambda\,,\ \mu\,,\ \nu\,,\ \sigma\,,\ \sigma\,,$$

with a sum over σ.

We must calculate the two expressions featuring in (4.35),

$$\sum_\tau \frac{\partial R_{\alpha\tau}}{\partial x_\tau}\,, \qquad \sum_\tau \frac{\partial R_{\tau\tau}}{\partial x_\alpha}\,.$$

The first is obtained by setting

$$\lambda = \alpha\,, \qquad \mu = \nu = \tau\,,$$

in (4.38) and summing over τ to yield

$$\sum_\tau \frac{\partial R_{\alpha\tau}}{\partial x_\tau} = -(\alpha\sigma)-(\tau\sigma)+(\alpha\tau)+(\sigma\sigma)\,.$$

The five indices are α, σ, σ, τ, and τ. We can exchange the two dummy indices σ and τ in the sums:

$$(\alpha\sigma) = (\alpha\tau)\,.$$

Cancelling like terms, we now arrive at

$$\sum_\tau \frac{\partial R_{\alpha\tau}}{\partial x_\tau} = -(\tau\sigma)+(\sigma\sigma)\,. \tag{4.39}$$

The second expression is found by setting

$$\nu = \alpha\,, \qquad \lambda = \mu = \tau\,,$$

in (4.38), then summing over τ to yield

$$\sum_\tau \frac{\partial R_{\tau\tau}}{\partial x_\alpha} = -(\tau\sigma) - (\tau\sigma) + (\tau\tau) + (\sigma\sigma)\,.$$

The five indices are still α, σ, σ, τ, and τ. We have

$$(\sigma\sigma) = (\tau\tau)\,.$$

Hence,

$$\sum_\tau \frac{\partial R_{\tau\tau}}{\partial x_\alpha} = -2(\tau\sigma) + 2(\sigma\sigma)\,. \tag{4.40}$$

Combining (4.39) and (4.40), equation (4.35) can thus be written

$$c_1 \sum_\tau \frac{\partial R_{\alpha\tau}}{\partial x_\tau} + c_2 \sum_\tau \frac{\partial R_{\tau\tau}}{\partial x_\alpha} = (c_1 + 2c_2)\big[-(\tau\sigma) + (\sigma\sigma)\big] = 0\,.$$

This can be satisfied identically by setting

$$c_1 + 2c_2 = 0\,. \tag{4.41}$$

With this condition, the equations (4.31) are a consequence of equation (4.30) when we use the particular coordinate system chosen for this proof. The same will be true in any coordinate system because, when all the components of a tensor vanish at a point in one coordinate system, they will likewise all vanish at this point for any other coordinate system.

The gravitational equations must therefore take the form

$$T_{\mu\nu} = c_1 \left(R_{\mu\nu} - \frac{1}{2} g_{\mu\nu} R\right) + c_3 g_{\mu\nu}\,.$$

These equations can be solved for $R_{\mu\nu}$. Indeed, noting that

$$\sum_\mu \sum_\nu g_{\mu\nu} g^{\mu\nu} = \sum_\mu g_\mu^\mu = 4\,,$$

we have

$$T = \sum_\mu \sum_\nu g^{\mu\nu} T_{\mu\nu} = c_1 (R - 2R) + 4c_3 = -c_1 R + 4c_3\,.$$

Then,

$$R_{\mu\nu} = -\kappa\left(T_{\mu\nu} - \frac{1}{2}g_{\mu\nu}T\right) + \lambda g_{\mu\nu}\,, \tag{4.42}$$

where we have set

$$\kappa = -\frac{1}{c_1}\,, \qquad \lambda = \frac{c_3}{c_1}\,.$$

This is the form in which the general equations of gravitation are usually written. As we have seen, they imply the general dynamical equations

$$f^\mu \equiv \sum_\sigma \frac{\partial \mathscr{T}^{\mu\sigma}}{\partial x_\sigma} + \sum_\sigma\sum_\tau \left\{ {\sigma\,\tau \atop \mu} \right\} \mathscr{T}^{\sigma\tau} = 0\,. \tag{4.43}$$

In the next section, we shall examine the consequences of the gravitational equations when we set $\lambda = 0$. Later, we shall see what changes are introduced when this constant is not zero.

4.4 Applications to Astronomy[5]

It is always possible to choose a coordinate system such that at a point M $\mathrm{d}s^2$ reduces to

$$\mathrm{d}s^2 = -\mathrm{d}x^2 - \mathrm{d}y^2 - \mathrm{d}z^2 + \mathrm{d}t^2\,, \tag{4.44}$$

the derivatives of the $g_{\mu\nu}$ are zero, and the d'Alembertians of the $g_{\mu\nu}$ reduce to

$$\Box g_{\mu\nu} = 2R_{\mu\nu}\,. \tag{4.45}$$

These conditions can always be realised to within a given approximation in some domain containing M.

If we imagine that they are indeed realised throughout the Solar System, we obtain an approximation which corresponds to classical mechanics. By examining a second approximation, we can explain the motion of the perihelion of Mercury's orbit, something that the classical theory failed to account for, without making any further assumptions.

Equation (4.44) expresses the fact that the geometry is Euclidean and that the coordinates x, y, z, can be considered as Cartesian coordinates relative to three orthogonal axes. The time t is the time indicated by a stationary clock and can be measured, for example, by the Earth's rotation relative to the distant stars. The unit of time is chosen such that the unit of speed is the speed of the propagation of light in vacuum. This is the time required for light to propagate a unit length.

[5] For the original text of this section in French see p. 217.

The derivatives of $g_{\mu\nu}$ are approximately zero. The smaller they are, the greater the region where equation (4.44) and its consequences are acceptable. We shall assume that equation (4.44) and also the vanishing of the derivatives of the $g_{\mu\nu}$ are exact in the limit as we move infinitely far from any mass.

Equation (4.45) is calculated using (4.13),

$$g_{\mu\nu}(x,y,z,t) = -\frac{1}{2\pi}\iiint \frac{R_{\mu\nu}(\xi,\eta,\zeta,t-r)}{r}\,\mathrm{d}\xi\,\mathrm{d}\eta\,\mathrm{d}\zeta\,,$$

and taking into account the gravitational equations

$$R_{\mu\nu} = -\kappa\left(T_{\mu\nu} - \frac{1}{2}g_{\mu\nu}T\right), \tag{4.46}$$

whence

$$g_{\mu\nu} = \frac{\kappa}{2\pi}\iiint \frac{T_{\mu\nu} - \frac{1}{2}g_{\mu\nu}T}{r}\,\mathrm{d}\xi\,\mathrm{d}\eta\,\mathrm{d}\zeta\,. \tag{4.47}$$

We may also ignore the delay in the calculation of the potentials because the speeds of the celestial bodies are small compared to the unit speed (the speed of light).

By definition, we have

$$\begin{aligned} T_{\mu\nu} &= \sum_\sigma\sum_\tau g_{\mu\sigma}g_{\nu\tau}T^{\sigma\tau} \\ &= \frac{\rho}{\sqrt{-g}}\sum_\sigma\sum_\tau g_{\mu\sigma}g_{\nu\tau}\frac{\mathrm{d}x_\sigma}{\mathrm{d}s}\frac{\mathrm{d}x_\tau}{\mathrm{d}s}\,. \end{aligned}$$

As the celestial bodies move slowly, we may set

$$\frac{\mathrm{d}x_1}{\mathrm{d}s} = v_x\,, \qquad \frac{\mathrm{d}x_2}{\mathrm{d}s} = v_y\,, \qquad \frac{\mathrm{d}x_3}{\mathrm{d}s} = v_z\,, \qquad \frac{\mathrm{d}x_4}{\mathrm{d}s} = \sqrt{1-v^2}\,.$$

In particular, when the velocities v_x, v_y, v_z of these bodies are zero or negligible, $T_{\mu\nu}$ reduces to

$$T_{\mu\nu} = \frac{\rho}{\sqrt{-g}}g_{\mu 4}g_{\nu 4}\,.$$

When solving (4.47), if κ is a small quantity, as we shall indeed soon observe, we may replace $g_{\mu\nu}$ under the integral sign by the approximate values extracted from (4.44), which we denote by $\delta_{\mu\nu}$, whence

$$\delta_{\mu\nu} = \begin{array}{cccc} -1 & 0 & 0 & 0 \\ 0 & -1 & 0 & 0 \\ 0 & 0 & -1 & 0 \\ 0 & 0 & 0 & 1 \end{array}$$

All the $T_{\mu\nu}$ are then zero except for

$$T_{44} = \rho \, .$$

The invariant T is equal to

$$T = \frac{\rho}{\sqrt{-g}} \approx \rho \, .$$

Defining

$$\varpi = \frac{\kappa}{4\pi} \iiint \frac{\rho}{r} \mathrm{d}\xi \, \mathrm{d}\eta \, \mathrm{d}\zeta \, , \tag{4.48}$$

the solution of (4.47) is found to be

$$g_{\mu\nu} = \delta_{\mu\nu} + g_{\mu}^{\nu} \varpi \, , \tag{4.49}$$

where g_{μ}^{ν} still denotes

$$g_{\mu}^{\nu} = \begin{cases} 1 & \text{if } \mu = \nu \, , \\ 0 & \text{if } \mu \neq \nu \, . \end{cases}$$

The determinant of the $g_{\mu\nu}$ will thus be

$$g = \begin{vmatrix} -1+\varpi & 0 & 0 & 0 \\ 0 & -1+\varpi & 0 & 0 \\ 0 & 0 & -1+\varpi & 0 \\ 0 & 0 & 0 & 1+\varpi \end{vmatrix} \approx -1 + 2\varpi \, .$$

The $g^{\mu\nu}$ are equal to the minors of the corresponding elements in this determinant, divided by the value of the determinant. Since

$$\frac{1}{g} \approx -1 - 2\varpi \, ,$$

they are given by the table

$$g^{\mu\nu} = \begin{matrix} -1-\varpi & 0 & 0 & 0 \\ 0 & -1-\varpi & 0 & 0 \\ 0 & 0 & -1-\varpi & 0 \\ 0 & 0 & 0 & 1-\varpi \end{matrix}$$

that is,

$$g^{\mu\nu} = \delta_{\mu\nu} - g_{\mu}^{\nu} \varpi \, . \tag{4.50}$$

Finally,

$$\sqrt{-g} = 1 - \varpi \,. \tag{4.51}$$

The equations of motion for a free point are calculated using (2.15),

$$\frac{d^2 x_i}{dx_4^2} + \sum_\mu \sum_\nu \left(\left\{ \begin{matrix} \mu\ \nu \\ i \end{matrix} \right\} - \left\{ \begin{matrix} \mu\ \nu \\ 4 \end{matrix} \right\} \frac{dx_i}{dx_4} \right) \frac{dx_\mu}{dx_4} \frac{dx_\nu}{dx_4} = 0 \qquad (i = 1,2,3) \,, \tag{4.52}$$

where the coordinates x_1, x_2, x_3, x_4 should be replaced by x, y, z, t, respectively.

The Christoffel symbols are calculated using (2.13), which yields

$$\begin{aligned} \left\{ \begin{matrix} \sigma\ \sigma \\ \tau \end{matrix} \right\} &= -\frac{1}{2} \frac{1}{g_{\tau\tau}} \frac{\partial g_{\sigma\sigma}}{\partial x_\tau} \approx -\frac{1}{2} \delta_{\tau\tau} \frac{\partial \varpi}{\partial x_\tau} \qquad (\sigma \neq \tau) \,, \\ \left\{ \begin{matrix} \sigma\ \tau \\ \sigma \end{matrix} \right\} &= \frac{1}{2} \frac{1}{g_{\sigma\sigma}} \frac{\partial g_{\sigma\sigma}}{\partial x_\tau} \approx \frac{1}{2} \delta_{\sigma\sigma} \frac{\partial \varpi}{\partial x_\tau} \,. \end{aligned} \tag{4.53}$$

The equations of motion will thus be[6]

$$\frac{d^2 x}{dt^2} + \frac{1}{2} (1 - 3v_x^2 + v_y^2 + v_z^2) \frac{\partial \varpi}{\partial x} - v_x v_y \frac{\partial \varpi}{\partial y} - v_x v_z \frac{\partial \varpi}{\partial z} - \frac{1}{2} v_x \frac{\partial \varpi}{\partial t} = 0 \,, \tag{4.54}$$

and similar equations. These are the approximate equations of motion when the masses producing the field are stationary and when we adopt the approximation indicated at the beginning of this section.

The acceleration is a function of the velocity of the moving body. This velocity is expressed taking the speed of light as a unit and is therefore very small. The main term is therefore the one that does not depend on the velocity:

$$\frac{d^2 x}{dt^2} + \frac{1}{2} \frac{\partial \varpi}{\partial x} = 0 \,,$$

whence ϖ is proportional to the Newtonian potential. This is equivalent to the equation (4.1). We deduce that ϖ is twice the Newtonian potential. To obtain this, the constant κ must be chosen in such a way that

$$\frac{\kappa}{4\pi} = 2 \frac{K}{c^2} \,,$$

where K is the gravitational constant introduced in (4.2) when the unit of time is chosen as we have done here. We will thus have

$$\kappa = \frac{8\pi K}{c^2} = 1.87 \times 10^{-27} \,, \tag{4.55}$$

[6] Correction by G. Lemaître: "formula (4.54), read
$$\frac{d^2 x}{dt^2} + \frac{1}{2} (1 - 3v_x^2 + v_y^2 + v_z^2) \frac{\partial \varpi}{\partial x} - 2v_x v_y \frac{\partial \varpi}{\partial y} - 2v_x v_z \frac{\partial \varpi}{\partial z} - \frac{1}{2} v_x (3 - v^2) \frac{\partial \varpi}{\partial t} = 0 \,."$$

which is indeed a very small quantity. To obtain the equations of motion in C.G.S. units, the equations of motion should be multiplied by c^2.

Equation (4.54) is an approximation. To obtain a better approximation, we may take the one already obtained for all the terms in (4.52) containing the small quantities v_x, v_y, v_z, and the only one for which we need a better approximation is therefore

$$\begin{Bmatrix} 4\,4 \\ i \end{Bmatrix} = \sum_\sigma g^{i\sigma} \begin{bmatrix} 4\,4 \\ \sigma \end{bmatrix} \approx (1+\varpi)\left(-\frac{\partial g_{4i}}{\partial x_4} + \frac{1}{2}\frac{\partial g_{44}}{\partial x_i}\right) .$$

Here, $\partial g_{4i}/\partial x_4$ will be zero to a first approximation since g_{4i} vanishes to a first approximation. The time rates of change of the potentials should be small since the speeds of the celestial bodies are assumed to be small. If we consider the standard case of a quasi-stationary field, we will be able to ignore the time derivatives of the potentials in the present calculation. It thus remains to find a better approximation for $\partial g_{44}/\partial x_i$, ignoring the derivatives with respect to x_4. By (4.36), we have

$$\begin{aligned} R_{44} &= \sum_\sigma \left(-\frac{\partial}{\partial x_\sigma}\begin{Bmatrix} 4\,4 \\ \sigma \end{Bmatrix} + \frac{\partial}{\partial x_4}\begin{Bmatrix} 4\,\sigma \\ \sigma \end{Bmatrix}\right) \\ &\quad + \sum_\sigma \sum_\tau \left(\begin{Bmatrix} 4\,\sigma \\ \tau \end{Bmatrix}\begin{Bmatrix} 4\,\tau \\ \sigma \end{Bmatrix} - \begin{Bmatrix} 4\,4 \\ \tau \end{Bmatrix}\begin{Bmatrix} \sigma\,\tau \\ \sigma \end{Bmatrix}\right) . \end{aligned}$$

The second term is zero if we assume that the derivatives with respect to x_4 vanish.

Let us calculate the others, treating ϖ as a small quantity of the first order and ignoring terms of order higher than the second. We will have

$$\frac{\partial}{\partial x_4} = 0 , \qquad \begin{Bmatrix} 4\,4 \\ \sigma \end{Bmatrix} = -\frac{1}{2}\sum_\tau g^{\sigma\tau}\frac{\partial g_{44}}{\partial x_\tau} ,$$

and hence,

$$-\sum_\sigma \frac{\partial}{\partial x_\sigma}\begin{Bmatrix} 4\,4 \\ \sigma \end{Bmatrix} = \frac{1}{2}\sum_\sigma\sum_\tau g^{\sigma\tau}\frac{\partial^2 g_{44}}{\partial x_\sigma \partial x_\tau} + \frac{1}{2}\sum_\sigma\sum_\tau \frac{\partial g^{\sigma\tau}}{\partial x_\sigma}\frac{\partial g_{44}}{\partial x_\tau} .$$

To a first approximation, we will have

$$\begin{cases} g^{\sigma\tau} = -g_\sigma^\tau(1+\varpi) & (\sigma,\tau \neq 4) , \\ g^{44} = 1+\varpi , \end{cases}$$

and we may therefore write

$$-\sum_\sigma \frac{\partial}{\partial x_\sigma}\begin{Bmatrix} 4\,4 \\ \sigma \end{Bmatrix} = -\frac{1}{2}(1+\varpi)\sum_\sigma \frac{\partial^2 g_{44}}{\partial x_\sigma^2} - \frac{1}{2}\sum_\sigma\left(\frac{\partial \varpi}{\partial x_\sigma}\right)^2 .$$

To calculate the terms in the second line of the expression for R_{44}, we must replace the Christoffel symbols by their values obtained to the first order approximation as given in (4.53). We note that all the

$$\left\{ \begin{matrix} 4\,\sigma \\ \tau \end{matrix} \right\}$$

are zero except for

$$\left\{ \begin{matrix} 4\,4 \\ \tau \end{matrix} \right\} = \frac{1}{2} \frac{\partial \varpi}{\partial x_\tau} \qquad (\tau \neq 4)$$

and

$$\left\{ \begin{matrix} 4\,\sigma \\ 4 \end{matrix} \right\} = \frac{1}{2} \frac{\partial \varpi}{\partial x_\sigma} \qquad (\sigma \neq 4)\,.$$

It follows that

$$\sum_\sigma \sum_\tau \left\{ \begin{matrix} 4\,\sigma \\ \tau \end{matrix} \right\} \left\{ \begin{matrix} 4\,\tau \\ \sigma \end{matrix} \right\} = \frac{1}{2} \sum_\sigma \left(\frac{\partial \varpi}{\partial x_\sigma} \right)^2 .$$

Furthermore, from (2.12),

$$\sum_\sigma \left\{ \begin{matrix} \sigma\,\tau \\ \sigma \end{matrix} \right\} = \frac{1}{\sqrt{-g}} \frac{\partial \sqrt{-g}}{\partial x_\tau} = -\frac{\partial \varpi}{\partial x_\tau}\,,$$

whence

$$-\sum_\sigma \sum_\tau \left\{ \begin{matrix} \sigma\,\tau \\ \sigma \end{matrix} \right\} \left\{ \begin{matrix} 4\,4 \\ \tau \end{matrix} \right\} = -\frac{1}{2} \sum_\sigma \left(\frac{\partial \varpi}{\partial x_\sigma} \right)^2 .$$

Finally, we obtain

$$R_{44} = -\frac{1}{2}(1+\varpi) \sum_\sigma \frac{\partial^2 g_{44}}{\partial x_\sigma^2} - \frac{1}{2} \sum_\sigma \left(\frac{\partial \varpi}{\partial x_\sigma} \right)^2 .$$

We solve this equation for

$$\sum_\sigma \frac{\partial^2 g_{44}}{\partial x_\sigma^2} = \Box g_{44}\,.$$

In the present approximation, we will have

$$\Box g_{44} = -2(1+\varpi) R_{44} + \sum_\sigma \left(\frac{\partial \varpi}{\partial x_\sigma} \right)^2 .$$

Replacing R_{44} by the expression

$$R_{44} = \kappa \left(T_{44} - \frac{1}{2} T \right) = -\frac{1}{2} \kappa \rho\,,$$

we then obtain

$$\Box g_{44} = -\kappa(1-\varpi)\rho + \sum_\sigma \left(\frac{\partial \varpi}{\partial x_\sigma} \right)^2 .$$

We will thus obtain g_{44} in the form

$$g_{44} = 1 + \varpi$$

by replacing the density ρ by a fictitious density

$$\rho' = (1 - \varpi)\rho - \frac{1}{\kappa}\sum_{\sigma}\left(\frac{\partial \varpi}{\partial x_\sigma}\right)^2$$

in the calculation of ϖ from (4.48). If V denotes the Newtonian potential in C.G.S. units and g the corresponding acceleration

$$g^2 = \left(\frac{\partial V}{\partial x}\right)^2 + \left(\frac{\partial V}{\partial y}\right)^2 + \left(\frac{\partial V}{\partial z}\right)^2 ,$$

then the expression for ρ' can be written

$$\rho' = \left(1 - \frac{2V}{c^2}\right)\rho^2 - \frac{1}{\kappa c^4}g^2 . \tag{4.56}$$

The constant factor takes the value

$$\frac{1}{\kappa c^4} = 0.66 \times 10^{-15} .$$

The term in g^2 is the only one we have encountered so far in which the non-homogeneous nature of the gravitational equations shows up. The fact that these equations are not homogeneous means that the acceleration of a moving body in the field of several massive objects is not exactly the resultant of the accelerations that each of those massive objects would communicate to it were it acting alone. The difference will be extremely small, but it can contribute to certain phenomena, such as the motion of the perihelion of a planetary orbit.

We have thus found the equations of motion of the celestial bodies in an approximation that is widely suitable for astronomical applications. These are (4.54) and their analogues, where ϖ is calculated using the fictitious density ρ' instead of the real density ρ.

We have carried out this calculation assuming that we may ignore the influence of the velocities and rotations of the attracting masses. It is straightforward to take this influence into account. The tensor $T_{\mu\nu}$ can be written[7]

$$T_{\mu\nu} = \begin{matrix} \rho v_x^2 & \rho v_x v_y & \rho v_x v_z & \rho v_x \\ \rho v_y v_x & \rho v_y^2 & \rho v_y v_z & \rho v_y \\ \rho v_z v_x & \rho v_z v_y & \rho v_z^2 & \rho v_z \\ \rho v_x & \rho v_y & \rho v_z & \rho(1-v^2) \end{matrix}$$

The integral (4.47) can be calculated for each celestial body. If v_{0x}, v_{0y}, v_{0z} denote the velocity components of the centre of gravity of the body and ω_x, ω_y, ω_z the components of the angular velocity of the body, we can write

$$\begin{aligned} v_x &= v_{0x} + z\omega_y - y\omega_z \,, \\ v_y &= v_{0y} + x\omega_z - z\omega_x \,, \\ v_z &= v_{0z} + y\omega_x - x\omega_y \,, \end{aligned}$$

where x, y, z are the coordinates of the points relative to axes passing through the centre of gravity.

Let $\mathrm{d}\Pi$ be the volume element. Then,

$$\iiint \rho v_x^2 \mathrm{d}\Pi = M v_{0x}^2 + \frac{I}{2}(\omega_y^2 + \omega_z^2) \,,$$

where M is the mass of the body and I its moment of inertia about an axis passing through the centre of gravity. (We shall assume that the body is spherically symmetric.) Similarly,

$$\iiint \rho v_x v_y \mathrm{d}\Pi = M v_{0x} v_{0y} - \frac{I}{2}\omega_x \omega_y \,.$$

The values of

$$\int T_{\mu\nu} \mathrm{d}\Pi$$

are then given for each body in the following table, where we have dropped the subscript zero on the velocity components:

[7] Correction by G. Lemaître: "for this formula, read

$$T_{\mu\nu} = \begin{matrix} \rho v_x^2 & , \cdots & , \cdots & , -\rho v_x \\ \cdots & , \cdots & , \cdots & , -\rho v_y \\ \cdots & , \cdots & , \cdots & , -\rho v_z \\ -\rho v_x & , -\rho v_y & , -\rho v_z & , \rho(1+v^2) \end{matrix}$$ ".

$$
\begin{array}{cccc}
Mv_x^2+\frac{I}{2}(\omega_y^2+\omega_z^2) & Mv_yv_x-\frac{I}{2}\omega_y\omega_x & Mv_zv_x-\frac{I}{2}\omega_z\omega_x & Mv_x \\
Mv_xv_y-\frac{I}{2}\omega_x\omega_y & Mv_y^2+\frac{I}{2}(\omega_x^2+\omega_z^2) & Mv_zv_y-\frac{I}{2}\omega_z\omega_y & Mv_y \\
Mv_xv_z-\frac{I}{2}\omega_x\omega_z & Mv_yv_z-\frac{I}{2}\omega_y\omega_z & Mv_z^2+\frac{I}{2}(\omega_x^2+\omega_y^2) & Mv_z \\
Mv_x & Mv_y & Mv_z & M(1-v^2)-I\omega^2
\end{array}
$$

For example, we will have

$$
g_{11}=-1+\varpi+\frac{\kappa}{2\pi}\sum\frac{Mv_x^2+\frac{I}{2}(\omega_y^2+\omega_z^2)}{r}\,,
$$

$$
g_{12}=\frac{\kappa}{2\pi}\sum\frac{Mv_xv_y-\frac{I}{2}\omega_x\omega_y}{r}\,,
$$

$$
g_{14}=\frac{\kappa}{2\pi}\sum\frac{Mv_x}{r}\,,
$$

$$
g_{44}=1+\varpi-\frac{\kappa}{2\pi}\sum\frac{Mv^2+I\omega^2}{r}\,.
$$

Analogous expressions are formed by permuting the subscripts x, y, z.

Geometric Consequences

If we remain in the situation where we may ignore the influence of the velocities of the attracting masses, the fundamental form can be written

$$
\mathrm{d}s^2=(-1+\varpi)(\mathrm{d}x^2+\mathrm{d}y^2+\mathrm{d}z^2)+(1+\varpi)\mathrm{d}t^2\,.
$$

The corresponding geometry is characterised by the distance element

$$
\mathrm{d}\sigma^2=(1-\varpi)(\mathrm{d}x^2+\mathrm{d}y^2+\mathrm{d}z^2)\,.
$$

We can observe that this geometry is not Euclidean.

A map of the space can be constructed in a Euclidean space. By considering x, y, z as the Cartesian coordinates of the Euclidean space in which we construct the map, we observe that the scale will be independent of the direction of the lengths and equal to

$$
\frac{1}{\sqrt{1-\varpi}}\approx 1+\frac{1}{2}\varpi\,,
$$

or again, given the interpretation of ϖ as $\varpi = 2V/c^2$,

$$1+\frac{V}{c^2},$$

where V is the Newtonian potential in C.G.S. units.

This result can be interpreted by saying that the geometry is Euclidean but that rulers suffer a deformation proportional to

$$\sqrt{1-\varpi} \approx 1-\frac{V}{c^2}.$$

Naturally, this could be interpreted in other ways by using, instead of a conformal map as we have done here, any other map.

Hence in the case where there is only one attracting mass,

$$\varpi = -2K\frac{m}{r},$$

and using polar coordinates, we will have

$$\mathrm{d}\sigma^2 = \left(1+\frac{2Km}{r}\right)\left[\mathrm{d}r^2 + r^2\left(\sin^2\theta\,\mathrm{d}\varphi^2 + \mathrm{d}\theta^2\right)\right].$$

By applying a coordinate transformation of the form

$$\rho^2 = r^2\left(1+\frac{2Km}{r}\right),$$

which implies

$$\rho\,\mathrm{d}\rho = r\left(1+\frac{2Km}{r}\right)\mathrm{d}r$$

and

$$\left(1+\frac{2Km}{r}\right)\mathrm{d}r^2 \approx \frac{\rho^2}{r^2}\mathrm{d}\rho^2 \approx \frac{\mathrm{d}\rho^2}{1-2Km/\rho},$$

it follows that

$$\mathrm{d}\sigma^2 = \frac{\mathrm{d}\rho^2}{1-2Km/\rho} + \rho^2\left(\sin^2\theta\,\mathrm{d}\varphi^2 + \mathrm{d}\theta^2\right).$$

We see that ρ is the length element in a direction normal to the radius vector.

We may say that the geometry is Euclidean but that our rulers suffer a dilation in the radial direction proportional to

$$\frac{1}{\sqrt{1-2Km/\rho}}.$$

In the case of a single mass at rest, Schwarzschild found the exact value of $\mathrm{d}s^2$ to be

$$\mathrm{d}s^2 = -\frac{\mathrm{d}r^2}{1+\varpi} - r^2(\sin^2\theta\,\mathrm{d}\varphi^2 + \mathrm{d}\theta^2) + (1+\varpi)\mathrm{d}t^2\,, \qquad \varpi = -2\frac{K}{c^2}\frac{m}{r}\,.$$

If we assume that the mass moves away along the negative x axis, we obtain in the limit as r and m tend to infinity,

$$\mathrm{d}s^2 = -\frac{\mathrm{d}x^2}{1+\varpi} - \mathrm{d}y^2 - \mathrm{d}z^2 + (1+\varpi)\mathrm{d}t^2\,, \qquad \varpi = \frac{2gx}{c^2}\,, \quad g = \lim\frac{Km}{r^2}\,.$$

This is the form found in (3.9) when we studied the artificial gravitational field obtained by a uniformly accelerating motion. We can now observe that this field is strictly the same as the field produced by an infinite mass, albeit infinitely far away.

Gravitational Deflection of Light Rays

Light rays follow geodesics along which $\mathrm{d}s$ is zero. We can observe that their velocity varies from point to point, independently of the direction, and is equal to

$$\frac{\mathrm{d}\sigma}{\mathrm{d}t} = \sqrt{1+\varpi} \approx 1 + \frac{K}{c^2}V\,.$$

It follows that light rays do not propagate along straight lines. Let us compare the acceleration along the x axis of a light ray with velocity

$$v_x = 0\,, \qquad v_y = 1\,, \qquad v_z = 0\,,$$

with the acceleration in the same direction of a stationary test mass. Equation (4.54) gives

$$\frac{\mathrm{d}^2x}{\mathrm{d}t^2} + \frac{\partial\varpi}{\partial x} = 0$$

for the light ray and

$$\frac{\mathrm{d}^2x}{\mathrm{d}t^2} + \frac{1}{2}\frac{\partial\varpi}{\partial x} = 0$$

for the test mass. The acceleration of the light ray perpendicular to its direction of propagation is therefore twice the acceleration of a mass at rest at the same point. We know that observations made during the eclipse of 29 May 1919 confirm these conclusions.

Redshift of Rays in the Solar Spectrum

The proper time on different celestial bodies is equal to

$$ds = \sqrt{1+\varpi}\,dt \approx \left(1+\frac{V}{c^2}\right)dt\,.$$

Identical clocks indicate the same lapse of time between the beginning and end of any of the similar periods by which they measure time. The corresponding time interval dt will thus differ in places where the potential is different.

If we assume that chemical elements have the same constitution in the Sun and on Earth, we must conclude that the characteristic wavelengths of these elements will be *longer* on the Sun by a factor equal to

$$1+\frac{V_E - V_S}{c^2}\,.$$

It is extremely difficult to check this conclusion experimentally owing to certain anomalies in the observation of the Sun's rays, and above all because it is difficult to determine the influence of pressure on the observed shifts. However, even though they may not be definitive, the results of research carried out so far are clearly favourable to the conclusions of the new theory.

4.5 Fixed Stars[8]

When we try to apply Newton's theory of gravity to the general distribution of stars, we encounter difficulties for which the new theory provides a perfectly satisfactory solution.

Let us suppose to begin with that the stars are finite in number. The density of the stellar universe will then tend to zero as we move away indefinitely. This idea raises several difficulties. The light emitted by the stars will move away and never return. The energy of the stars will gradually dissipate to infinity. Moreover, it is not obvious how to account for the stability of the universe. We may compare the multitude of stars to the molecules in a gas. The central forces acting between the stars are analogous to those assumed to act between the molecules in the kinetic theory of gases. But we cannot assume the stability of a gas that is not contained within walls and whose density tends to zero as we move toward infinity. Likewise, we cannot assume a similar solution for the general distribution of stars. We could abandon Newton's laws and assume that the potential increases without limit at infinity. Those stars with a tendency to drift away for ever would thus be brought back toward the others. We should therefore observe that the speeds of the more distant stars are much greater than those of nearby stars. But observation reveals no such thing.

[8] For the original text of this section in French see p. 229.

If on the other hand we assume that, as we move away indefinitely, we encounter a mean density of stars equal to the mean density of stars observed in the region accessible to our observations, we must necessarily assume that there are infinitely many stars. Newton's law does not apply and we should observe an infinite force directed toward the centre of gravity of the general distribution of stars. On the other hand, it is not obvious what would be the centre of gravity of an infinite homogeneous mass, and there is no way we could accept the idea of these infinite forces.

The latter problems can be avoided by modifying Newton's law. We need only replace the Newtonian potential V satisfying Poisson's equation

$$\Box V = 4\pi K\rho \, , \tag{4.57}$$

where K is the constant of attraction and ρ the density of the matter, by a potential V satisfying the equation

$$\Box V - \lambda V = 4\pi K\rho \, , \tag{4.58}$$

where λ is a universal constant.

When we consider a region containing a very large number of stars, the density can be treated as a constant ρ_0 and the modified Poisson equation then has a constant solution

$$V = -\frac{4\pi K}{\lambda}\rho_0 \, .$$

We can therefore conceive of an infinite universe in which the average density is constant. But the difficulty involved in assuming an infinite number of stars remains.

What modifications must be made to the tensor equations of gravity to ensure that they reduce to the theory we have just proposed by adopting the usual coordinates? When matter is at rest, the equations

$$R_{\mu\nu} = -\kappa\left(T_{\mu\nu} - \frac{1}{2}g_{\mu\nu}T\right)$$

reduce to

$$R_{\mu\nu} = -\kappa g_\mu^\nu \frac{\rho}{\sqrt{-g}} \, ,$$

and using special coordinates, we may write

$$\Box g_{\mu\nu} = \frac{R_{\mu\nu}}{2} = -\frac{\kappa}{2}g_\mu^\nu \rho \, .$$

These are equivalent to Poisson's equation in its usual form (4.57). We obtain to first approximation

$$g_{\mu\nu} = \delta_{\mu\nu} + 2V \, . \tag{4.59}$$

We have seen that the gravitational equations can be put in the more general form (4.42),

$$R_{\mu\nu} - \lambda g_{\mu\nu} = -\kappa\left(T_{\mu\nu} - \frac{1}{2}g_{\mu\nu}T\right) . \tag{4.60}$$

We observe that the approximate solution (4.59) for $g_{\mu\nu}$ will reduce these equations to the new form of Poisson's equation

$$\Box V - \lambda V = 4\pi K\rho .$$

We shall now integrate these new equations.

In a universe of constant density,

$$\rho = \rho_0 .$$

As usual, let $\delta_{\mu\nu}$ denote coefficients of the form

$$\mathrm{d}s^2 = -\mathrm{d}x^2 - \mathrm{d}y^2 - \mathrm{d}z^2 + \mathrm{d}t^2 .$$

We shall show that the equations (4.60) possess the solution

$$g_{44} = 1 , \qquad g_{i4} = 0 , \qquad g_{ij} = -\gamma_{ij} \qquad (i, j = 1, 2, 3) , \tag{4.61}$$

where the γ_{ij} are the metric potentials for a spherical space of constant curvature R.

These potentials are obtained by eliminating x_4 from the equations

$$\mathrm{d}\sigma^2 = \mathrm{d}x_1^2 + \mathrm{d}x_2^2 + \mathrm{d}x_3^2 + \mathrm{d}x_4^2$$

and

$$R^2 = x_1^2 + x_2^2 + x_3^2 + x_4^2 .$$

By defining

$$r^2 = x_1^2 + x_2^2 + x_3^2 , \tag{4.62}$$

we find

$$\mathrm{d}\sigma^2 = \sum_{\mu=1}^{3}\sum_{\nu=1}^{3} \gamma_{\mu\nu}\mathrm{d}x_\mu \mathrm{d}x_\nu ,$$

with

$$\gamma_{\mu\nu} = g_\mu^\nu + \frac{x_\mu x_\nu}{R^2 - r^2} . \tag{4.63}$$

In the current situation, all points must be equivalent. The solution (4.61) satisfies this condition. This follows from the way we calculate the $\gamma_{\mu\nu}$.

It suffices therefore to establish this for an arbitrary point. We choose

$$x_1 = x_2 = x_3 = x_4 = 0\,.$$

The first derivatives of the $g_{\mu\nu}$ vanish at this point. The second derivatives are all zero at this point as well except for

$$\frac{\partial^2 g_{ij}}{\partial x_i \partial x_j} = -\frac{1}{R^2} \qquad (i, j = 1, 2, 3)\,.$$

The $R_{\mu\nu}$ then reduce to

$$R_{\mu\nu} = \sum_{\sigma=1}^{3} \frac{\partial}{\partial x_\sigma} \begin{bmatrix} \mu\ \nu \\ \sigma \end{bmatrix} - \sum_{\sigma=1}^{3} \frac{\partial}{\partial x_\nu} \begin{bmatrix} \mu\ \sigma \\ \sigma \end{bmatrix}\,.$$

For $\mu = \nu = 1, 2, 3$, we find

$$R_{\mu\mu} = \frac{1}{2} \frac{1}{R^2}\,,$$

whereas $R_{\mu\nu}$ is zero for $\mu = 4 = \nu$ and for $\mu \neq \nu$.

Then for $\mu = \nu = 1, 2, 3$, (4.60) becomes

$$\frac{2}{R^2} + \lambda = -\frac{\kappa\rho}{2}\,,$$

and for $\mu = 4 = \nu$,

$$-\lambda = -\frac{\kappa\rho}{2}\,.$$

These equations are satisfied if we identify

$$\lambda = \frac{\kappa\rho}{2} = \frac{1}{R^2}\,. \tag{4.64}$$

The geometry of stationary bodies in the field is characterised by the distance element (4.63).

The $\gamma_{\mu\nu}$ are the metric potentials of the spherical geometry (Riemannian geometry in the narrow sense). Although unbounded, the space nevertheless has a finite volume. Hence, we no longer need to assume an infinite number of stars, as we did previously.

Let us calculate the total volume of the space. We know that it is equal to the integral of the volume element (1.8) over the whole space,

$$\iiint \sqrt{\gamma}\, \mathrm{d}x_1\, \mathrm{d}x_2\, \mathrm{d}x_3\,,$$

where γ is the determinant of the potentials $\gamma_{\mu\nu}$:

$$\gamma = \begin{vmatrix} 1+\dfrac{x_1^2}{R^2-r^2} & \dfrac{x_2x_1}{R^2-r^2} & \dfrac{x_3x_1}{R^2-r^2} \\ \dfrac{x_1x_2}{R^2-r^2} & 1+\dfrac{x_2^2}{R^2-r^2} & \dfrac{x_3x_2}{R^2-r^2} \\ \dfrac{x_1x_3}{R^2-r^2} & \dfrac{x_2x_3}{R^2-r^2} & 1+\dfrac{x_3^2}{R^2-r^2} \end{vmatrix} .$$

This determinant is equal to

$$\gamma = \frac{R^2}{R^2-r^2} .$$

According to (4.62), we can write

$$\mathrm{d}x_1\,\mathrm{d}x_2\,\mathrm{d}x_3 = 4\pi r^2 \mathrm{d}r .$$

Hence the total volume of the space is

$$\int_0^\infty \frac{4\pi R r^2 \mathrm{d}r}{\sqrt{R^2-r^2}} = 2\pi R^3 .$$

The total mass of an unbounded universe of constant density ρ is therefore

$$M = 2\pi\rho R^3 , \tag{4.65}$$

so that, by (4.62),

$$M = 4\pi\frac{R}{\kappa} = \sqrt{\frac{32\pi^2}{\kappa^3\rho}} . \tag{4.66}$$

Chapter 5
Electric Charges[1]

The definition of the electromagnetic field by the motion it communicates to electrically charged particles was investigated in Section 2.2. The laws describing the action of the fields on these particles were thus expressed in a form that was independent of the frames of reference used to study these phenomena. It is a straightforward matter to show that the quantities introduced in this way must be tensors.

In addition to the invariant ds, we considered another invariant,

$$\varphi_1 \mathrm{d}x_1 + \varphi_2 \mathrm{d}x_2 + \varphi_3 \mathrm{d}x_3 + \varphi_4 \mathrm{d}x_4 = \sum_\alpha \varphi_\alpha \mathrm{d}x_\alpha \,.$$

Given a change of coordinates, the electromagnetic potentials φ_α transform according to equations of the form

$$\varphi'_\sigma = \sum_\alpha \frac{\partial x_\alpha}{\partial x'_\sigma} \varphi_\alpha \,. \tag{5.1}$$

This implies that φ_σ is a covariant tensor. The electromagnetic field

$$F_{\mu\nu} = \frac{\partial \varphi_\mu}{\partial x_\nu} - \frac{\partial \varphi_\nu}{\partial x_\mu} \tag{5.2}$$

is also a tensor. Indeed, we have

$$F'_{\mu\nu} = \frac{\partial \varphi'_\mu}{\partial x'_\nu} - \frac{\partial \varphi'_\nu}{\partial x'_\mu} \,,$$

and according to (5.1),

$$\frac{\partial \varphi'_\mu}{\partial x'_\nu} = \sum_\beta \frac{\partial x_\beta}{\partial x'_\nu} \frac{\partial}{\partial x_\beta} \sum_\alpha \frac{\partial x_\alpha}{\partial x'_\mu} \varphi_\alpha \,.$$

[1] For the original text of this chapter in French see p. 234.

G. Lemaître, *Learning the Physics of Einstein with Georges Lemaître*,
https://doi.org/10.1007/978-3-030-22030-3_5

By carrying out the differentiation,

$$\frac{\partial \varphi'_\mu}{\partial x'_\nu} = \sum_\alpha \sum_\beta \frac{\partial x_\alpha}{\partial x'_\mu} \frac{\partial x_\beta}{\partial x'_\nu} \frac{\partial \varphi_\alpha}{\partial x_\beta} + \sum_\alpha \sum_\beta \frac{\partial x_\beta}{\partial x'_\nu} \frac{\partial}{\partial x_\beta} \left(\frac{\partial x_\alpha}{\partial x'_\mu} \right) \varphi_\alpha \,.$$

The last term can be written

$$\sum_\alpha \frac{\partial^2 x_\alpha}{\partial x'_\mu \partial x'_\nu} \varphi_\alpha \,,$$

which is symmetric under the exchange of μ and ν.

We thus obtain, as claimed,

$$F'_{\mu\nu} = \sum_\alpha \sum_\beta \frac{\partial x_\alpha}{\partial x'_\mu} \frac{\partial x_\beta}{\partial x'_\nu} F_{\alpha\beta} \,.$$

The electric current and electric charge have been represented by

$$\mathcal{J}^\alpha = e \frac{\mathrm{d}x_\alpha}{\mathrm{d}s} \,, \tag{5.3}$$

where e is the charge density. This quantity is not a tensor. To obtain a tensor, we must do as we did previously in the case of material masses, whence

$$J^\alpha = \frac{e}{\sqrt{-g}} \frac{\mathrm{d}x_\alpha}{\mathrm{d}s}$$

will be a tensor. Then $\mathcal{J}^\alpha$ is the corresponding tensor density. Finally, the electromagnetic force

$$f_\mu = -\sum_\sigma F_{\mu\sigma} \mathcal{J}^\sigma \tag{5.4}$$

is also a tensor density.

The quantity f_μ in (2.29) is the product of m with the left-hand side of the equation (2.7) obtained when studying the motion of a free point. Moreover, f^μ is the product of m with the left-hand side of (2.10) of the same section.[2] The way in which we proceed from one form to the other shows that f_μ and f^μ are associated tensor densities. We thus obtain

$$f^\mu = \sum_\alpha g_{\mu\alpha} f^\alpha \,.$$

[2] Correction by G. Lemaître: "rather than 'f^μ is the product of m' read 'f^μ is the product of $-m$'. Consequently one has to change the sign for the expressions of f_μ as a function of $\mathcal{T}_\mu^\nu$ (in the four equations hereafter and in (5.11)); the second term in formula (5.16) then changes sign. The text that follows that formula is then correct."

We expressed f^α as a function of the material energy tensor in (4.27),

$$f^\alpha = \sum_\sigma \frac{\partial \mathcal{T}^{\alpha\sigma}}{\partial x_\sigma} + \sum_\sigma \sum_\tau \left\{ \begin{matrix} \sigma\ \tau \\ \alpha \end{matrix} \right\} \mathcal{T}^{\sigma\tau} .$$

This means that f_μ is expressed in terms of the material energy tensor by the equation

$$f_\mu = \sum_\alpha \sum_\sigma g_{\mu\alpha} \frac{\partial \mathcal{T}^{\alpha\sigma}}{\partial x_\sigma} + \sum_\alpha \sum_\sigma \sum_\tau g_{\alpha\mu} \left\{ \begin{matrix} \sigma\ \tau \\ \alpha \end{matrix} \right\} \mathcal{T}^{\sigma\tau} .$$

This expression can be transformed by introducing the tensor density

$$\mathcal{T}_\mu^\sigma = \sum_\alpha g_{\mu\alpha} \mathcal{T}^{\alpha\sigma}$$

associated with $\mathcal{T}^{\alpha\sigma}$.

By noting from (2.18) that

$$\sum_\alpha g_{\alpha\mu} \left\{ \begin{matrix} \sigma\ \tau \\ \alpha \end{matrix} \right\} = \left[\begin{matrix} \sigma\ \tau \\ \mu \end{matrix} \right] ,$$

and by using the symmetry of $\mathcal{T}^{\sigma\tau}$, we may write

$$f_\mu = \sum_\sigma \frac{\partial \mathcal{T}_\mu^\sigma}{\partial x_\sigma} - \sum_\alpha \sum_\sigma \frac{\partial g_{\mu\alpha}}{\partial x_\sigma} \mathcal{T}^{\alpha\sigma} + \sum_\tau \sum_\sigma \frac{\partial g_{\tau\mu}}{\partial x_\sigma} \mathcal{T}^{\sigma\tau} - \frac{1}{2} \sum_\sigma \sum_\tau \frac{\partial g_{\sigma\tau}}{\partial x_\mu} \mathcal{T}^{\sigma\tau} .$$

The second and third terms cancel one another, as can be seen immediately by changing the dummy index α to τ. This leaves

(5.5)
$$f_\mu = \sum_\sigma \frac{\partial \mathcal{T}_\mu^\sigma}{\partial x_\sigma} - \frac{1}{2} \sum_\sigma \sum_\tau \frac{\partial g_{\sigma\tau}}{\partial x_\mu} \mathcal{T}^{\sigma\tau} .$$

Equations (5.4) and (5.5) are the equations of motion of electrical charges expressed in terms of the material energy tensor.

The laws just outlined provide a complete picture of the way the fields act on electric charges. We must now consider the way these charges act on the fields. We begin by looking at what experience can tell us about this.

The law which tells us how the scalar potential depends on the charge distribution is analogous to Newton's law of gravity. If we use a suitable physical unit for the electric charge (Heaviside units), it can be written in the form

$$\varphi = \sum \frac{1}{4\pi} \frac{e}{r} ,$$

or alternatively, using Poisson's equation,

(5.6)
$$\Box \varphi = -e .$$

These laws are equivalent to Coulomb's law.

The action of currents or material particles on the field is expressed by Laplace's law. Using the electromagnetic potential

$$-\varphi_x\,,\ -\varphi_y\,,\ -\varphi_z\,,$$

we may write

$$-\varphi_x = \frac{1}{4\pi}\sum\frac{ev_x}{r}\,,$$

and two analogous equations. Alternatively, using Poisson's equation, we have

$$\text{(5.7)}\qquad \begin{cases} \Box\,\varphi_x = ev_x\,,\\ \Box\,\varphi_y = ev_y\,,\\ \Box\,\varphi_z = ev_z\,. \end{cases}$$

The electromagnetic potential would not be fully determined by the equations we have just listed. To remove this lack of determination, another equation is required. Following Maxwell, it can be written in the form

$$\text{(5.8)}\qquad \frac{\partial\varphi_x}{\partial x}+\frac{\partial\varphi_y}{\partial y}+\frac{\partial\varphi_z}{\partial z}-\frac{\partial\varphi}{\partial t}=0\,.$$

The term $\partial\varphi/\partial t$ is necessary here when we use retarded potentials.

We now seek a tensor equation which reduces to the above equations when we use the coordinates adopted there. These coordinates are those for which the derivatives of the $g_{\mu\nu}$ are zero (at least approximately) and where $\mathrm{d}s^2$ reduces to

$$\mathrm{d}s^2 = -\mathrm{d}x^2-\mathrm{d}y^2-\mathrm{d}z^2+\mathrm{d}t^2\,.$$

We found in (4.29) that

$$\sum_\sigma \frac{\partial\mathscr{T}^{\mu\sigma}}{\partial x_\sigma}$$

is a tensor expression if

$$\mathscr{T}^{\mu\sigma} = -\mathscr{T}^{\sigma\mu}\,.$$

The electromagnetic field $F_{\alpha\beta}$ corresponds to the tensor density $\mathscr{F}_{\alpha\beta}$. The associated contravariant density is

$$\mathscr{F}^{\mu\nu} = \sum_\alpha\sum_\beta g^{\mu\alpha}g^{\nu\beta}\mathscr{F}_{\alpha\beta}\,.$$

The expression

$$\sum_\sigma \frac{\partial\mathscr{F}^{\mu\sigma}}{\partial x_\sigma}$$

is therefore a tensor.

When we adopt the coordinates used in (5.6)–(5.8), this expression simplifies. Indeed,

$$g^{11} = g^{22} = g^{33} = -1\,, \qquad g^{44} = 1\,, \qquad \sqrt{-g} = 1\,,$$

while all the other $g_{\mu\nu}$ and all first order derivatives are zero. We obtain

$$\sum_\sigma \frac{\partial \mathscr{F}^{\mu\sigma}}{\partial x_\sigma} = \sum_\sigma \pm \frac{\partial \mathscr{F}_{\mu\sigma}}{\partial x_\sigma}\,,$$

where we must take the $+$ sign if neither of the two indices μ and σ is equal to 4 and the $-$ sign otherwise. Replacing $F_{\mu\sigma}$ by its expression (5.2) in terms of the φ_μ, this expression becomes

$$\sum_\sigma \pm \frac{\partial^2 \varphi_\mu}{\partial x_\sigma^2} - (\pm)\frac{\partial}{\partial x_\mu}\sum_\sigma \frac{\partial \varphi_\sigma}{\partial x_\sigma}\,,$$

which yields, for $\mu = 1$,

$$\frac{\partial^2 \varphi_x}{\partial x^2} + \frac{\partial^2 \varphi_x}{\partial y^2} + \frac{\partial^2 \varphi_x}{\partial z^2} - \frac{\partial^2 \varphi_x}{\partial t^2} + \frac{\partial}{\partial x}\left(\frac{\partial \varphi_x}{\partial x} + \frac{\partial \varphi_y}{\partial y} + \frac{\partial \varphi_z}{\partial z} - \frac{\partial \varphi}{\partial t}\right) = \Box\, \varphi_x\,,$$

where we have used the extra Maxwell equation (5.8).

For $\mu = 2, 3$, we would have analogous expressions in y and in z. For $\mu = 4$, we find

$$-\Box\, \varphi\,.$$

Moreover, for velocities typical of those encountered in experiments,

$$e\,,\ ev_x\,,\ ev_y\,,\ ev_z\,,$$

can be represented by the tensor density

$$\mathscr{J}^\mu = e\frac{\mathrm{d}x_\mu}{\mathrm{d}s}\,.$$

Equations (5.6) and (5.7) can be put in the tensor form

$$\sum_\sigma \frac{\partial \mathscr{F}^{\mu\sigma}}{\partial x_\sigma} = \mathscr{J}^\mu\,. \tag{5.9}$$

This is the general law describing the way electric charges act on electromagnetic fields.

Gravitational Effects of Electromagnetic Fields

Taking into account the law just obtained, the equation of motion of electrically charged bodies is found by substituting the above expression for $\mathscr{J}^{\mu}$ in (5.4) to yield

$$f_{\mu}+\sum_{\sigma}\sum_{\tau}F_{\mu\sigma}\frac{\partial \mathscr{F}^{\sigma\tau}}{\partial x_{\tau}}=0\,. \tag{5.10}$$

The quantity f_{μ} can be expressed in terms of the material energy tensor using (5.5). We have seen that the gravitational equations imply that f^{μ} is identically zero. It follows that f_{μ} is also identically zero. The equations we have just found therefore tell us that the gravitational equations are no longer applicable. They must be modified when the electromagnetic action on a charged particle is not zero. In the gravitational equations, we must add a further term to the material energy tensor, namely, the electromagnetic energy tensor.

The expression

$$f_{\mu}=\sum_{\sigma}\frac{\partial \mathscr{T}_{\mu}^{\sigma}}{\partial x_{\sigma}}-\frac{1}{2}\sum_{\sigma}\sum_{\tau}\frac{\partial g_{\sigma\tau}}{\partial x_{\mu}}\mathscr{T}^{\sigma\tau} \tag{5.11}$$

vanishes when $\mathscr{T}_{\mu}^{\sigma}$ represents the material energy tensor alone. If we replace $\mathscr{T}_{\mu}^{\sigma}$ by the electromagnetic energy tensor we are looking for, f_{μ} must reduce to (5.10). We obtain this result by imposing the tensor equation

$$\mathscr{T}_{\mu}^{\sigma}=-\sum_{\alpha}F_{\mu\alpha}\mathscr{F}^{\sigma\alpha}+\frac{1}{4}g_{\mu}^{\sigma}\sum_{\alpha}\sum_{\beta}F_{\alpha\beta}\mathscr{F}^{\alpha\beta}\,. \tag{5.12}$$

We only need to show this at a point M, by using a particular set of coordinates. We shall assume therefore that the derivatives of the $g_{\mu\nu}$ vanish at this point and that the $g_{\mu\nu}$ reduce to

$$g_{\mu\nu}=g_{\mu}^{\nu}\,.$$

In this case, (5.10)–(5.12) reduce respectively to

$$f_{\mu}+\sum_{\sigma}\sum_{\tau}F_{\mu\sigma}\frac{\partial F_{\sigma\tau}}{\partial x_{\tau}}=0\,, \tag{5.13}$$

$$f_{\mu}=\sum_{\sigma}\frac{\partial T_{\mu}^{\sigma}}{\partial x_{\sigma}}\,, \tag{5.14}$$

and

$$T_{\mu}^{\sigma}=-\sum_{\alpha}F_{\mu\alpha}F_{\sigma\alpha}+\frac{1}{4}g_{\mu}^{\sigma}\sum_{\alpha}\sum_{\beta}F_{\alpha\beta}^{2}\,. \tag{5.15}$$

Substituting the expression for T_μ^σ given in the last expression into (5.14) and carrying out the differentiation, we deduce that

$$f_\mu = -\sum_\sigma\sum_\alpha \frac{\partial F_{\mu\alpha}}{\partial x_\sigma}F_{\sigma\alpha} - \sum_\sigma\sum_\alpha F_{\mu\alpha}\frac{\partial F_{\sigma\alpha}}{\partial x_\sigma} + \frac{1}{4}2\sum_\alpha\sum_\beta F_{\alpha\beta}\frac{\partial F_{\alpha\beta}}{\partial x_\mu}\,. \tag{5.16}$$

The second term is exactly the term we were looking for, so we must establish that the other two terms cancel. Using (5.2) to replace $F_{\alpha\beta}$ by its expression in terms of the φ_α, the first term becomes

$$-\sum_\sigma\sum_\alpha\left(\frac{\partial^2\varphi_\mu}{\partial x_\alpha\partial x_\sigma} - \frac{\partial^2\varphi_\alpha}{\partial x_\mu\partial x_\sigma}\right)\left(\frac{\partial\varphi_\sigma}{\partial x_\alpha} - \frac{\partial\varphi_\alpha}{\partial x_\sigma}\right)$$

$$= -\sum_\sigma\sum_\alpha\left(\frac{\partial^2\varphi_\mu}{\partial x_\alpha\partial x_\sigma}\frac{\partial\varphi_\sigma}{\partial x_\alpha} - \frac{\partial^2\varphi_\mu}{\partial x_\alpha\partial x_\sigma}\frac{\partial\varphi_\alpha}{\partial x_\sigma} - \frac{\partial^2\varphi_\alpha}{\partial x_\mu\partial x_\sigma}\frac{\partial\varphi_\sigma}{\partial x_\alpha} + \frac{\partial^2\varphi_\alpha}{\partial x_\mu\partial x_\sigma}\frac{\partial\varphi_\alpha}{\partial x_\sigma}\right)\,.$$

The first two terms on the right-hand side cancel because they have the same absolute value when we exchange the summation indices.

For the last term of (5.16), we obtain

$$\frac{1}{2}\sum_\alpha\sum_\beta\left(\frac{\partial\varphi_\alpha}{\partial x_\beta} - \frac{\partial\varphi_\beta}{\partial x_\alpha}\right)\left(\frac{\partial^2\varphi_\alpha}{\partial x_\beta\partial x_\mu} - \frac{\partial^2\varphi_\beta}{\partial x_\alpha\partial x_\mu}\right)\,.$$

By expanding and simplifying those terms that differ only through the letters used for the summation indices, this reduces to

$$\sum_\alpha\sum_\beta\left(\frac{\partial\varphi_\alpha}{\partial x_\beta}\frac{\partial^2\varphi_\alpha}{\partial x_\beta\partial x_\mu} - \frac{\partial\varphi_\alpha}{\partial x_\beta}\frac{\partial^2\varphi_\beta}{\partial x_\mu\partial x_\alpha}\right)\,.$$

Replacing the dummy index β by σ, we see that all the terms cancel.

We have used covariant tensors to express the gravitational equations. We must therefore write the electromagnetic energy tensor in covariant form. We can do this by using the tensor associated with $\mathscr{T}_\mu^\sigma$,

$$T_{\mu\nu} = \sum_\sigma \frac{g_{\sigma\nu}}{\sqrt{-g}}\mathscr{T}_\mu^\sigma\,.$$

In the general gravitational equations

$$R_{\mu\nu} = -\kappa\left(T_{\mu\nu} - \frac{1}{2}g_{\mu\nu}T\right)\,,$$

we must take $T_{\mu\nu}$ to be the sum of the material energy tensor and the electromagnetic energy tensor. The contracted electromagnetic energy tensor is zero. This can be seen immediately by calculating it using (5.12):

$$T = \frac{1}{\sqrt{-g}}\sum_\sigma \mathscr{T}_\sigma^\sigma = 0\,.$$

La Physique d'Einstein

Préambule

Les Archives Georges Lemaître conservent un exemplaire[1] – à savoir une copie – du manuscrit original dactylographié de 131 pages du mémoire de Georges Lemaître intitulé "*La Physique d'Einstein*" et daté du 31 mai 1922. Ce tapuscrit unique constitue un précieux témoin historique à plus d'un titre.

Ce travail est sans aucun doute la prémisse des réflexions ayant conduit Georges Lemaître, dès 1927 avec son célèbre article[2] paru dans les *Annales de la Société Scientifique de Bruxelles*, et ceci contre l'avis sans appel d'Albert Einstein lui-même, à proposer une théorie de l'expansion de l'Univers, suivie seulement quelques années plus tard encore, en 1931, de sa proposition publiée dans la revue *Nature* d'une origine quantique au temps et à l'espace,[3] tous deux ayant surgi d'un "atome primitif" en "cet instant unique, qui n'avait pas d'hier". Mais aussi, on découvre dans ce mémoire, en compagnie du jeune Georges Lemaître lui-même tout fraîchement diplômé en sciences mathématiques – et ceci tout en poursuivant durant les années fin 1920 à fin 1923 ses autres études en tant que séminariste au Grand Séminaire de Malines/Mechelen en vue de son sacerdoce de prêtre – la manière dont l'architecte visionnaire de la cosmologie physique moderne a appréhendé, étudié, approfondi et finalement compris et reconstruit, dans une approche fondamentalement personnelle, les développements des théories alors toutes récentes de la relativité restreinte et de la relativité générale d'Albert Einstein.

---

[1] Archives de l'Université catholique de Louvain | Archives Georges Lemaître, BE A4006 FG LEM-447.

[2] G. Lemaître, *Un univers homogène de masse constante et de rayon croissant rendant compte de la vitesse radiale des nébuleuses extra-galactiques*, *Annales de la Société Scientifique de Bruxelles*, série A : *Sciences Mathématiques*, 1ère partie : *Comptes rendus des Séances*, t. 47, 1927, pp. 49-59.

[3] G. Lemaître, *The beginning of the world from the point of view of quantum theory*, *Nature* 127, 9 May 1931, p. 706.

Ce mémoire a déjà fait l'objet une première fois d'un travail d'édition original, en 1996, pour connaître une première publication, dans sa version française, dans

Mgr. G. Lemaître, savant et croyant.
Actes du colloque commémoratif du centième anniversaire de sa naissance, Louvain-la-Neuve, le 4 novembre 1994, 1996 (J.-Fr. Stoffel, ed.),
Réminiscences 3, Centre Interfacultaire d'Étude en Histoire des Sciences, Université catholique de Louvain.

À l'occasion de la publication du présent ouvrage, et au bénéfice de l'histoire des sciences et de ses chercheurs, ce travail d'édition est poursuivi ici en profondeur et avec force détails sur base de la version originale française du texte de Lemaître, pour produire la nouvelle version critique proposée et commentée dans ce chapitre. En outre, depuis 1996 et avec la numérisation des Archives Georges Lemaître (voir le site web `https://archives.uclouvain.be/`), des documents additionnels ont été mis au jour. Ceux-ci éclairent de quelle manière Lemaître lui-même a commenté la première version de son mémoire avec des notes complémentaires et des corrections de quelques erreurs, en réponse à des échanges épistolaires avec le Professeur Maurice Alliaume, ce dernier agissant vraisemblablement en tant que Secrétaire du Jury, constitué pour l'année 1923, du Concours des Bourses de voyage du Gouvernement belge pour lequel Lemaître avait rédigé cette dissertation.[4] Ces notes ainsi que cette correspondance entre Maurice Alliaume et Georges Lemaître sont reproduites dans leur contenu original en français dans le chapitre à la suite de celui-ci, qui constitue cette nouvelle édition critique du mémoire "*La Physique d'Einstein*".

Outre ces documents additionnels ainsi que le manuscrit original qui sont donc accessibles pour consultation sous forme numérisée dans tous leurs détails *via* le site web du Service des Archives de l'Université catholique de Louvain (UCLouvain), à savoir `https://archives.uclouvain.be/`, le sont encore de cette manière quelques autres notes fragmentaires et de brouillon que Lemaître a travaillées en vue de la préparation des notes finales transmises à M. Alliaume. Bien que celles-ci présentent également leur propre intérêt historique, ces notes fragmentaires ne sont donc pas reprises dans le présent ouvrage.

Les principes d'édition critique qui ont prévalu pour la préparation de la version éditée publiée avec ce chapitre sont repris ci-dessous.

Principes d'édition critique

Le travail d'édition étant basé sur le texte initial, c'est-à-dire la version originale la plus ancienne (à savoir A), on signale en note de bas de page (avec le sigle B) toutes les corrections et modifications apportées dans un second temps (y compris la correction manuscrite des simples coquilles), ainsi que (avec le sigle C) les corrections signalées dans l'Errata rédigé par Lemaître lui-même et reproduit dans le chapitre qui suit celui-ci.

[4] Ce même mémoire lui a permis d'obtenir encore une autre bourse, à savoir "a Graduate Fellowship of the Commission for Relief in Belgium – Educational Foundation".

Modalités

- L'appel de note se fait tout de suite après le mot concerné ou le dernier mot du groupe de mots concernés ; lorsque c'est un ajout, l'appel de note se fait entre les mots concernés sans être collé à l'un ou à l'autre.
- En note, le sigle est d'abord renseigné (B ou C) puis, selon les cas, "ajout", "modification", "suppression" (pour B) ou "correction" (pour C exclusivement).

Modifications nécessaires, mais non signalées par Lemaître

- Accentuation des lettres majuscules.
- Ajout des ligatures (par exemple : œuf).
- Signalement par un [*sic*] des *fautes d'orthographe*, des fautes dans les *noms propres*, de mots manquants ou d'autres erreurs qui ne sont *pas corrigées*.

Modifications par Lemaître qui sont signalées

- Ses *ajouts* qui ont un sens.
- Ses *suppressions* qui ont un sens.
- Ses modifications du texte ou de la ponctuation qui sont susceptibles de *modifier ou de préciser le sens* (essentiellement, lorsqu'il s'agit d'un autre mot et non de la correction d'un mot); cependant "ou" ou "où" changent le sens et sont donc signalés.

Modifications par Lemaître qui sont intégrées sans être signalées

- Ses corrections de fautes de frappe et de fautes d'orthographe qui *ne changent pas le sens.*
- Ses ajouts pour des *manques manifestes* (typiquement, un article ou un verbe [par exemple, être] qui manquait au point de rendre la phrase *grammaticalement* incorrecte ; ne sont donc pas visés ici les manques qui rendraient la phrase incompréhensible).
- Sa suppression du *doublon* d'un groupe de mots répété par inadvertance.
- Ses corrections dans les *noms propres.*
- Sa suppression de *mots entamés en fin de ligne* sans espace suffisant et repris en début de ligne.
- Ses *ajouts prévus* (la plupart du temps des équations, mais parfois aussi des mots), c'est-à-dire qui viennent s'insérer dans un espace prévu à cet effet.

G Lemaître
Maison St Rombaut
84 rue de Merode
Malines

Table des matières

La Physique d'Einstein

[ ][1]

[p. i] Nous pourrions concevoir un univers où vivraient des hommes semblables à nous, mais où régnerait un génie malfaisant, maître des lois de la nature et qui s'amuserait à les changer dès que quelque savant serait sur le point de les découvrir. Quelle pourrait bien être la physique de ces infortunés ? Chaque génération dénouerait[2] les erreurs des anciens ; les expériences effectuées jadis, donneraient aujourd'hui des résultats totalement différents de ceux que décrivent les mémoires de leurs auteurs ; peut-être un jour l'humanité renoncerait-elle à connaître la vérité en devinant le maléfice dont elle est la victime impuissante.

Cette fiction montre clairement la première condition que suppose l'existence d'une science expérimentale : nos expériences et celles de nos ancêtres doivent nous servir ; nous devons pouvoir nous adapter à l'univers où nous vivons : cela n'est possible que si nos sens peuvent reconnaître une série de sensations analogues à d'autres dont la mémoire garde le souvenir et dont nous attendons la succession régulière dès que nous en reconnaissons le commencement.

De temps en temps, une modification se produit dans l'ordre habituel ; elle attire notre attention, nous en cherchons la cause, nous cherchons à la rattacher à quelque fait nouveau qui nous avait échappé : petit-à-petit nous organisons notre connaissance de l'univers, connaissance vulgaire d'abord, connaissance scientifique ensuite, poursuivant plus attentivement l'œuvre de notre éducation, en utilisant l'expérience accumulée pendant de longs siècles et transmise précieusement de génération en génération.

La constance du monde physique ne suffirait pourtant pas à notre adaptation à l'univers et au développement de notre science expérimentale.

Il faut encore que des relations semblables se produisent assez [p. ii] souvent dans la suite des phénomènes et soient assez simples pour que nous puissions les remarquer. Ceci est nécessaire, non seulement au début de notre éducation, mais à chaque étape du développement de la science. Nous devons procéder par étapes successives : les anomalies qui surgissent dans le vaste champ des phénomènes expliqués, doivent être assez simples et assez fréquentes pour pouvoir être formulées en quelque loi nouvelle. Celle-ci étendra le domaine des faits que nous comprenons, que nous pouvons prévoir ou produire par[3] volonté.

Toutes ces lois empiriques s'accumulent ; leur ensemble finit par devenir aussi complexe et inextricable que dut apparaître à nos yeux d'enfant le monde sensible que nous débrouillons si facilement maintenant. Il faut synthétiser ces lois éparses : sinon nous ne pourrons même plus les connaître, ni les retenir, et le progrès de la science sera bien vite arrêté.

De nouveau, il faut que cela soit possible. Il faut qu'une simplicité supérieure domine la variété des phénomènes et la multiplicité des lois. C'est au génie du sa-

---

[1] B ajout : "Introduction".

[2] B modification : "dénoncerait".

[3] B modification : "à".

vant à deviner le point de vue d'où la multiplicité se résoudra dans une synthèse plus vaste. Dans cet effort d'unification il devra généralement sacrifier la simplicité de certaines lois partielles à la simplicité plus profonde d'une loi générale. Les lois simples de Mariotte et de Gay-Lussac, premières étapes de notre connaissance des propriétés des gaz, disparaissent devant la simplicité plus compréhensive de la théorie cinétique, qui nous fait mieux connaître les propriétés des gaz en même temps qu'elle nous permet d'atteindre celles des vapeurs et des liquides.

Le progrès scientifique est la découverte d'une simplicité plus compréhensive, d'un point de vue augmentant l'étendue du domaine réduit à l'unité. Les succès passés nous donnent confiance dans l'avenir de la science : nous prenons de plus en plus confiance[4] que l'univers est intelligible. [p. iii] Les formes infiniment variées des phénomènes se laissent comprendre dans quelques énoncés clairs, dans quelques formules qui en découvrent le cours régulier et permettent d'en prévoir la succession lorsque les conditions initiales sont connues.

Nous allons parcourir les grandes étapes de cette emprise de l'intelligence humaine sur l'univers si merveilleusement adapté à toutes les formes de notre activité. Nous verrons mieux ainsi la place qu'occupe l'œuvre d'Einstein dans le développement de notre connaissance et le pas qu'il nous fait franchir vers l'unité.

Les premières notions que se forme l'enfant, du monde où il commence à vivre, sont sans doute d'ordre géométrique. Nous sommes constamment en contact avec des objets qui se présentent sous le même aspect : nous nous habituons à les revoir et à les reconnaître : nous remarquons l'identité des mouvements nécessaires pour les atteindre ou les palper. Leur forme doit être une des premières notions que nous en abstrayons.

Quelle que soit la valeur de cette opinion sur la genèse de notre connaissance sensible, c'est par l'établissement de la géométrie qu'on[5] a débuté la connaissance scientifique de l'humanité. La géométrie, définitivement systématisée par Euclide, domine toute la science grecque. La constance de la forme des corps solides est une des plus immédiates que nous présente la nature. La notion de distance s'en abstrait immédiatement et les relations entre les distances sont assez simples pour avoir été découvertes les premières.

L'astronomie des grecs [*sic*] était toute [*sic*] entière dominée par le besoin de simplicité géométrique : il fallait que le mouvement des astres soit un mouvement simple ou une combinaison de mouvements simples. En fait, ils parvinrent à un échafaudage fort compliqué de mouvements relatifs [p. iv] circulaires et uniformes qui dut vraisemblablement les satisfaire assez peu. Le triomphe du point de vue géométrique dans l'explication des phénomènes astronomiques fut la découverte du mouvement elliptique des planètes : les lois de Képler font décrire aux planètes un mouvement d'une simplicité idéale qui aurait sans doute ravi l'esprit géométrique des grecs [*sic*].

---

[4] B modification : "conscience".

[5] B suppression : "on".

Comment pouvons-nous considérer comme un progrès, dans l'intelligence de la simplicité du monde, le remplacement de ces trajectoires idéales par la courbe infiniment compliquée que l'observation plus précise nous a révélée et qui résulte des calculs des perturbations que Newton nous a fait connaître ? N'avons-nous pas fait un pas en arrière ? Nous avons perdu la simplicité du mouvement Keplerien [*sic*], mais c'est pour atteindre une simplicité plus profonde et infiniment plus féconde, celle de la mécanique et des lois d'attraction de Newton. Les corps s'attirent, leur mouvement peut se calculer quand on en connaît les données initiales ; il suffit d'intégrer les équations générales de la mécanique.

La mécanique a dominé la physique comme la géométrie avait dominé l'astronomie des grecs [*sic*], on chercha partout des actions à distance. La loi de Coulomb calquée sur la loi de l'attraction universelle exprima l'action réciproque des masses électriques. On espéra même un moment pouvoir ramener toute la physique à la mécanique, mais les phénomènes électriques se montrèrent rebelles à cette tentative. Sous l'influence des idées Newtonnieuses[6], toute l'attention s'était portée d'abord sur les masses électriques et les conducteurs du courant. Faraday montra bientôt l'influence des diélectriques ; le champ apparut comme une réalité répandue dans le milieu ; les masses électriques ou les aiguilles aimantées décelaient sa présence mais il existait en dehors d'eux. Maxwell découvrit les équations différentielles qui [p. v] rendent compte des propriétés des champs. Ses idées furent définitivement comprimées[7] par la découverte des ondes hertziennes.

La physique s'est ainsi développée en trois grandes étapes : la mécanique utilisant la géométrie, l'électricité utilisant la géométrie et la mécanique. Les développements nouveaux n'avaient en rien modifié les parties plus anciennes de la science, et celles-ci s'étaient développées de leur côté d'une manière aussi indépendante que possible. Aussi, les trois étages de la science présentent-ils des caractères très différents, presqu'opposés. La géométrie apparaît comme une réalité indépendante de toutes les vicissitudes de la matière : les corps sont dans l'espace ; il semble que l'espace existe de tout temps, attendant les corps qui vont pénétrer dans telle ou telle de ses parties, puis la quitter sans rien changer à ses propriétés.

Les actions électriques se transmettant avec une vitesse fixée[8], la mécanique classique nous parle encore des[9] actions à distance instantanées. Pouvons-nous conserver l'ancienne théorie du potentiel de gravitation à côté des potentiels retardés de Lorentz ? La mécanique ne devrait-elle pas subir le contre-coup, des transformations de l'électricité ? Les développements récents ne doivent-ils pas réagir sur les branches plus anciennes ? L'ensemble ne gagnera-t-il pas ainsi la[10] simplicité ?

C'est cette unification qu'a tentée Einstein. Il montre que les lois physiques prennent une forme plus profondément simple dans leur ensemble lorsqu'on admet

[6] B modification : "Newtoniennes".

[7] B modification : "comfirmées" [*sic*].

[8] B modification : "finie".

[9] B modification : "d' ".

[10] B modification : "en".

une certaine influence des phénomènes électriques sur les phénomènes mécaniques, et l'influence des uns et des autres sur les propriétés des règles matérielles. Les synthèses partielles se fondent dans une synthèse plus vaste ; la simplicité de la géométrie euclidienne[11] et de la mécanique classique disparaissent [*sic*] ; elles doivent être sacrifiées[12] à la simplicité de l'ensemble de la physique, comme la simplicité des lois de Képler a été jadis abandonnée sans regret devant la synthèse de Newton.

[p. vi] La méthode d'Einstein est fort simple.

Dans l'étude d'un phénomène, nous sommes bien forcés d'adjoindre au fait que nous étudions des éléments subjectifs qui nous permettent de le connaître. À côté de l'objet, nous devons considérer un observateur qui repère à chaque instant la position des divers points et les modifications qui s'y présentent. En cherchant à résumer les observations en une formule intelligible, nous ne pouvons distinguer dans la simplicité de la loi, le caractère plus ou moins subjectif de cette simplicité. Elle est peut-être tout-à-fait artificielle. Elle provient non de l'objet mais de l'observateur particulier que nous avons dû lui adjoindre et du mode de repérage qu'il a employé.

Einstein dénonce la relativité de l'observateur. [13] Un phénomène peut-être[14] observé par un observateur quelconque. Un objet peut être repéré d'une infinité de manières. La simplicité d'une loi sera vraiment objective lorsqu'elle subsiste, quel que soit le mode de repérage employé.

Les équations[15] qui expriment une loi physique doivent garder leur forme algébrique, lorsqu'on fait un changement arbitraire de coordonnées.

C'est le principe de relativité.

Les propriétés qu'il atteint sont débarrasées [*sic*], aussi complètement qu'il est possible, de tout caractère subjectif. Aussi, l'instrument mathématique qui permet d'en réaliser le programme porte-t-il le nom de calcul différentiel absolu (on l'appelle aussi calcul tensoriel). Il détermine les diverses formes possibles des lois physiques en accord avec le principe de relativité. Il reste alors à voir ce que deviennent ces lois lorsqu'on emploie le mode de repérage des événements utilisé par les expérimentateurs et à choisir, parmi les diverses lois possibles, celles qui rendent compte de leurs observations.

On obtient ainsi des lois vraiment objectives, et on peut légitimement espérer qu'elles rendront plus parfaitement compte des faits observés et en feront découvrir de nouveaux.

---

[11] Notons que l'adjectif "euclidien" a, presque systématiquement, été mal dactylographié, exigeant de ce fait une correction manuscrite.

[12] On s'attendrait à un singulier pour "elles" et "sacrifiées", puisque le "elles " renvoie normalement au sujet de la première partie de la phrase, à savoir "la simplicité".

[13] B modification : symbole requérant le début d'un nouveau paragraphe.

[14] B modification : "peut être".

[15] Notons que le tapuscrit indique, le plus souvent, "équatations" au lieu de "équations".

Chapitre I – L'ESPACE ET LE TEMPS.

[§ 1] LA GÉOMÉTRIE GÉNÉRALE DE RIEMANN[16]

[p. 1] La géométrie dut vraisemblablement son origine à des besoins d'ordre expérimental, tels que la nécessité de mesurer des terrains, mais elle se dégagea bientôt de ce caractère empirique et nous apparaît déjà avec Euclide sous une forme très systématique. Elle s'est rendue indépendante des faits expérimentaux qui lui avaient donné naissance : des postulats nettement énoncés la séparent des faits contingents et toute la théorie s'en déduit sans nouvel appel à l'expérience. Depuis, ce caractère s'est accentué. Les discussions soulevées par les effets[17] de démonstration du fameux postulat des parallèles ont conduit à des géométries bien différentes de celle que nous suggérait notre intuition de l'espace. En modifiant l'un ou l'autre des postulats, on obtenait des géométries égales en valeur logique à celle d'Euclide et dont l'application au monde physique, quoique moins commode, restait encore possible.

L'expérience peut-elle décider entre ces différentes formes de la géométrie ? Est-il possible de réaliser une expérience géométrique ?

Il n'y a pas d'être qui soit purement géométrique : tous ont, à côté de leurs propriétés géométriques, d'autres propriétés physiques ; et les expériences vont dépendre des unes et des autres. Si on réalisait un triangle matériel dont la somme des angles fût plus grande que deux droits, il resterait à prouver que ses côtés sont rectilignes : on pourrait le faire, par exemple en les superposant deux à deux, puis en recommençant, en en retournant un. Mais, comment saurait-on s'ils ne se sont pas déformés pendant l'opération. On pourrait toujours prétendre que la géométrie est euclidienne. Le triangle dont la somme des angles est plus grande que deux droits est un triangle curviligne. Ses côtés se sont déformés pendant les opérations par lesquelles on voulait prouver qu'ils sont droits.

Tout ce que l'expérience peut nous faire vérifier c'est une géométrie et le reste de la physique et nous pouvons toujours choisir ce reste de manière [p. 2] à rendre n'importe quelle géométrie conforme à l'expérience. Nous pouvons, par exemple, choisir la géométrie la plus simple, la géométrie euclidienne et il sera possible d'établir la physique sur cette géométrie.

Mais sommes-nous bien sûrs qu'à la géométrie la plus simple correspondra la physique la plus simple ? Ce qui nous importe [18] la simplicité d'une partie de l'édifice, c'est la simplicité de l'ensemble. Ne serait-il pas plus sage de chercher à profiter de la forme arbitraire que peut prendre la géométrie, pour simplifier l'ensemble de la physique et ne pas craindre de compliquer quelque peu des par-

---

[16] Notons que tout au long du tapuscrit, le nom "Riemann" et l'adjectif "riemannien" ont mal été dactylographiés, exigeant de ce fait une correction manuscrite.

[17] B modification : "essais".

[18] B ajout : "n'est pas".

ties bien connues comme la géométrie, si nous pouvons faciliter ainsi l'exploration de régions nouvellement découvertes ?

La géométrie générale ou[19] de Riemann est la généralisation toute naturelle de la géométrie d'Euclide suivant le procédé usité dans toutes les branches de la physique mathématique. On l'obtient en admettant que la géométrie d'Euclide n'est généralement plus valable dans un domaine fini mais qu'elle subsiste à la limite dans un domaine infiniment petit. Autrement dit, il est toujours possible de trouver un domaine assez petit pour que les relations métriques de la géométrie d'Euclide y soient valables avec une approximation donnée.

Les notions géométriques courantes ne sont plus applicables qu'à la limite, c'est ainsi qu'il est généralement impossible de construire des figures superposables de dimensions finies. Lorsqu'on veut conserver la possibilité de superposer des figures géométriques, on obtient des cas particuliers remarquables de la géométrie pour lesquels l'espace est homogène. Leur ensemble est souvent appelé géométrie générale ; ce n'est pourtant qu'un cas particulier de la géométrie générale que nous étudions ici pour laquelle l'espace n'est pas nécessairement homogène. Il comprend la géométrie d'Euclide, la géométrie non-euclidienne proprement dite de [p. 3] Bolyai-Lobatschefsky que l'on obtient en répétant[20] le fameux postulat des parallèles et enfin la géométrie sphérique qui étend à l'espace les propriétés de la sphère et qu'on appelle souvent géométrie de Riemann, au sens étroit. Einstein et Weyl emploient toujours l'expression géométrie de Riemann au sens large pour désigner la géométrie générale telle que nous l'étudions ici. Lorsqu'ils veulent exprimer le sens étroit ils parlent d'espace ou de géométrie sphérique.

À première vue, il paraît bien étrange que l'on puisse construire une géométrie où la superposition des figures soit une chose impossible ; nous sommes habitués à définir l'égalité par une telle superposition et elle nous paraît une notion fondamentale. Nous savons pourtant que l'on peut mesurer la longueur d'un arc de courbe, bien qu'il soit impossible de [21] faire coïncider avec le mètre rectiligne qui sert d'unité, ni d'en faire coïncider exactement une partie aussi petite que l'on veut avec une partie du mètre. On peut concevoir la réalisation de cette mesure de la manière suivante : on se sert, par exemple, d'un compas dont on applique les pointes sur l'arc à mesurer, on déplace le compas, une pointe restant en un point, l'autre marquant un nouveau point ; on le porte ainsi, un certain nombre de fois sur l'arc. Les extrémités de l'arc n'auront généralement pas été toutes deux touchées par les pointes, mais il sera impossible de marquer sur l'arc de nouveaux points. On cherche de même combien de fois on peut porter le compas sur le mètre unité. Le rapport des deux nombres obtenus tend vers une limite lorsque l'ouverture du compas tend vers zéro ; cette limite mesure la longueur de l'arc.

---

[19] B suppression : "ou".

[20] B modification : "rejetant".

[21] B ajout : "le".

Il faut pour cela que le compas ne se déforme pas pendant les opérations, ou, si on préfère marquer[22] une série de compas ayant deux à deux une pointe en contact, il faut que ces compas puissent être superposables. Il suffit naturellement que les pointes soient superposables. Cette superposition ne doit d'ailleurs être définie qu'à la limite lorsque les deux pointes tendent l'une vers l'autre. Il suffit donc que les notions de la géométrie d'Euclide soit [*sic*] [p. 4] applicables à la limite dans un domaine infiniment petit.

Cette définition de la distance par superposition d'un instrument de mesure tel qu'un compas est parfaitement adaptée à la mesure des distances, mais elle ne nous apprend pas ce qu'est la distance. Il est bien clair que lorsqu'on introduit un compas dans un milieu, la distance entre les points de ce milieu existait avant qu'on introduise le compas au milieu[23] duquel on peut la mesurer. Il doit y avoir une réalité physique qui est mesurée par la distance, de même que la température est mesurée par un thermomètre mais subsiste là où il n'y a pas de thermomètre. D'une manière analogue l'aiguille [24] explore le champ magnétique et ne le crée pas.

Si nous [25] plaçons à ce point de vue, nous devons admettre qu'il y a une réalité physique répandue dans l'espace et dont la distance est une manifestation : il y a un champ métrique qui s'emploie[26] avec un compas, comme un champ magnétique s'explore avec une boussole et nous pouvons nous attendre à ce que cette réalité dont nous connaissons l'aspect métrique intervienne dans des phénomènes physiques d'un autre ordre.

La notion d'égalité des éléments de distance doit donc avoir un sens physique indépendant des instruments avec lesquels nous mesurons cette distance.

Pour pouvoir aborder l'étude de cette réalité physique, il faut que nous repérions la position des points. Nous assignerons à chaque point trois nombres ($x1$, $x2$, $x3$.) ses coordonnées. De plus à une suite continue de points doit correspondre une variation continue de leurs coordonnées. À part ces restrictions générales d'une[27] et de continuité, le mode de repérage employé est totalement indifférent.

L'égalité des distances nous apprend qu'une certaine grandeur physique caractérisée par chaque couple de points est égale pour chacun d'eux. Cette grandeur dépend pour chaque couple des coordonnées des deux points [p. 5], coordonnées qui doivent être considérées comme infiniment voisines puisque l'égalité n'est définie qu'à la limite lorsque chaque distance tend vers zéro. C'est donc une fonction des coordonnées et des différentielles des coordonnées.

Quelle peut-être [*sic*] la forme de cette fonction ?

Il est clair qu'elle doit être homogène par rapport aux différentielles. De plus, elle ne doit pas changer lorsqu'on change le signe des différentielles, car, elle ne peut

---

[22] B modification : "imaginer".

[23] B modification : "moyen".

[24] B ajout : "aimantée".

[25] B ajout : "nous".

[26] B modification : "s'explore".

[27] B modification : "univocité".

dépendre de l'ordre dans lequel on considère les points. Enfin, elle doit pouvoir être employée quelque soit [*sic*] la manière dont on repère les points, elle doit donc garder sa forme algébrique lorsqu'on fait un changement arbitraire des coordonnées.

La fonction la plus simple qui vérifie ces conditions est une forme quadratique des différentielles des coordonnées.

~~“ dx,2 + 2V12 dx1dx2 + 2V13dx 1dx 3~~ [28]

$$\begin{aligned}\gamma_{11}dx_1^2 + 2\gamma_{12}dx_1dx_2 &+ 2\gamma_{13}dx_1dx_3 \\ + \quad \gamma_{22}dx_2^2 &+ 2\gamma_{23}dx_2dx_3 \\ &+ \gamma_{33}dx_3^2\end{aligned}$$

Les couples de points équidistants sont alors ceux dont les coordonnées font prendre une même valeur à cette fonction.

La distance mesurée sur un arc de courbe peut être utilisée comme une des coordonnées repérant les points de cet arc. Il faut donc que d'élément [*sic*] de distance $d\sigma$ soit [29] infiniment plus[30] petit de même ordre que les différentielles des coordonnées.
On devra donc poser

$$d\sigma^2 = \sum_\mu \sum_\nu \gamma_{\mu\nu}dx_\mu dx_\nu \qquad (\mu, \nu = 1,2,3) \tag{1}$$

avec

$$\gamma_{\mu\nu} = \gamma_{\nu\mu}$$

La longueur d'un arc de courbe s'exprimera alors par l'intégrale curviligne

$$\int d\sigma = \int \sqrt{\sum_\mu \sum_\nu \gamma_{\mu\nu} \frac{dx_\mu}{d\lambda}\frac{dx_\nu}{d\lambda}} \cdot d\lambda \tag{2}$$

où les dérivées se calculent d'après les équations paramétriques de la courbe sur laquelle se fait la mesure

$$x_\mu = \varphi_\mu(\lambda) \qquad (\mu = 1,2,3)$$

[p. 6] Les propriétés géométriques du milieu sont donc définies par la valeur en chaque point des six fonctions des coordonnées $\gamma_{\mu\nu}$; ce sont des[31] potentiels du champ métrique. Ces fonctions sont naturellement supposées continues. Ceci veut

---

[28] B suppression : cette seule et unique équation dactylographiée est supprimée. Dorénavant, elles seront toutes écrites à la main.

[29] B ajout : “un”.

[30] B suppression : “plus”.

[31] B modification : “les”.

dire qu'on peut toujours déterminer autour d'un point un domaine assez petit de telle sorte que, dans ce domaine, les potentiels diffèrent de leur valeur en ce point de moins qu'une quantité donnée. Nous allons montrer que lorsque les potentiels sont constants la géométrie est euclidienne. On pourra donc conclure que les relations de la géométrie euclidienne sont valables à une approximation donnée dans un domaine suffisamment petit. Cette propriété caractérise la géométrie générale de Riemann.

Une forme quadratique à coefficients constants peut, par une transformation linéaire des coordonnées, être mise sous la forme d'une somme algébrique de carrés. Le calcul est le même que celui de la réduction des quadriques. On peut en effet considérer dx_1, dx_2, dx_3 comme les coordonnées cartésiennes d'un point : l'équation (1) représente alors pour une valeur donnée de $d\sigma$ une quadrique dont l'équation réduite est de la forme

$$d\sigma^2 = dx^2 + dy^2 + dz^2 \tag{3}$$

où on a posé

$$dx = \sum_\sigma a_\sigma dx_\sigma$$
$$dy = \sum_\sigma b_\sigma dx_\sigma$$
$$dz = \sum_\sigma c_\sigma dx_\sigma$$

$a_\sigma, b_\sigma, c_\sigma$ désignant des constantes convenables.
L'équation (3) exprime le théorème de Pythagore étendu à l'espace ; elle définit l'élément de distance de la géométrie euclidienne.

Les diverses notions géométriques de droite, d'angle etc, peuvent facilement être étendues à la géométrie générale. Il suffit de trouver [p. 7] une définition qui soit indépendante d'un choix particulier du mode de repérage des points et qui coïncide avec les notions euclidiennes lorsque $d\sigma^2$ se réduit à la forme (3).

La droite est une ligne définie par deux de ces[32] points, c'est le plus court chemin entre ces deux points ; on en obtient l'équation en appliquant les méthodes du calcul des variations à l'équation

$$\delta \int d\sigma = 0 \tag{4}$$

où les limites sont supposées constantes.

La droite est définie par les coordonnées d'un de ces[33] points et les différentielles des coordonnées en ce point.

La notion de perpendiculaire peut se définir de la manière suivante.

---

[32] B modification : "ses".

[33] B modification : "ses".

Considérons deux directions

$$\delta x_1,\ \delta x_2,\ \delta x_3$$

et

$$dx_1,\ dx_2,\ dx_3$$

L'ensemble de ces deux directions s'appelle un angle.

Les deux directions sont dites perpendiculaires si l'angle qu'elles forment est identique à l'angle qu'on obtient en remplaçant une des deux directions par la direction opposée.

Considérons la forme bilinéaire associée à la forme fondamentale (1)

$$\sum_\mu \sum_\nu \gamma_{\mu\nu}\, dx_\mu \delta x_\nu$$

Cette forme est invariante ; si on fait un changement quelconque des coordonnées

$$dx_\mu = \sum_\sigma \frac{\partial x_\mu}{\partial x'_\sigma} dx'_\sigma$$

$$\delta x_\nu = \sum_\tau \frac{\partial x_\nu}{\partial x'_\tau} \delta x'_\tau$$

elle se transforme en

$$\sum_\sigma \sum_\tau \gamma'_{\sigma\tau}\, dx'_\sigma \delta x'_\tau$$

ou[34]

(5) $$\gamma'_{\sigma\tau} = \sum_\mu \sum_\nu \frac{\partial x_\mu}{\partial x'_\sigma} \frac{\partial x_\nu}{\partial x'_\tau} \gamma_{\mu\nu}$$

Les coefficients se transforment de la même façon que ceux de la forme quadratique qu'on obtiendrait en remplaçant δx_ν par dx_ν.

Pour que les deux directions soient perpendiculaires, la forme [p. 8] bilinéaire ne doit pas changer lorsqu'on change le signe des δx_ν. Il faut pour cela qu'elle soit nulle. La condition de perpendicularité est donc

(6) $$\sum_\mu \sum_\nu \gamma_{\mu\nu}\, dx_\mu \delta x_\nu = 0$$

Dans le cas de la géométrie euclidienne (3), on retrouve l'équation bien connue

$$dx\,\delta x + dy\,\delta y + dz\,\delta z = 0$$

[34] B modification : "où".

L'angle des[35] deux directions s'exprime par la formule

$$(7) \qquad \cos\theta = \frac{\sum_\mu\sum_\nu \gamma_{\mu\nu}\, dx_\mu \delta x_\nu}{\sqrt{\sum_\mu\sum_\nu \gamma_{\mu\nu}\, dx_\mu dx_\nu}\ \sqrt{\sum_\mu\sum_\nu \gamma_{\mu\nu}\, \delta x_\mu \delta x_\nu}}$$

Cette expression est une combinaison d'invariants et elle se réduit à la forme classique dans le cas de la géométrie euclidienne.

Enfin l'élément de volume s'écrit

$$(8) \qquad \sqrt{\gamma}dx_1dx_2dx_3$$

ou[36] γ représente le déterminant symétrique formé au moyen des potentiels. Cette expression se réduit au produit des différentielles dans le cas de la géométrie euclidienne. C'est un invariant : en effet, d'après (5) et les propriétés des déterminants,

$$\gamma' = |\gamma'_{\sigma\tau}| = \left|\frac{\partial x_\mu}{\partial x'_\sigma}\right|^2 \gamma$$

mais

$$dx_1dx_2dx_3 = \left|\frac{\partial x_\mu}{\partial x'_\sigma}\right| dx'_1dx'_2dx'_3$$

On a donc bien

$$(8)^{37} \qquad \sqrt{\gamma}dx_1dx_2dx_3 = \sqrt{\gamma'}\, dx'_1dx'_2dx'_3$$

On peut se faire une représentation assez simple des conséquences de l'adoption de la géométrie générale pour l'étude de la physique, en la comparant avec la géométrie euclidienne des surfaces. Soit[38]

$$x_1 = f_1(u_1,u_2)$$
$$x_2 = f_2(u_1,u_2)$$
$$x_3 = f_3(u_1,u_2)$$

les équations paramétriques d'une surface ; à tout couple de valeurs des paramètres u_1 et u_2 correspond un point sur la surface, ces paramètres déterminent donc les points de la surface et peuvent être considérés comme leurs coordonnées. L'élément de distance euclidienne [p. 9] s'exprime en fonction des x_i par l'expression (3).

---

[35] B modification : "de".

[36] B modification : "où".

[37] B modification : "(9)".

[38] B modification : "Soient".

Différentions les équations de la surface

$$dx_i = \frac{\partial f_i}{\partial u_1} du_1 + \frac{\partial f_i}{\partial u_2} du_2 \qquad (i = 1,2,3)$$

et portons ces valeurs dans (3), nous obtenons une expression de la forme

(9)[39] $$d\sigma^2 = \gamma_{11} du_1{}^2 + 2\gamma_{12} du_1 du_2 + \gamma_{22} du_2{}^2$$

où γ_{11}, γ_{12}, γ_{22} sont des fonctions des u_1 et u_2. Cette expression analogue à (1) montre que la géométrie euclidienne sur une surface quelconque est équivalente à la géométrie générale de Riemann à deux dimensions.

Il devait d'ailleurs en être ainsi ; car la géométrie de Riemann suppose simplement que la géométrie euclidienne est valable à la limite dans un domaine infiniment petit et il est clair que la géométrie de figures tracées sur une portion de surface dont l'aire tend vers zéro se confond à la limite avec celle des projections de ces figures sur le plan tangent à la surface.

La géométrie de la sphère nous est bien connue. On peut en effectuant uniquement des mesures sur la sphère se rendre compte de la courbure de la surface ; on ne pourra pourtant pas savoir ainsi si la surface sur laquelle on se trouve est concave ou convexe, mais on constatera simplement qu'elle est courbe. Des hommes qui ne sauraient pas que la lumière se propage en ligne droite et pour qui toutes les mesures géométriques se réduiraient à arpenter la surface de la terre, pourraient bien constater que la terre est ronde, ils pourraient en faire le tour, ils pourraient tracer le plus court chemin entre trois villes et constater que la somme des angles du triangle ainsi déterminé est supérieure à deux droits, mais ils ne pourraient savoir s'ils se trouvent sur une sphère pleine ou dans une sphère creuse de même rayon. Les propriétés spatiales qu'ils pourraient découvrir se rattachent à la valeur de la courbure totale de la surface, ou courbure de Gauss qui dans le cas de la sphère est égale à [p. 10] l'inverse du carré du rayon.

Sur la sphère la courbure est constante ; aussi, les figures qu'on y trace sont-elles superposables ; on peut étudier l'égalité des triangles sphériques de la même manière que celle des triangles plans. Si nous étudions la géométrie euclidienne d'une surface à courbure variable, une éllipsoïde [*sic*] à trois axes inégaux, par exemple, une pareille méthode serait inutilisable, il serait impossible de déplacer une figure sur l'éllipsoïde [*sic*] sans la déformer. Les figures qu'on y trace ont pourtant encore des propriétés géométriques, longueurs, angles, surfaces etc. On peut les étudier directement par les méthodes de la géométrie générale en se servant de l'expression de $d\sigma$ au moyen de coordonnées quelconques.

La géométrie générale à trois dimensions est appelée par comparaison avec la géométrie des surfaces quelconques, une géométrie d'un espace à courbure variable ; on peut chercher les grandeurs dont dépendent les propriétés géométriques de cet espace et qui généralisent la courbure de Gauss des surfaces. Le mouve-

[39] B modification : "(10)".

ment sans déformation d'un solide y est aussi possible[40] que le déplacement sans déformation d'une figure sans[41] un ellipsoïde à axes inégaux, et cependant les grandeurs géométriques, longueur, droite, angle et[42] volume etc. peuvent y être définies et peuvent être étudiées tout aussi bien que dans un espace euclidien.

Remarquons enfin que de même qu'on peut faire la carte d'une sphère ou d'un ellipsoïde sur un plan, on peut faire la carte d'un espace Riemannien dans un espace euclidien ; il suffit de considérer les coordonnées x_1, x_2, x_3, comme des coordonnées cartésiennes de cet espace. La forme $d\sigma^2$ (1) déterminera alors l'échelle de la carte dans chaque direction et jouera le même rôle que l'équation (10) dans l'étude des cartes géographiques.

[p. 11] On dira alors que tout se passe comme si l'étalon de longueur se déformait lorsqu'on change sa direction et l'on pourra décrire le caractère de l'espace en chaque point par la forme (1) de l'ellipsoïde des échelles. Bien entendu une pareille description dépendra en partie de la manière dont on a fait la carte. On pourra considérer des cartes conformes, pour lesquelles l'éllipsoïde [*sic*] des échelles se réduit à une sphère de rayon variable d'un point à l'autre, qui conservent les angles ou des cartes équivalentes, où le volume de l'éllipsoïde [*sic*] est constant et qui conservent les volumes (Il faut pour cela comme dans le cas des cartes planes que le déterminant des $\gamma_{\mu\nu}$ soit égal à une constante. Nous voyons donc qu'il n'y a aucune difficulté à se représenter un espace Riemannien ; il suffit d'en faire la carte dans un espace euclidien dont l'expérience courante nous a donné l'intuition.

Toutes les formes de la géométrie générale sont applicables à l'étude du monde physique ; on peut choisir la forme de la géométrie qui se prête le mieux à l'étude de la physique, comme on choisit le système de coordonnées qui se prête le mieux à l'étude d'un problème de géométrie analytique. Toutes sont d'ailleurs aussi commodes au point de vue pratique du moment que la géométrie euclidienne est valable à l'approximation de nos mesures dans le domaine bien petit où s'étendent nos laboratoires et nos instruments.

---

[40] B modification : "impossible".

[41] B modification : "sur".

[42] B suppression : "et".

§ 2 LE TEMPS ET LA MÉCANIQUE :

[p. 12] Que voulons-nous exprimer lorsque nous disons qu'il s'écoule le même temps, entre le passage de l'aiguille de notre montre, d'un chiffre quelconque du cadran au chiffre voisin ? Qu'est-ce qu'un chronomètre ? Quelle définition idéale avons-nous en vue lorsque nous disons que tel chronomètre est meilleur qu'un autre ?

Pour nous en rendre compte, considérons une horloge rudimentaire : un sablier ; nous estimons que le sable met un temps toujours le même pour s'écouler d'un réservoir à l'autre : pourquoi ? Parce que nous ne voyons aucune différence dans la situation du sablier chaque fois que nous le renversons, le même phénomène va se passer : nous affirmons qu'il va durer le même temps.

Supposons qu'on ait cuit un œuf mollet en le laissant dans l'eau bouillante pendant que le sable s'écoule d'un réservoir à l'autre. On recommence l'opération avec un œuf semblable : nous serions bien étonnés si on obtenait un œuf dur. Nous nous demanderions si le premier œuf n'était pas plus gros que le second, si l'écoulement du sable a été bien régulier, que sais-je ? Mais si aucune des explications que nous pourrions imaginer ne se montrait suffisante, nous n'aurions aucune raison de décider s'il vaut mieux mesurer la durée par le temps qu'il faut pour cuire un œuf mollet ou par celui qui est nécessaire pour que le sable s'écoule d'un réservoir à l'autre. Nous prétendrons toujours que, s'il y a une différence dans les résultats des mesures par l'un et l'autre procédé, c'est qu'une circonstance a changé dans la reproduction des phénomènes que nous croyions identiques.

Mais est-il possible de réaliser deux fois le même phénomène ? Strictement, non. Si on a renversé une première fois un sablier, on aura[43] jamais plus un sablier qui n'a jamais été renversé. Il n'est [p. 13] naturellement pas difficile de faire abstraction de différences de cette nature. Mais, même parmi les différences physiques, on pourra faire abstraction de certains facteurs dont un dispositif convenable pourra compenser l'action. On jugera généralement que l'indication d'une bonne horloge doit être indépendante des circonstances de température, de champ électrique ou gravifique etc. Mais on [44] peut faire abstraction de tout ; il faut bien admettre qu'il y a une réalité physique qui est la même entre le commencement et la fin des diverses secondes de tous les chronomètres.

Pour pouvoir étudier la réalité physique que mesurent les chronomètres nous devons adopter un mode de repérage des divers événements.

Nous ferons correspondre à chaque événement quatre nombres x_1, x_2, x_3, x_4, ses coordonnées. Le mode de repérage est complètement indifférent. Il faut pourtant que deux événements ne soient pas désignés par les mêmes coordonnées et que des événements voisins soient représentés par des nombres voisins.

L'égalité de deux intervalles de temps, nous apprend qu'une certaine grandeur caractérisée par les deux couples d'événements est égale pour chacun d'eux. Cette grandeur dépend pour chaque couple des coordonnées des événements. Pour des

---

[43] B modification : "n'aura".

[44] B ajout : "ne".

intervalles de temps infiniment petits, c'est une fonction des coordonnées et des différentielles de ces coordonnées.

Cette fonction doit être homogène par rapport aux différentielles. De plus, elle doit pouvoir être employée quel que soit le mode de repérage des événements, elle doit donc garder sa forme algébrique lorsqu'on fait un changement arbitraire de coordonnées.

L'intervalle de temps peut servir à repérer les divers événements ; il doit donc être du même ordre de grandeur que les différentielles des coordonnées.

Les formes les plus simples que l'on emploie[45] sont, des formes [p. 14] linéaires ou quadriques[46] des différentielles des coordonnées. On posera alors

$$ds = \sum_{\sigma} a_{\sigma}\, dx_{\sigma} \qquad (\sigma = 1,2,3,4)$$

ou

$$ds^2 = \sum_{\mu}\sum_{\nu} g_{\mu\nu}\, dx_{\mu} dx_{\nu} \qquad (\mu,\nu = 1,2,3,4) \tag{1}$$

avec

$$g_{\mu\nu} = g_{\nu\mu} \tag{2}$$

Remarquons que la première forme est un cas particulier de la seconde, celui ou[47] la forme quadrique[48] est un carré parfait.

Le temps mesuré par un chronomètre est, en tout cas, égal à l'intégrale

$$\int ds$$

calculée en suivant le chronomètre ou, comme on dit, sur sa ligne d'univers, en appelant univers le contenu[49] à quatre dimensions des phénomènes physiques.

L'observation des phénomènes[50], nous fait donc atteindre des fonctions $g_{\mu\nu}$ des coordonnées, les potentiels d'univers, analogues à quatre variables des potentiels métriques de la géométrie.

Les potentiels métriques suffisent à déterminer dans l'espace les droites (ou à deux dimensions les géodésiques des surfaces). Ces lignes sont complètement définies par leurs conditions initiales : un point et la direction en ce point.

[45] B modification : "peut employer".
[46] B modification : "quadratiques".
[47] B modification : "où".
[48] B modification : "quadratique".
[49] B modification : "continu".
[50] B modification : "chronomètres".

Il existe dans l'univers un fait analogue. Il y a des lignes d'univers qui sont complètement déterminées par les coordonnées d'un de leurs points et des[51] différentielles de ces coordonnées. Ce sont les trajectoires que suivent des points matériels abandonnés à eux-mêmes. Il y a une propriété du milieu qui les oblige à suivre un chemin déterminé ; cette propriété s'appelle, suivant les circonstances, l'inertie ou le champ de gravitation. Ces réalités physiques qui tracent l'orbite des planètes ne seraient-elles pas les mêmes que celles qui déterminent le mouvement des horloges ? Celles-ci sont généralement basées sur la reproduction de phénomènes mécaniques [p. 15]. La réalité qu'elles mesurent est donc bien de même nature que celle qui se manifeste dans la dynamique des points libres.

La forme (1) définit généralement des lignes d'univers particulièrementi[52] qui sont déterminées par deux de leurs points. Il suffit d'appliquer les formules du calcul des variations à l'équation

$$\delta \int ds = 0$$

prise entre ces deux points. La trajectoire d'un point libre se calculera ainsi comme les droites de la géométrie générale ou les géodésiques d'une surface.

Mais pour que ce procédé soit applicable il est nécessaire que ds ne soit pas une différentielle exacte. Si la droite ou la géodésique peut-être[53] définie comme étant le plus court chemin entre deux points, c'est parce que la longueur d'un arc de courbe dépend de la forme de la courbe et n'est pas la même pour tous les arcs qui ont mêmes extrémités.

Pour que les potentiels $g_{\mu\nu}$ définissent la trajectoire des points libres, il faut donc que l'intervalle de temps entre deux événements dépende de la ligne d'univers suivant laquelle on passe d'un événement à l'autre. En d'autres termes, il faut renoncer à l'idée familière d'un temps absolu. Des chronomètres identiques et identiquement réglés ne marqueront généralement plus le même temps entre deux rencontres. Leur indication dépendra des lignes d'univers qu'ils ont suivies de l'une à l'autre. Il suffit d'ailleurs que le temps défini sur les diverses lignes d'univers différent [*sic*][54] l'un de l'autre de quantités extrêmement petites. Il faut simplement que le temps ne soit pas rigoureusement une différentielle exacte ; il pourra alors être plus grand ou plus petit, sur une trajectoire, que sur toutes les trajectoires voisines de mêmes extrémités ; il y aura des géodésiques d'univers, les trajectoires des points matériels abandonnés à eux-mêmes.

[p. 16] Voici, par exemple, comment ces conditions peuvent être réalisées.

Lorsqu'on emploie des coordonnées cartésiennes x, y, z et un temps t défini d'une manière convenable, les équations du mouvement d'un point libre sous l'influence des seules forces de gravitation peuvent s'écrire, d'après la mécanique classique

---

[51] B modification : "les".

[52] B modification : "particulières".

[53] B modification : "peut être".

[54] On s'attendrait à lire "diffère" alors que le tapuscrit porte "différents" qui a été corrigé en "différent".

$$\frac{dv_x}{dt} + \frac{dV}{dx} = 0$$
$$\frac{dv_y}{dt} + \frac{dV}{dy} = 0$$
$$\frac{dv_z}{dt} + \frac{dV}{dz} = 0$$

où v_x, v_y, v_z désignent les composantes de la vitesse du mobile, et V le potentiel du champ de gravitation.

Ces équations peuvent se mettre sous la forme d'Hamilton

$$\delta \int_{t_0}^{t_1} \left(\frac{v^2}{2} - V\right) dt = 0 \tag{3}$$

Soit c un nombre très grand par rapport aux vitesses qui se présentent dans les expériences sur lesquelles ces équations sont basées, ce sera, par exemple, la vitesse de la lumière.

Considérons la forme quadratique

$$ds^2 = -dx^2 - dy^2 - dz^2 + (c^2 + 2V)\, dt^2 \tag{4}$$

Nous aurons alors

$$\int ds = \int_{t_0}^{t_1} \sqrt{-\left(\frac{dx}{dt}\right)^2 - \left(\frac{dy}{dt}\right)^2 - \left(\frac{dz}{dt}\right)^2 + c^2 + 2V} \cdot dt$$
$$= c \int_{t_0}^{t_1} \sqrt{1 - \frac{1}{c^2}(v^2 - 2V)} \cdot dt$$

Puisque c^2 est grand par rapport à v^2 et par conséquent aussi par rapport à V qui est du même ordre de grandeur, nous pourrons écrire, très approximativement

$$\int ds = c \int_{t_0}^{t_1} \left[1 - \frac{1}{c^2}\left(\frac{v^2}{2} - V\right)\right] dt \tag{5}$$

L'équation des géodésiques

$$\delta \int ds = 0$$

sera alors équivalente à l'équation classique (3).

[p. 17] On pourrait naturellement obtenir le même résultat en se servant d'autres formes que (4) ; en particulier, en y prenant tous les termes positifs. Nous verrons plus tard les raisons qui ont conduit à ce choix des signes.

Voyons maintenant les conséquences qu'entraine l'équation (5), le temps des chronomètres ne sera plus une différentielle exacte, il y aura de petits écarts dans les mesures de l'intervalle de temps entre les événements suivant la trajectoire des horloges qui le mesurent.

Le terme principal de (5) est

$$c\int_{t_0}^{t_1} dt$$

C'est à un facteur près le temps employé par la mécanique classique.

Le temps en secondes sera donc

$$\frac{1}{c}\int ds = \int_{t_0}^{t_1}\left[1-\frac{1}{c^2}\left(\frac{v^2}{2}-V\right)\right]dt$$

Une montre immobile en un point où le potentiel est nul

$$v=0 \qquad V=0$$

marque le temps coordonnée

$$T_1 = \int_{t_0}^{t_1} dt = t_1 - t_0$$

Une deuxième montre en mouvement sur une surface équipotentielle

$$V=0$$

marque le temps

$$T_2 = \int_{t_0}^{t_1}\left(1-\frac{1}{c^2}\frac{v^2}{2}\right)dt = \left(1-\frac{1}{c^2}\frac{\overline{v^2}}{2}\right)T_1$$

où $\overline{v^2}$ est le carré moyen de la vitesse.

$$\overline{v^2} = \frac{1}{t_1-t_0}\int_{t_0}^{t_1} v^2\,dt$$

Considérons enfin une troisième montre, identique aux deux précédentes, lancée verticalement au temps t_0 avec une vitesse v_0 et retombant auprès des deux autres à l'instant t_1.

Puisque les équations[55] (5) n'est qu'une équation approchée nous pouvons nous entendre[56] de calculer la vitesse et le potentiel suivant [p. 18] sa[57] trajectoire en adoptant les équations classiques

$$x = v_0 t - \frac{gt^2}{2}$$

[55] B modification : "l'équation".

[56] B modification : "contenter".

[57] B modification : "la".

et

$$V = gx$$

Nous aurons alors

$$T_1 = t_1 - t_0 = \frac{2v_0}{g}$$

et le temps marqué par la troisième montre sera

$$T_3 = \int_{t_0}^{t_1} \left\{ 1 + \frac{1}{c^2} \left[g \left(v_0 t - \frac{gt^2}{2} \right) - \frac{1}{2} (v_0 - gt)^2 \right] \right\} dt$$

en[58] effectuant l'intégration, on trouve

$$T_3 = \left(1 + \frac{1}{6} \frac{v_0^2}{c^2} \right) T_1$$

La montre T_3 en mouvement libre marquera donc entre les mêmes événements (les rencontres des montres) un temps plus grand que chacune des deux autres.

Il est facile de voir que lorsqu'on adopte la forme (4) de l'élément de temps, la trajectoire d'un point libre est toujours une trajectoire de temps maximum. On peut toujours, en effet, imaginer une trajectoire virtuelle, aussi voisine que l'on veut de la trajectoire réelle, pour laquelle le temps est moindre. Il suffit de concevoir un mobile décrivant une spirale serrée autour de la trajectoire réelle. Le potentiel V aurait alors même valeur sur les deux trajectoires ; la vitesse v sera plus grande sur la trajectoire virtuelle et, par conséquent, le temps (5) sera moindre sur cette dernière.

Il ne faudrait pourtant pas conclure que le temps sur la trajectoire d'un mobile soit nécessairement un maximum absolu. C'est seulement un maximum relativement aux trajectoires voisines. Ainsi la deuxième montre dont nous venons de parler pourrait, sans quitter la surface de niveau, revenir à son point de départ en décrivant un grand cercle autour de la terre, avec une vitesse (7,9 Km.sec [*sic*]) telle que la force centrifuge neutralise exactement la pesanteur. [p. 19] Elle décrirait donc une géodésique, une trajectoire de point libre. Son temps sera plus grand que celui de toutes les trajectoires virtuelles voisines de mêmes extrémités, et cependant il serait plus petit que le temps (T_1')[59] marqué entre ces mêmes extrémités, par la montre immobile[60]. La différence serait, dans ce cas, en prenant

$$c = 3 \cdot 10^{10} \text{ cm. sec. } [sic]$$

de deux millionnièmes [*sic*] de secondes pour un trajet de 1 ^{h} 24 ^{m}.

[58] B modification : "En".

[59] B suppression : parenthèses supprimées.

[60] B suppression : souligné supprimé.

§ 3 LA SIMULTANÉITÉ ET L'ESPACE.

[p. 20] Les définitions que nous avons données de la longueur d'une ligne et de la durée d'un phénomène se réduisent l'une à[61] l'autre à la réalisation de phénomènes périodiques successifs ; points équidistants marqués sur la ligne, secondes identiques d'un chronomètre. Dans le cas de la mesure du temps cette périodicité est réalisée suivant la ligne d'univers suivie par le chronomètre. Sur quelle ligne d'univers est-elle réalisée dans le cas de la mesure des longueurs ?

Nous avons supposé jusqu'ici que l'objet à mesurer ne se déformait pas ; nous avions alors tout le temps de marquer des points équidistants et de vérifier l'égalité de leurs intervalles. Mais si nous voulons étudier la longueur d'un objet qui se déforme, il faut que nous réalisions toutes les opérations que comporte la mesure, au même instant. Si nous nous trompons dans l'appréciation de la simultanéité, notre mesure sera faussée. La périodicité des longueurs doit être réalisée sur une ligne de points simultanés.

Lorsque nous marquons deux points sur un objet, chacun de ces points décrit une ligne d'univers. Représentons par

$$dx_1 \quad dx_2 \quad dx_3 \quad dx_4$$

les différences des coordonnées des deux points et par

$$\delta x_1 \quad \delta x_2 \quad \delta x_3 \quad \delta x_4$$

les variations des coordonnées suivant les lignes d'univers voisines que décrivent les deux points. La condition de simultanéité ne peut dépendre que des coordonnées x_σ ($\sigma = 1, 2, 3, 4$) et des deux sortes de variations (dx_σ)[62] et (δx_σ)[63]. Cette condition doit être symétrique par rapport aux deux points, elle ne doit pas changer quand on en invertit l'ordre, c'est-à-dire quand on change le signe de tous les dx_σ. En effet, la notion de simultanéité, tout comme la notion des distances[64], est réciproque ; elle est indépendante de l'ordre dans lequel on [p. 21] considère les points.
La solution du problème de la simultanéité est analogue à celle du problème géométrique des directions perpendiculaires.
Considérons la forme bilinéaire associée à la forme quadratique ds^2

$$\sum_\mu \sum_\nu g_{\mu\nu}\, dx_\mu \delta x_\nu$$

C'est une fonction invariante des coordonnées et des deux sortes de variations. Pour qu'elle soit symétrique par rapport aux deux événements simultanés, il faut que sa valeur ne change pas, qu'[65] on change le signe des dx_μ ; elle doit donc s'annuler.

---

[61] B modification : "et".

[62] B suppression : parenthèses supprimées.

[63] B suppression : parenthèses supprimées.

[64] B modification : "de distance".

[65] B modification : "quand".

La condition de simultanéité

(1) $$\sum_{\mu}\sum_{\nu} g_{\mu\nu}\, dx_{\mu}\, \delta x_{\nu} = 0$$

doit donc être réalisée entre les points d'un objet dont on mesure la longueur. On peut utiliser cette relation pour éliminer de l'expression

(2) $$ds^2 = \sum_{\mu}\sum_{\nu} g_{\mu\nu}\, dx_{\mu} dx_{\nu}$$

une des coordonnées dx_{μ}, on obtiendra alors une forme quadratique

(3) $$ds^2 = \sum_{\mu}^{3}\sum_{\nu}^{3} \gamma_{\mu\nu}\, dx_{\mu} dx_{\nu}$$

à trois variables, où les $\gamma_{\mu\nu}$ sont des fonctions des potentiels d'univers $g_{\mu\nu}$ et de la vitesse δx_{ν} de l'objet. Cette expression doit être égale pour deux couples de points équidistants. En effet, superposer deux longueurs revient à effectuer un changement de coordonnées de telle sorte que les coordonnées des points ainsi que les diverses grandeurs qui caractérisent une des longueurs, se transforment dans les coordonnées et les grandeurs correspondantes de l'autre longueur. C'est alors seulement que nous pourrons dire qu'elles sont identiques, elles ne diffèrent que par le choix des coordonnées de repérage. Mais la forme bilinéaire de l'équation de simultanéité et l'expression de ds sont des expressions invariantes, indépendantes du choix des coordonnées. Leurs valeurs doivent donc être les mêmes pour deux couples de points équidistants. ds est donc l'élément de distance. Lorsque la forme quadratique est négative on la posera égale à $-d\sigma^2$, $d\sigma$ représentera alors l'élément [p. 22] de distance.

Nous obtenons ainsi la géométrie sous la forme générale de Riemann (§ 1-1). La longueur d'un arc se calculera par la formule (§ 1-2).

Les potentiels métriques sont des fonctions des potentiels d'univers $g_{\mu\nu}$ et de la vitesse δx_{σ} des divers points des objets que l'on mesure.

Considérons par exemple la forme que nous avons obtenue au § précédent

$$ds^2 = -dx^2 - dy^2 - dz^2 + (c^2 + 2V)\, dt^2$$

l'équation de simultanéité de points immobiles

$$\delta x = \delta y = \delta z = 0$$

sera

$$(c^2 + 2V)\, \delta t\, dt = 0$$

ou

$$dt = 0$$

La géométrie de ces points sera donc caractérisée par

$$d\sigma^2 = dx^2 + dy^2 + dz^2$$

qui dépeint[66] l'élément de distance de la géométrie euclidienne.

Si nous réglons des horloges immobiles par simultanéité – c'est-à-dire, dans le cas actuel, si nous leur faisons marquer le temps t – ces horloges ne pourront généralement plus être considérées comme identiques et ne seront plus interchangeables. Des horloges identiques immobiles marqueraient, en effet, le temps

$$\frac{1}{c}\int \sqrt{c^2 + 2V} \cdot dt$$

La définition de la simultanéité sur un corps dépend non seulement de la vitesse δx_σ du corps, mais elle peut aussi dépendre [p. 23] de la ligne suivant laquelle on la définit de proche en proche. Nous rencontrerons plus tard un exemple de ce cas.

* * *

Ces diverses complications de notre notion habituelle du temps disparaissent à la limite dans un domaine infiniment petit.

En géométrie générale on peut toujours trouver un domaine assez petit pour que les relations de la géométrie d'Euclide y soient valables avec une approximation donnée. On peut de même dans la mécanique d'Einstein adopter approximativement au voisinage d'un point une forme réduite du ds^2 analogue à l'élément de distance de la géométrie d'Euclide.

Dans un domaine petit, on peut remplacer les potentiels $g_{\mu\nu}$ par leurs valeurs en un point M. ds^2 est alors une forme quadratique à coefficients constants. Elle se réduit par un changement linéaire des coordonnées, à une somme algébrique de carrés.

On évitera généralement les coordonnées imaginaires en choisissant les signes de la manière suivante :

$$ds^2 = -d\xi_1^2 - d\xi_2^2 - d\xi_3^2 + d\xi_4^2$$

Cette réduction est possible de plusieurs manières.
Si

$$\delta x_1 \quad \delta x_2 \quad \delta x_3 \quad \delta x_4$$

représente la vitesse d'un point d'un corps, on pourra imposer aux fonctions

$$\xi_\sigma = \xi_\sigma(x_1, x_2, x_3, x_4) \quad (\sigma = 1,2,3,4)$$

66 B modification : "définit".

par lesquelles s'effectue la réduction, la condition suivante[67]

$$\frac{\partial \xi_4}{\partial x_\nu} = K \sum_\nu g_{\mu\nu}\, \delta x_\nu \tag{4}$$

[p. 24] Cette relation ne doit d'ailleurs être vérifiée exactement qu'en M (K est une constante quelconque)

L'équation de simultanéité (1) sera alors sur le corps

$$\sum_\mu \frac{\partial \xi_4}{\partial x_\mu} dx_\mu = d\xi_4 = 0$$

Les longueurs sur le corps ne dépendent pas de ξ_4, la géométrie y est définie par l'élément de distance

$$d\sigma^2 = d\xi_1^2 + d\xi_2^2 + d\xi_3^2$$

ξ_1, ξ_2, ξ_3 sont des coordonnées cartésiennes.
L'équation de simultanéité s'écrit dans le nouveau système de coordonnées

$$-d\xi_1\, \delta\xi_1 - d\xi_2\, \delta\xi_2 - d\xi_3\, \delta\xi_3 + d\xi_4\, \delta\xi_4 = 0$$

Pour que cette équation se réduise à

$$d\xi_4 = 0$$

il faut que

$$\delta\xi_1 = \delta\xi_2 = \delta\xi_3 = 0$$

Les[68] variations $\delta\xi_\sigma$ étant prises suivant les lignes d'univers des points du corps. C'est suivant ces lignes d'univers que se calcule le temps sur le corps, ce sera

$$\int d\xi_4$$

Les nouvelles coordonnées donnent donc immédiatement le résultat des mesures spatio-temporelles effectuées en un point d'un corps. C'est ce que nous appellerons les coordonnées propres du corps en ce point. Nous emploierons généralement les notations

$$ds^2 = -dx^2 - dy^2 - dz^2 + c^2\, dt^2$$

où c est une constante ; x, y, z sont des coordonnées cartésiennes, t le temps de chronomètres immobiles.

Lorsque cette forme du ds peut être adoptée rigoureusement, nous dirons que nous avons affaire à un champ de Galilée.

---

[67] Dans l'équation qui suit, il faut lire bien sûr : $\frac{\partial \xi_4}{\partial x_\mu} = K \sum_\nu g_{\mu\nu}\, \delta x_\nu$.

[68] B modification : "les".

Ce sont ces champs qu'Einstein a d'abord étudiés dans son mémoire [p. 25] de 1905 sur l'électrodynamique des corps en mouvement. Il est arrivé à la forme définitive de sa théorie (1915) en adoptant le procédé par lequel Riemann avait généralisé la géométrie d'Euclide, c'est-à-dire en admettant que la forme primitive de sa théorie n'était valable qu'à la limite dans un domaine infiniment petit. C'est le mode habituel de généralisation usité en physique.

* * *

§ 4 – Les Mesures Indirectes d'Espace et de Temps

[p. 26] Nous avons vu comment se mesurent les dimensions d'un corps par arpentage au moyen de mètres identiques portés bout à bout sur ce corps. Le résultat des mesures dépend du mouvement du corps. Celui-ci peut se déformer au cours du mouvement. Nous supposons seulement que la vitesse de ses points varie d'une façon continue d'un point à l'autre. Les dimensions du corps sont alors parfaitement déterminées lorsqu'on connaît les équations des lignes d'univers suivies par chacun de ses points.

Le résultat des mesures ne dépend pas du système de coordonnées adopté, ce sont des mesures absolues.

Il est souvent impossible de procéder par mesure directe en portant les instruments de mesure sur l'objet qu'on [69] propose de mesurer. Il faut alors déduire les longueurs de mesures indirectes effectuées au moyen des appareils dont on dispose.

Si nous voulons mesurer la longueur d'un obus à sa sortie du canon, nous ne disposons pas d'un mètre qui l'accompagne dans son mouvement. Mais nous pourrions[70] par exemple le photographier. Il faudra alors déduire des dimensions de l'épreuve les résultats que nous obtiendrions si nous pouvions suivre l'obus dans son mouvement et le mesurer à la manière ordinaire.

Quel que soit le procédé employé, on pourra le ramener à celui-ci. Nous marquerons au même instant sur un corps qui porte nos instruments, les points qui coïncident avec ceux du mobile. Nous obtiendrons ainsi une image accessible du mobile.

Nous devrons déduire des dimensions de l'image, les dimensions de l'objet.

[p. 27] Nous marquons les points au même instant par rapport à nos instruments de mesure ; c'est la seule simultanéité que nous pouvons songer à réaliser.

Représentons par ξ, η, ζ, τ les coordonnées propres du corps qui porte les instruments à l'endroit où se fait la mesure ; les points simultanés seront ceux pour lesquels les valeurs de τ seront égales.

Désignons par x, y, z, t les coordonnées propres de l'objet au même point. On doit avoir

$$ds^2 = -d\xi^2 - d\eta^2 - d\zeta^2 + c^2 d\tau^2 \tag{1}$$

et

$$ds^2 = -dx^2 - dy^2 - dz^2 + c^2 dt^2 \tag{2}$$

Pour un choix convenable de l'orientation des deux systèmes d'axes, on pourra poser

---

[69] B ajout : "se".

[70] B modification : "pourrons".

$$(3)\qquad \begin{cases} d\xi = \dfrac{1}{\sqrt{1-\frac{v^2}{c^2}}}(dx+v\,dt) \\ d\eta = dy \\ d\zeta = dz \\ d\tau = \dfrac{1}{\sqrt{1-\frac{v^2}{c^2}}}\left(\frac{v}{c^2}dx+dt\right) \end{cases}$$

Cette transformation est compatible, quel que soit v avec les équations (1) et (2).

La vitesse de l'objet s'obtient en y posant

$$dx = dy = dz = 0$$

On trouve

$$\frac{d\xi}{d\tau} = v, \quad \frac{d\eta}{d\tau} = 0, \quad \frac{d\zeta}{d\tau} = 0$$

Les équations (3) portent le nom de transformations de Lorentz. Elles peuvent aussi s'écrire

$$\begin{cases} dx = \dfrac{1}{\sqrt{1-\frac{v^2}{c^2}}}(d\xi - v\,d\tau) \\ dy = d\eta \\ dz = d\zeta \\ dt = \dfrac{1}{\sqrt{1-\frac{v^2}{c^2}}}(-\frac{v}{c^2}d\xi + d\tau) \end{cases}$$

[p. 28] En procédant par mesures indirectes, nous obtenons des différences de coordonnées

$$d\xi, \quad d\eta, \quad d\zeta$$

entre des points liés par la condition de simultanéité.

$$d\tau = 0$$

Les vraies longueurs

$$dx, \quad dy, \quad dz$$

sont donc

$$\begin{cases} dx = \dfrac{1}{\sqrt{1-\frac{v^2}{c^2}}}d\xi \\ dy = d\eta \\ dz = d\zeta \end{cases}$$

Les dimensions de l'image sont différentes de celles de l'objet. Le procédé par mesure indirecte ne donne pas immédiatement les dimensions de l'objet. Les dimensions normales à la vitesse sont conservées mais les mesures effectuées dans la direction de cette vitesse doivent être corrigées. L'objet paraît contracté dans le sens de son mouvement. Ses longueurs parallèles à la vitesse paraissent réduites dans le rapport

$$\sqrt{1-\frac{v^2}{c^2}}$$

On peut aussi procéder par mesures indirectes à la mesure du temps.

On se propose de mesurer l'intervalle de temps dt entre deux événements qui se produisent en un point d'un mobile. Il est impossible de le mesurer directement au moyen d'un chronomètre qui accompagnerait ce mobile. Mais on dispose de chronomètres placés sur un corps accessible et réglés par simultanéité.

On note la différence $d\tau$ des indications des deux chronomètres devant lesquels se produisent les deux événements.

Il s'agit de déduire de la mesure indirecte $d\tau$ le temps dt que mesurerait un chronomètre accompagnant le mobile.

[p. 29] Puisque les événements ont lieu au même point que l'objet, on doit poser

$$dx = 0$$

dans les équations de Lorentz, on trouve donc

$$d\tau = \frac{1}{\sqrt{1-\frac{v^2}{c^2}}}dt$$

On peut interpréter cette correction à apporter aux mesures indirectes en disant que les phénomènes qui se passent sur le mobile paraissent retardés dans le rapport :

$$\frac{1}{\sqrt{1-\frac{v^2}{c^2}}}$$

Nous avons vu comment des grandeurs absolues, durée d'un phénomène et longueur d'un corps peuvent être déterminées par des mesures indirectes relatives à un autre corps. Nous allons aborder maintenant l'étude d'une grandeur qui est essentiellement relative : la vitesse.

Nous avons parlé à plusieurs reprises de la vitesse d'un point, elle est caractérisée par les différentielles

$$dx_1, \quad dx_2, \quad dx_3, \quad dx_4$$

des coordonnées. Elle dépend essentiellement du mode de repérage.

Lorsqu'on emploie des appareils de mesures situés sur un corps, on entendra par vitesse d'un mobile par rapport à ce corps les différentielles

$$dx, \quad dy, \quad dz, \quad dt$$

des coordonnées propres du corps au point où se trouve le mobile. On appellera

$$\frac{dx}{dt}, \quad \frac{dy}{dt}, \quad \frac{dz}{dt}$$

les composantes de la vitesse [71]

$$\sqrt{\left(\frac{dx}{dt}\right)^2 + \left(\frac{dy}{dt}\right)^2 + \left(\frac{dz}{dt}\right)^2}$$

la grandeur de la vitesse.

[p. 30] On considère parfois, sous le nom de vitesse propre, la vitesse dont les composantes sont

$$\frac{dx}{ds}, \quad \frac{dy}{ds}, \quad \frac{dz}{ds}$$

La transformation de Lorentz permet de calculer les composantes

$$\frac{d\xi}{d\tau}, \quad \frac{d\eta}{d\tau}, \quad \frac{d\zeta}{d\tau}$$

de la vitesse d'un point, par rapport à un corps de coordonnées propres ξ, η, ζ, τ, lorsqu'on connaît la vitesse du même point par rapport à un autre corps de coordonnées propres x, y, z, t, animé par rapport au premier d'une vitesse v suivant l'axe des ξ.

On trouve immédiatement

$$\begin{cases} \dfrac{d\xi}{d\tau} = \dfrac{\frac{dx}{dt} + v}{1 + \frac{v}{c^2}\frac{dx}{dt}} \\[2ex] \dfrac{d\eta}{d\tau} = \dfrac{\sqrt{1 - \frac{v^2}{c^2}}}{1 + \frac{v}{c^2}\frac{dx}{dt}} \dfrac{dy}{dt} \\[2ex] \dfrac{d\zeta}{d\tau} = \dfrac{\sqrt{1 - \frac{v^2}{c^2}}}{1 + \frac{v}{c^2}\frac{dx}{dt}} \dfrac{dz}{dt} \end{cases}$$

Ces formules donnent en première approximation (lorsqu'on néglige les termes contenant $1/c^2$) les formules ordinaires de composition des vitesses. Ces formules classiques restent appliquables [*sic*] lorsque les vitesses que l'on combine sont suffisamment petites.

Si celles-ci sont très grandes, la différence entre les équations nouvelles et les anciennes pourra devenir accessible à l'expérience.

Certains phénomènes résultent de la différence des composantes de la vitesse de la lumière suivant le mouvement du corps par rapport auquel on la mesure.

Supposons que les composantes de la vitesse mesurée soient de l'ordre de grandeur de c, la vitesse relative v des systèmes de comparaison étant petite par rapport à c. La première des [p. 31] équations précédentes s'écrit approximative-

[71] B ajout : ", et".

ment

$$\frac{d\xi}{d\tau} = \frac{dx}{dt} + v\left[1 - \frac{1}{c^2}\left(\frac{dx}{dt}\right)^2\right]$$

ou, en posant

$$\frac{c}{\frac{dx}{dt}} = n$$

$$\frac{d\xi}{d\tau} - \frac{dx}{dt} = v\left(1 - \frac{1}{n^2}\right)$$

C'est la formule trouvée par Fizeau et vérifiée expérimentalement avec une grande précision. Fizeau a fait interférer des rayons lumineux d'une même source se propageant dans un courant d'eau de vitesse v, l'un dans le sens du courant, l'autre [72] le sens contraire. Le déplacement des franges d'interférence lorsqu'on change le sens du courant résulte d'une vitesse d'entrainement

$$v\left(1 - \frac{1}{n^2}\right)$$

au lieu de v comme on devrait avoir d'après la loi classique de composition des vitesses. n est l'indice de réfraction du milieu, c'est à dire [*sic*] le rapport de la vitesse dans le vide et de la vitesse dans le milieu.

La nouvelle loi de composition des vitesses rend immédiatement compte de l'expérience de Fizeau si on admet que c est la vitesse de la lumière dans le vide. En suivant un rayon lumineux, on aura donc

$$ds = 0$$

Il en résulte que la grandeur de la vitesse de la lumière est constante en un point, quels que soient la direction du rayon et le mouvement du système par rapport auquel on la mesure.

Mr. Michelson a essayé vainement de mettre en évidence une différence de vitesse de rayons lumineux se propageant dans des directions différentes ; il pensait ainsi pouvoir déceler le mouvement de la terre par rapport au milieu transmetteur de la lumière. Le résultat négatif de son expérience a joué un grand rôle dans l'établissement [p. 32] et la diffusion des théories d'Einstein.

Le fait que la vitesse de la lumière est indépendante de la direction du rayon lumineux donne un moyen commode de réaliser la simultanéité des chronomètres. Deux rayons qui se croisent dans deux directions opposées doivent être considérés comme symétriques l'un de l'autre.

Considérons deux rayons lumineux se propageant dans les deux sens sur la droite joignant deux points A et B. Le premier signal partira de A au moment où un chronomètre placé en A marque le temps t'_A, passe devant des chronomètres intermédiaires qui marquent le temps t' et parvient en B lorsque le chronomètre de B

[72] B ajout : "dans".

marque t'_B. Le signal envoyé de B en t''_B passe devant les chronomètres intermédiaires lorsqu'ils marquent t'' et parvient en A en t''_A.

Les indications

$$\frac{t'_A+t''_A}{2}, \quad \frac{t'+t''}{2}, \quad \frac{t'_B+t''_B}{2}$$

des chronomètres respectifs sont simultanées.

Cela résulte simplement de ce que la vitesse de la lumière est la même dans les deux sens. La vitesse peut varier d'un point à l'autre, il faut seulement qu'elle reste la même en chaque point pendant le temps qui sépare le passage des deux signaux en ce point.

Ce procédé peut être employé sur une ligne polygonale quelconque ; il suffit de disposer des miroirs qui réfléchissent les rayons lumineux suivant les côtés du polygone. On peut ainsi comparer expérimentalement la réalisation de la simultanéité entre deux points suivant des trajets différents.

Si, par exemple, le polygone que suivent les signaux est fermé, [p. 33] les montres de A et B sont au même endroit, leur simultanéité réalisée directement revient à leur équivalence, elles marquent la même heure lorsqu'un événement a lieu près d'elles. On peut d'ailleurs n'employer qu'une seule montre. [73] Lorsqu'on définit la simultanéité suivant le polygone, on doit considérer comme simultanées les indications[74]

$$\frac{t'_A+t''_A}{2}$$

du chronomètre supposé en A, et l'indication

$$\frac{t'_B+t''_B}{2}$$

du même chronomètre considéré comme étant en B.

Si les deux signaux partent au même moment

$$t'_A=t''_B$$

la différence des instants de leur retour

$$t'_B-t''_A$$

sera égale au double de l'écart Δt résultant de la définition de la simultanéité suivant le polygone.

On pourra réaliser l'expérience en divisant un [75] rayons qui décrivent en sens opposés un même polygone et se réunissent à nouveau. Le retard d'un rayon sur l'autre se déterminera par des mesures d'interférence.

---

[73] B modification : symbole requérant le début d'un nouveau paragraphe.

[74] B modification : "l'indication".

[75] B ajout : "rayon lumineux en deux".

Cette expérience a été réalisée par Mr. Sagnac dans le cas d'un champ centrifuge dont l'équation en coordonnées polaires (r étant la distance de[76] l'axe de rotation) peut s'écrire (§ 3-1[77])

$$ds^2 = -dr^2 - r^2(d\theta - \omega\, dt)^2 - dz^2 + c^2 dt^2$$

l'équation de simultanéité pour un corps immobile

$$\delta r = \delta\theta = \delta z = 0$$

est

$$r^2(d\theta - \omega\, dt)\, \omega\, \delta t + c^2\, dt\, \delta t = 0$$

ou

$$\omega\, r^2 d\theta + (c^2 - \omega^2 r^2)\, dt = 0$$

La variation du temps t sur un contour fermé de points simultanés est

$$\Delta t = -\frac{\omega}{c^2}\int_0^{2\pi} \frac{r^2\, d\theta}{1 - \frac{\omega^2 r^2}{c^2}} \cong -\frac{2\,\omega S}{c^2}$$

ou[78]

$$S = \int_0^{2\pi} \frac{r^2\, d\theta}{2}$$

[p. 34] représente l'axe[79] de la projection du polygone sur le plan normal à l'axe de rotation.

Le retard des rayons doit donc être égal à

$$2\Delta t = \frac{4\,\omega S}{c^2}$$

C'est ce qui a été vérifié expérimentalement.

Chapitre II – LES CHAMPS DE FORCE :

La loi de l'attraction universelle et la notion de forces réciproques exercées d'un corps sur l'autre suggèrent l'idée que la force est une réalité ayant son siège dans l'une et l'autre masse ; il ne semble pas qui[80] lui corresponde quelque chose en de-

[76] B modification : "à".
[77] Il faut lire bien sûr : § 3-2.
[78] B modification : "où".
[79] B modification : "l'aire".
[80] B modification : "qu'il".

hors des masses. Le rôle du milieu en[81] est réduit au minimum. Il apparaît pourtant en ceci : la force est en raison inverse du carré de la distance des corps ; elle dépend de la quantité d'espace qui les sépare.

Cette conception d'action à distance avait bien causé quelqu'inquiétude à Newton ; tout se passe comme si les corps s'attirent, disait-il ; mais dans l'enthousiasme que provoquaient avec raison les succès de la théorie, ses successeurs perdirent de vue les prudentes réserves du début et l'explication des divers phénomènes physiques fut cherchée dans des actions à distance.

Le rôle du milieu ne tarda pas à s'imposer de nouveau. Pour Newton, la lumière était une émission rapide de particules matérielles. Cette conception qui devait trouver plus tard son application dans la théorie des rayons cathodiques, se montra tout à fait insuffisante pour rendre compte des lois de l'optique. Il fallut faire appel à un milieu transmetteur, l'éther. On le dota naturellement, suivant les idées de l'époque, de propriétés mécaniques ; il devait être plus rigide que l'acier et plus léger que l'air ; c'était un peu étrange ; mais enfin, cela entrait dans le schéma général des corps solides, dont les propriétés s'interprétaient [p. 35] par des actions, fonctions de la distance.

Le rôle du milieu dans la transmission des forces s'affirma plus nettement en électricité.

L'action réciproque des masses s'exprime par une loi tout à fait analogue à celle de l'attraction universelle la loi de Coulomb. Des masses électriques de même nature se repoussent en raison directe du produit des masses et en raison inverse du carré des distances. Mais la répulsion ne dépend plus seulement de la quantité d'espace qui séparent [*sic*] les masses, elle dépend en outre de la nature du milieu qui les sépare. La force varie lorsqu'on intercale entre elles une lame de verre par exemple. L'attention se détourna ainsi petit à petit des conducteurs portant les masses électriques et se porta d'avantage [*sic*] sur les phénomènes qui se passent dans les diélectriques.

La théorie se développant, il apparut bientôt que les champs induits par des courants doivent se propager avec une vitesse finie, égale au rapport des grandeurs électriques mesurées dans les deux systèmes d'unités électrostatiques et électromagnétiques. Les courants n'agissent pas l'un sur l'autre instantanément ; il y a un retard entre la cause et l'effet ; il faut donc bien que l'action se transmette entre les corps dans le milieu qui les sépare.

La vitesse de cette transmission, rapport des unités des deux systèmes est justement égale à la vitesse de la lumière. Il y a donc une relation entre ces phénomènes si dissemblables. Maxwell montra que la propagation d'une onde d'induction a toutes les propriétés de la lumière. Ces ondes ont été réalisées par Hertz qui a ainsi confirmé d'une manière définitive la théorie de Maxwell. On sait l'importance pratique inatendue [*sic*] qu'ont trouvé ces méthodes[82] théoriques dans la télégraphie sans fil.

Ces résultats nous imposent une nouvelle conception des actions réciproques des corps. Ils n'agissent pas directement l'un sur l'autre [p. 36] mais ils sont dans un

---

[81] B modification : "y".

[82] B modification : "études".

milieu qu'ils modifient et dont ils subissent l'action ; nous pouvons étudier l'état du milieu en observant le mouvement qu'il impose à des particuliers[83] neutres ou électrisées ou encore à des aiguilles aimantées. Nous explorons ainsi des[84] champs d'inertie et de gravitation ou les champs électriques et magnétiques.

Un second problème consiste à étudier la réaction que la matière excerce sur le champ. Une théorie complète devra synthétiser ces deux points de vue ; elle rendra compte de la production des champs par les masses et de l'exploration de ces champs par l'observation du mouvement de petits corps d'épreuve.

Nous allons étudier, dans ce chapitre, le premier problème : l'exploration des champs au moyen de corps d'épreuve et la définition des grandeurs qui caractérisent l'état du milieu. Nous verrons, dans les chapitres suivants, les divers modes de production des champs, par mouvements relatifs[85] (champ de force centrifuge) par exemple, ou par la présence de masses matérielles et électriques. Nous aboutirons enfin aux lois générales qui interprètent les [86] ordres de phénomènes, l'action et la réaction des champs sur les masses.

Les champs se divisent immédiatement en deux grandes classes : ceux qui agissent de la même manière sur une particule quelle que soit sa nature, sa charge électrique etc, et ceux dont l'action dépend de l'état électrique du corps d'épreuve. Nous considérerons d'abord les champs de la première classe, champ d'inertie et de gravitation ; nous en avons déjà dit quelques mots au chapitre précédent, § 2. Nous allons reprendre leur théorie et en développer les calculs.

[83] B modification : "particules".

[84] B modification : "les".

[85] B modification : "mouvement relatif".

[86] B ajout: "deux".

§ 5. LE CHAMP D'INERTIE ET DE GRAVITATION.

[p. 37] Nous avons vu que les propriétés spatio-temporelles du milieu peuvent être représentées par la forme fondamentale à quatre variables

$$ds^2 = \sum_\mu \sum_\nu g_{\mu\nu} \, dx_\mu dx_\nu \tag{1}$$

Des lignes d'univers particulières, les géodésiques, définissent le mouvement libre de particules matérielles. Nous avons vu qu'il était possible de choisir la forme (1) de telle sorte que les équations des points libres diffèrent très peu de celles que vérifie l'expérience. Nous allons maintenant calculer ces équations lorsque les potentiels sont des fonctions quelconques des coordonnées.

Le calcul est naturellement le même que celui par lequel on détermine les géodésiques d'une surface, ou les droites de la géométrie générale lorsqu'on connaît l'expression de l'élément de distance pour un système de coordonnées.

Nous avons à comparer la valeur de

$$\int ds$$

sur la trajectoire réelle d'un point libre et sur des trajectoires virtuelles de mêmes extrémités. Pour calculer les intégrales curvilignes nous prendrons comme variable la valeur λ, que prend en chaque point une fonction convenable des coordonnées.

Les intégrales curvilignes se transforment alors en des intégrales définies de mêmes limites.

$$\int_{\lambda_1}^{\lambda_2} F \, d\lambda \tag{2}$$

ou[87]

$$F = \sqrt{\sum_\mu \sum_\nu g_{\mu\nu} \frac{dx_\mu}{d\lambda} \frac{dx_\nu}{d\lambda}} \tag{3}$$

Les dérivées, que nous représentons par

$$u^\sigma = \frac{dx_\sigma}{d\lambda} \qquad (\sigma = 1,2,3,4) \tag{4}$$

doivent être calculées sur les trajectoires respectives.

[p. 38] Soit[88]

$$x_\sigma = x_\sigma(\lambda) \qquad (\sigma = 1,2,3,4)$$

les équations de la trajectoire réelle.

---

[87] B modification : "où".

[88] B modification : "Soient".

Les équations

$$x_\sigma = x_\sigma(\lambda) + \alpha\,\varpi_\sigma(\lambda) \qquad (\sigma = 1,2,3,4)$$

représenteront des trajectoires virtuelles, si les fonctions continues $\varpi_\sigma(\lambda)$ s'annulent aux limites.

Lorsque α tend vers zéro, les trajectoires virtuelles deviennent aussi voisines que l'on veut de la trajectoire réelle. Pour que celle-ci soit une géodésique, il faut que la différence de l'intégrale (2) sur la trajectoire réelle et sur la trajectoire virtuelle, ne dépende pas du signe de α lorsque α tend vers zéro.

Cette différence peut s'écrire

$$\int \left[F(x_\sigma, u^\sigma) - F\left(x_\sigma + \alpha\,\varpi_\sigma, u^\sigma + \alpha \frac{d\varpi_\sigma}{d\lambda} \right) \right] d\lambda$$

dont le terme principal, lorsque α tend vers zéro, est

$$\alpha \int \sum_\sigma \left[\frac{\partial F}{\partial x_\sigma} \varpi_\sigma + \frac{\partial F}{\partial u^\sigma} \frac{d\varpi_\sigma}{d\lambda} \right] d\lambda$$

Ce terme doit être nul.

Transformons le en intégrant par parties. On a

$$\int_{\lambda_1}^{\lambda_2} \frac{\partial F}{\partial u^\sigma} \frac{d\varpi_\sigma}{d\lambda} d\lambda = \left[\frac{\partial F}{\partial u^\sigma} \varpi^\sigma \right]_{\lambda_1}^{\lambda_2} - \int_{\lambda_1}^{\lambda_2} \varpi_\sigma \frac{d}{d\lambda} \left(\frac{\partial F}{\partial u^\sigma} \right) d\lambda$$

Et puisque ϖ_σ s'annule aux limites

$$\int \sum_\sigma \left[\frac{\partial F}{\partial x_\sigma} - \frac{d}{d\lambda} \left(\frac{\partial F}{\partial u^\sigma} \right) \right] \varpi_\sigma\, d\lambda = 0$$

Cette équation doit être vérifiée quelles que soient les fonctions ϖ_σ continues et admettant une dérivée continue.

Il en résulte qu'en tout point [89] la trajectoire réelle doit avoir [*sic*]

$$(5) \qquad \frac{\partial F}{\partial x_\sigma} - \frac{d}{d\lambda} \left(\frac{\partial F}{\partial u^\sigma} \right) = 0 \qquad (\sigma = 1,2,3,4)$$

Si le premier membre d'une de ces équations était différent de zéro en un point, on pourrait facilement trouver des valeurs de ϖ_σ pour lesquelles l'intégrale ne s'annulerait pas.

On pourrait en effet trouver, au voisinage de ce point, un intervalle

$$\ell < \lambda < L$$

[89] B ajout : "de".

ou[90] une des expressions entre crochets ne change pas de signe. On peut [p. 39] facilement trouver une valeur du ϖ_σ correspondant qui ne change pas de signe dans cet intervalle.

$$\varpi_\sigma = (\ell - \lambda)^2 (L - \lambda)^2$$

On peut alors annuler ϖ_σ en dehors de cet intervalle. ϖ_σ sera ainsi une fonction continue admettant une dérivée continue.

En annulant les autres ϖ_σ, la quantité sous le signe intégrale aurait partout même signe sans s'annuler partout et l'intégrale ne pourrait s'annuler.

Pour développer l'équation (5) il est commode de choisir la fonction λ de telle sorte que, sur la trajectoire réelle, on ait

$$d\lambda = ds$$

On pourra alors, après avoir calculé les dérivées partielles de

$$F = \sqrt{\sum_\mu \sum_\nu g_{\mu\nu}\, u^\mu\, u^\nu}$$

remplacer la quantité sous le radical par sa valeur 1 sur la trajectoire réelle.

On a ainsi

$$\frac{\partial F}{\partial x_\sigma} = \frac{1}{2} \sum_\mu \sum_\nu \frac{\partial g_{\mu\nu}}{\partial x_\sigma}\, u^\mu\, u^\nu$$

et

$$\frac{\partial F}{\partial u^\sigma} = \frac{1}{2} \left[\sum_\nu g_{\sigma\nu}\, u^\nu + \sum_\mu g_{\mu\sigma}\, u^\mu \right]$$

Les équations (5) des géodésiques sont alors

$$\frac{1}{2} \sum_\mu \sum_\nu \frac{\partial g_{\mu\nu}}{\partial x_\sigma} \frac{dx_\mu}{ds} \frac{dx_\nu}{ds} - \frac{1}{2} \frac{d}{ds} \left[\sum_\nu g_{\sigma\nu} \frac{dx_\nu}{ds} + \sum_\mu g_{\mu\sigma} \frac{dx_\mu}{ds} \right] = 0$$

Effectuons alors la dérivation, en remarquant que

$$\frac{d}{ds} g_{\sigma\nu} = \sum_\mu \frac{\partial g_{\sigma\nu}}{\partial x_\mu} \frac{dx_\mu}{ds}$$

et

$$\frac{d}{ds} g_{\mu\sigma} = \sum_\nu \frac{\partial g_{\mu\sigma}}{\partial x_\nu} \frac{dx_\nu}{ds}.$$

[90] B modification : "où".

Il vient

$$\frac{1}{2}\sum_{\mu}\sum_{\nu}\left(\frac{\partial g_{\mu\nu}}{\partial x_\sigma}-\frac{\partial g_{\sigma\nu}}{\partial x_\mu}-\frac{\partial g_{\mu\sigma}}{\partial x_\nu}\right)\frac{dx_\mu}{ds}\frac{dx_\nu}{ds}$$
$$-\frac{1}{2}\left(\sum_{\nu} g_{\sigma\nu}\frac{d^2x_\nu}{ds^2}+\sum_{\mu} g_{\mu\sigma}\frac{d^2x_\mu}{ds^2}\right)=0$$

Les deux termes de la dernière parenthèse sont égaux par suite de la symétrie des $g_{\mu\nu}$ (§ 2-2)

$$g_{\sigma\nu}=g_{\nu\sigma}$$

[p. 40] En introduisant la notation suivante due à Christoffel

$$\text{(6)}\qquad \begin{bmatrix}\mu\,\nu\\ \sigma\end{bmatrix}=\frac{1}{2}\left(\frac{\partial g_{\mu\sigma}}{\partial x_\nu}+\frac{\partial g_{\nu\sigma}}{\partial x_\mu}-\frac{\partial g_{\mu\nu}}{\partial x_\sigma}\right)$$

Les équations des géodésiques s'écrivent

$$\text{(7)}\qquad \sum_{\mu} g_{\mu\sigma}\frac{d^2x_\mu}{ds^2}+\sum_{\mu}\sum_{\nu}\begin{bmatrix}\mu\,\nu\\ \sigma\end{bmatrix}\frac{dx_\mu}{ds}\frac{dx_\nu}{ds}=0$$

Il est commode de résoudre ces quatre équations par rapport aux dérivées secondes. Cette résolution s'effectue aisément en introduisant des quantités $g^{\sigma\tau}$ définies par les équations

$$\text{(8)}\qquad \sum_{\sigma} g_{\mu\sigma}g^{\sigma\tau}=g_{\mu}^{\tau}\qquad(\mu,\tau=1,2,3,4)$$

ou[91]

$$g_{\mu}^{\tau}=\begin{cases}1 & \text{si}\quad \mu=\tau\\ 0 & \text{si}\quad \mu\neq\tau\end{cases}$$

Ces équations constituent pour chaque valeur de τ un système de quatre équations linéaires à quatre inconnues $g^{1\tau}$, $g^{2\tau}$, $g^{3\tau}$, $g^{4\tau}$, qui permet de les calculer. Multiplions la première des équations des géodésiques ($\sigma=1$) par $g^{1\tau}$ la seconde par $g^{2\tau}$ la troisième et la quatrième respectivement par $g^{3\tau}$ et $g^{4\tau}$ et sommons les expressions ainsi obtenues. Il vient

$$\sum_{\sigma}\sum_{\mu} g_{\mu\sigma}g^{\sigma\tau}\frac{d^2x_\mu}{ds^2}+\sum_{\sigma}\sum_{\mu}\sum_{\nu} g^{\sigma\tau}\begin{bmatrix}\mu\,\nu\\ \sigma\end{bmatrix}\frac{dx_\mu}{ds}\frac{dx_\nu}{ds}=0$$

Le premier terme s'écrit

$$\sum_{\mu} g_{\mu}^{\tau}\frac{d^2x_\mu}{ds^2}=\frac{d^2x_\tau}{ds^2}$$

le second se simplifie par l'introduction du symbole de Christoffel de deuxième espèce

---

[91] B modification : "où".

$$(9) \qquad \left\{ \begin{matrix} \mu\,\nu \\ \tau \end{matrix} \right\} = \sum_{\sigma} g^{\sigma\tau} \left[\begin{matrix} \mu\,\nu \\ \sigma \end{matrix} \right]$$

Les équations des géodésiques s'écrivent finalement

$$(10) \qquad \frac{d^2 x_\tau}{ds^2} + \sum_{\mu} \sum_{\nu} \left\{ \begin{matrix} \mu\,\nu \\ \tau \end{matrix} \right\} \frac{dx_\mu}{ds} \frac{dx_\nu}{ds} = 0$$

$$(\tau = 1,2,3,4)$$

[p. 41] Ce sont les équations du mouvement de la nouvelle mécanique.
Remarques 1) Les quantités $g^{\sigma\tau}$ que nous avons introduites dans le calcul sont dites associées aux potentiels $g_{\sigma\tau}$. Il est facile de trouver leur expression en fonction de ces potentiels.

Les équations par lesquelles nous les avons définies se décomposent en quatre systèmes ($\tau = 1,2,3,4$) de quatre équations à quatre inconnues

$$g_{\mu 1} g^{1\tau} + g_{\mu 2} g^{2\tau} + g_{\mu 3} g^{3\tau} + g_{\mu 4} g^{4\tau} = g_{\mu}^{\tau}$$

$$(\mu = 1,2,3,4)$$

Le déterminant de chacun de ces systèmes est le déterminant

$$g = |g_{\mu\nu}|$$

formé au moyen des potentiels.

Calculons par exemple g^{23} dans le système ($\tau = 3$)

$$g^{23} = \frac{1}{g} \begin{vmatrix} g_{11}\,, 0\,, g_{31}\,, g_{41} \\ g_{12}\,, 0\,, g_{32}\,, g_{42} \\ g_{13}\,, 1\,, g_{33}\,, g_{43} \\ g_{14}\,, 0\,, g_{34}\,, g_{44} \end{vmatrix}$$

g^{23} est le quotient par g du mineur du potentiel g_{23} de mêmes indices dans le déterminant g.
En général, $g^{\mu\nu}$ est le quotient par g du mineur de $g_{\mu\nu}$ dans le déterminant g.

Cette propriété peut être utilisée pour calculer la dérivée du déterminant des potentiels. La dérivée d'un déterminant s'obtient en dérivant chaque élément, en le multipliant par son mineur et en faisant la somme de tous les termes obtenus. Nous aurons ainsi

$$(11) \qquad \frac{\partial g}{\partial x_\mu} = g \sum_{\sigma} \sum_{\tau} g^{\sigma\tau} \frac{\partial g_{\sigma\tau}}{\partial x_\mu}$$

2) Symétrie des symboles
Il résulte du calcul des $g^{\mu\nu}$ que la symétrie des potentiels

$$g_{\mu\nu} = g_{\nu\mu}$$

[p. 42] entraîne la symétrie de leurs associés

$$g^{\mu\nu} = g^{\nu\mu}$$

La définition des symboles de Christoffel montre que ceux-ci sont symétriques par rapport aux indices supérieurs

$$\begin{bmatrix} \mu\,\nu \\ \sigma \end{bmatrix} = \begin{bmatrix} \nu\,\mu \\ \sigma \end{bmatrix}$$

et

$$\begin{Bmatrix} \mu\,\nu \\ \upsilon \end{Bmatrix} = \begin{Bmatrix} \nu\,\mu \\ \upsilon \end{Bmatrix}$$

La symétrie des symboles permet souvent de simplifier les sommations.

Ainsi dans l'expression

$$\sum_{\tau} \begin{Bmatrix} \mu\,\tau \\ \tau \end{Bmatrix} = \frac{1}{2} \sum_{\sigma} \sum_{\tau} g^{\sigma\tau} \left(\frac{\partial g_{\sigma\tau}}{\partial x_{\mu}} + \frac{\partial g_{\mu\sigma}}{\partial x_{\tau}} - \frac{\partial g_{\mu\tau}}{\partial x_{\sigma}} \right)$$

les deux derniers termes se détruisent par l'échange les [*sic*] indices de sommation σ et τ. On peut les supprimer à cause de la symétrie de $g^{\sigma\tau}$.

On aura alors à cause de l'équation (11)

$$\sum_{\tau} \begin{Bmatrix} \mu\,\tau \\ \tau \end{Bmatrix} = \frac{1}{2g} \frac{\partial g}{\partial x_{\mu}} \tag{12}$$

Dans la plupart des applications, ds^2 est de la forme

$$ds^2 = g_{11}\, dx_1^2 + g_{22}\, dx_2^2 + g_{33}\, dx_3^2 + g_{44}\, dx_4^2$$

le déterminant des $g_{\mu\nu}$ se réduit alors à sa diagonale principale, et tous les $g^{\mu\nu}$ sont nuls à l'exception de[92]

$$g^{\sigma\sigma} = \frac{1}{g^{\sigma\sigma}} \qquad (\sigma = 1, 2, 3, 4)$$

[93]

$$\begin{bmatrix} \sigma\,\sigma \\ \tau \end{bmatrix} = -\frac{1}{2} \frac{\partial g_{\sigma\sigma}}{\partial x_{\tau}} \qquad (\sigma \neq \tau)$$

$$\begin{bmatrix} \sigma\,\tau \\ \sigma \end{bmatrix} = \begin{bmatrix} \tau\,\sigma \\ \sigma \end{bmatrix} = \frac{1}{2} \frac{\partial g_{\sigma\sigma}}{\partial x_{\tau}}$$

pour les symboles de première espèce, et[94]

---

[92] Dans l'équation qui suit, il faut lire bien sûr : $g^{\sigma\sigma} = \frac{1}{g_{\sigma\sigma}}$.

[93] B ajout : "Les symboles de Christoffel sont nuls à l'exception de".

[94] Dans l'équation qui suit, il faut lire bien sûr : $\begin{Bmatrix} \sigma\,\sigma \\ \tau \end{Bmatrix} = -\frac{1}{2} \frac{1}{g_{\tau\tau}} \frac{\partial g_{\sigma\sigma}}{\partial x_{\tau}} \quad (\sigma \neq \tau)$.

$$(13)\qquad \begin{aligned} &\left\{\begin{matrix}\sigma\sigma\\ \tau\end{matrix}\right\} = -\frac{1}{2}\frac{1}{g_{\tau\tau}}\frac{\partial g_{\sigma\sigma}}{\partial g_\tau} \qquad (\sigma \neq \tau)\\ &\left\{\begin{matrix}\sigma\tau\\ \sigma\end{matrix}\right\} = \left\{\begin{matrix}\tau\sigma\\ \sigma\end{matrix}\right\} = \frac{1}{2}\frac{1}{g_{\sigma\sigma}}\frac{\partial g_{\sigma\sigma}}{\partial x_\tau}\end{aligned}$$

pour ceux de deuxième espèce.

Dans le cas plus particulier que nous avons envisagé au chapitre premier où

$$ds^2 = -dx_1^2 - dx_2^2 - dx_3^2 + g_{44}dx_4^2$$

(Nous avions alors $g_{44} = c^2 + 2V$)
[p. 43] Nous[95] aurons

$$(14)\qquad \begin{aligned} &\left\{\begin{matrix}44\\ i\end{matrix}\right\} = \frac{1}{2}\frac{\partial g_{44}}{\partial x_i} \qquad (i = 1,2,3)\\ &\left\{\begin{matrix}4\sigma\\ 4\end{matrix}\right\} = \left\{\begin{matrix}\sigma 4\\ 4\end{matrix}\right\} = \frac{1}{2}\frac{1}{g_{44}}\frac{\partial g_{44}}{\partial x_\sigma} \qquad (\sigma = 1,2,3,4)\end{aligned}$$

les autres symboles de seconde espèce étant nuls.

* * *

Nous avons trouvé les équations du mouvement en prenant comme variable indépendante suivant la trajectoire

$$\int ds$$

c'est-à-dire le temps marqué par un chronomètre qui suit le mobile.

Il est souvent commode de prendre comme variable principale une des coordonnées x_4 qui représentera généralement l'indication de chronomètres échelonnés le long de trajectoire [*sic*].

On a pour $(i = 1,2,3,4)$

$$\frac{d^2x_i}{ds^2} = \frac{d}{ds}\left(\frac{dx_i}{dx_4}\frac{dx_4}{ds}\right) = \frac{d^2x_i}{dx_4^2}\left(\frac{dx_4}{ds}\right)^2 + \frac{dx_i}{dx_4}\frac{d^2x_4}{ds^2};$$

d'où

$$\frac{d^2x_i}{dx_4^2} = \frac{d^2x_i}{ds^2}\left(\frac{ds}{dx_4}\right)^2 - \frac{d^2x_4}{ds^2}\frac{dx_i}{dx_4}\left(\frac{ds}{dx_4}\right)^2$$

Portons dans cette expression les valeurs de $\frac{d^2x_i}{ds^2}$ et $\frac{d^2x_4}{ds^2}$ tirées des équations du mouvement (10), il vient

$$(15)\qquad \frac{d^2x_i}{dx_4^2} + \sum_\mu\sum_\nu\left(\left\{\begin{matrix}\mu\nu\\ i\end{matrix}\right\} - \left\{\begin{matrix}\mu\nu\\ 4\end{matrix}\right\}\frac{dx_i}{dx_4}\right)\frac{dx_\mu}{dx_4}\frac{dx_\nu}{dx_4} = 0$$

$$(i = 1,2,3)$$

[95] B modification : "nous".

En particulier dans le cas où

$$ds^2 = -dx^2 - dy^2 - dz^2 + (c^2 + 2V)\,dt^2$$

On[96] aura, d'après les valeurs des symboles de Christoffel calculées plus haut,

$$\frac{d^2x}{dt^2} = -\frac{\partial V}{\partial x} + \frac{1}{c^2 + 2V}\frac{dx}{dt}\left[2\left(\frac{\partial V}{\partial x}\frac{dx}{dt} + \frac{\partial V}{\partial y}\frac{dy}{dt} + \frac{\partial V}{\partial z}\frac{dz}{dt}\right) + \frac{\partial V}{\partial t}\right] \tag{16}$$

et les équations analogues en y et en z.

[p. 44] Nous avions déjà vu que cette forme du ds^2 conduit [97] équations qui se confondent en première approximation avec les équations de la mécanique classique, lorsque les vitesses sont petites par rapport à c.

Lorsque ces vitesses ne sont pas négligeables, il faut adjoindre à l'expression classique de la force un terme complémentaire fonction de la vitesse.

* * *

Lorsque toutes les dérivées des potentiels sont nulles, il en est de même des symboles de Christoffel et les équations du mouvement se réduisent à

$$\frac{d^2x_\sigma}{ds^2} = 0$$

ou

$$\frac{d^2x_i}{dx_4^2} = 0$$

Nous dirons alors que le mouvement est rectiligne et uniforme.

Réciproquement, pour que le mouvement en un point soit rectiligne et uniforme quelle que soit la vitesse du mobile, il faut que les dérivées de tous les potentiels s'annulent en ce point.

Il faut, en effet, d'après les équations du mouvement, que tous les symboles de Christoffel, de seconde espèce s'annulent. On obtient en résolvant les équations qui les définissent par rapport aux dérivées des potentiels

$$\frac{\partial g_{\mu\nu}}{\partial x_\tau} = \sum_\sigma \left[g_{\mu\sigma}\begin{Bmatrix}\nu\,\tau\\ \sigma\end{Bmatrix} + g_{\nu\sigma}\begin{Bmatrix}\mu\,\tau\\ \sigma\end{Bmatrix}\right] \tag{17}$$

Cette équation s'établit facilement de la manière suivante.

D'après la définition des symboles (9) et des $g^{\mu\nu}$ (8)[98]

$$\sum_\sigma g_{\mu\sigma}\begin{Bmatrix}\nu\,\tau\\ \sigma\end{Bmatrix} = \sum_\sigma\sum_\rho g_{\mu\sigma}g^{\sigma\rho}\begin{bmatrix}\nu\,\tau\\ \rho\end{bmatrix} = \sum_\rho g_\mu^\rho\begin{bmatrix}\nu\,\tau\\ \rho\end{bmatrix} = \begin{bmatrix}\nu\,\tau\\ \mu\end{bmatrix} \tag{17'}$$

[96] B modification : "on".

[97] B ajout : "à des".

[98] Le manuscrit ne numérote pas l'équation qui suit mais s'y réfère par le numéro (17').

il reste à vérifier que

$$\frac{\partial g_{\mu\nu}}{\partial x_\tau} = \begin{bmatrix} \nu\,\tau \\ \mu \end{bmatrix} + \begin{bmatrix} \mu\,\tau \\ \nu \end{bmatrix} \tag{18}$$

Ceci résulte immédiatement de la définition des symboles de première espèce (6).

[p. 45] Il est possible de choisir des coordonnées ξ_1, ξ_2, ξ_3, ξ_4 de telle sorte que tout point libre passant en un point P ait un mouvement [99]. Les dérivées des potentiels s'annuleront alors en ce point.

Soit[100]

$$\xi_\tau = \xi_\tau(x_1, x_2, x_3, x_4) \qquad (\tau = 1,2,3,4)$$

les équations qui définissent le système de coordonnées dont nous voulons prouver l'existence. x_1, x_2, x_3, x_4 sont des coordonnées quelconques.

Développons ces fonctions en série de Taylor,

$$\xi_\tau = (\xi_\tau)_0 + \sum_\sigma \left(\frac{\partial \xi_\tau}{\partial x_\sigma}\right)_0 (x_\sigma - x_\sigma^0)$$
$$+ \frac{1}{2}\sum_\mu \sum_\nu \left(\frac{\partial^2 \xi_\tau}{\partial x_\mu \partial x_\nu}\right)_0 (x_\mu - x_\mu^0)(x_\nu - x_\nu^0) + \cdots$$

les expressions marquées de l'indice zéro désignent les valeurs des fonctions au point P de coordonnées $(x_1^0, x_2^0, x_3^0, x_4^0)$.

Nous devons montrer qu'il y a moyen de déterminer $\left(\frac{\partial \xi_\tau}{\partial x_\sigma}\right)_0$ et $\left(\frac{\partial^2 \xi_\tau}{\partial x_\mu \partial x_\nu}\right)_0$ de telle sorte que

$$\frac{d^2 \xi_\tau}{ds^2} = 0$$

pour tout point libre passant au point P.

Effectuons la dérivation

$$\frac{d^2 \xi_\tau}{ds^2} = \sum_\sigma \left(\frac{\partial \xi_\tau}{\partial x_\sigma}\right)_0 \frac{d^2 x_\sigma}{ds^2}$$
$$+ \frac{1}{2}\sum_\mu \sum_\nu \left(\frac{\partial^2 \xi_\tau}{\partial x_\mu \partial x_\nu}\right)_0 \left[(x_\mu - x_\mu^0)\frac{d^2 x_\nu}{ds^2} + 2\frac{dx_\mu}{ds}\frac{dx_\nu}{ds} + (x_\nu - x_\nu^0)\frac{d^2 x_\mu}{ds^2}\right] + \cdots$$

---

[99] B ajout : “rectiligne et uniforme”.

[100] B modification : “Soient”.

Au point P, tous les termes non écrits dans le développement s'annulent et il reste en remplaçant $\frac{d^2x_\sigma}{ds^2}$ par sa valeur tirée de l'équation du mouvement d'un point libre (10)[101]

$$\frac{d^2\xi_\tau}{ds^2} = \sum_\mu \sum_\nu \left[\frac{\partial^2 \xi_\tau}{\partial x_\mu \partial x_\nu} - \sum_\sigma \frac{\partial \xi_\tau}{\partial x_\nu} \left\{ \begin{matrix} \mu\nu \\ \sigma \end{matrix} \right\} \right] \frac{dx_\mu}{ds} \frac{dx_\nu}{ds}$$

Cette expression doit s'annuler quels que soient $\frac{dx_\mu}{ds}$ et $\frac{dx_\nu}{ds}$.
[p. 46] Il faut donc que au point P

$$(19) \qquad \frac{\partial^2 \xi_\tau}{\partial x_\mu \partial x_\nu} = \sum_\sigma \frac{\partial \xi_\tau}{\partial x_\sigma} \left\{ \begin{matrix} \mu\nu \\ \sigma \end{matrix} \right\}$$

quels que soient μ, ν et τ.

Nous pouvons donc nous donner arbitrairement les valeurs des dérivées premières des ξ_τ et calculer les dérivées secondes par la formule précédente.

* * *

Nous avons vu qu'il existait, pour chaque point d'un corps, un système de coordonnées qui mettait en évidence les mesures effectuées en ce point sur le corps. Les potentiels prenaient alors en ce point les valeurs exprimées par la forme

$$ds^2 = -d\xi_1^2 - d\xi_2^2 - d\xi_3^2 + c^2 d\xi_4^2$$

Nous avions appelé ce système : coordonnées propres du corps en ce point, et nous les avions obtenues par un changement linéaire de coordonnées, c'est-à-dire en disposant de la valeur des dérivées premières $\frac{\partial \xi_\tau}{\partial x_\sigma}$ (§ 3-4) des fonctions définissant le changement de coordonnées. Nous voyons maintenant que l'on peut en outre choisir les dérivées secondes des ξ_τ par la relation (19) de telle sorte que, dans le nouveau système de coordonnées ξ_τ, les dérivées premières des potentiels s'annulent en P.

Le mouvement d'un point libre y est alors rectiligne et uniforme.

Ces coordonnées s'appellent des coordonnées de Galilée. La relativité restreinte étudie le cas où elles peuvent être employées non seulement en un point P mais dans un domaine fini. On conclut de ce [p. 47] qui précède qu'il existe en un point quelconque un système de Galilée et qu'il est toujours possible de trouver autour de ce point un domaine suffisamment petit, pour que les formules de la relativité restreinte y soient valables, avec une approximation donnée.

---

[101] Dans l'équation qui suit, il faut lire bien sûr :

$$\frac{d^2\xi_\tau}{ds^2} = \sum_\mu \sum_\nu \left[\frac{\partial^2 \xi_\tau}{\partial x_\mu \partial x_\nu} - \sum_\sigma \frac{\partial \xi_\tau}{\partial x_\sigma} \left\{ \begin{matrix} \mu\nu \\ \sigma \end{matrix} \right\} \right] \frac{dx_\mu}{ds} \frac{dx_\nu}{ds}.$$

§ 6. Les champs électriques.

[p. 48] Les propriétés électro-magnétiques du milieu nous sont connues par l'action qu'il exerce sur des particules chargées, sur les conducteurs parcourus par un courant et enfin sur les aimants.

L'action du champ sur des particules électrisées se divise en deux effets nettement distincts attribués respectivement au champ électrique et au champ magnétique. Le premier communique aux diverses particules des accélérations parallèles indépendantes de leur vitesse. Les accélérations produites par le second sont normales aux vitesses et situées dans un même plan.

Le mouvement de la particule peut être représenté par les équations suivantes :

$$(1)\qquad \begin{cases} m\dfrac{d^2x}{dt^2} = X \\ m\dfrac{d^2y}{dt^2} = Y \\ m\dfrac{d^2z}{dt^2} = Z \end{cases}$$

ou[102]

$$(2)\qquad \begin{cases} X = e(e_x + h_z v_y - h_y v_z) \\ Y = e(e_y + h_x v_z - h_z v_x) \\ Z = e(e_z + h_y v_x - h_x v_y) \end{cases}$$

m, e, v_x, v_y, v_z représentent respectivement les masses matérielle et électrique et les composantes de la vitesse de la particule sur les[103] trois axes rectangulaires.

[p. 49] e_x, e_y, e_z sont les composantes d'un vecteur, caractérisant le champ électrique ; la force correspondante est parallèle à ce vecteur et indépendante de la direction ou de la vitesse de la particule.

Le vecteur h_x, h_y, h_z représente le champ magnétique ; la force correspondante est perpendiculaire à ce vecteur ainsi qu'à la vitesse de la particule et est proportionnelle à l'aire du parallélogramme construit sur ces deux vecteurs.

Les fils conducteurs du courant électrique doivent leurs propriétés à des particules d'électricité, les électrons, capables de subir les actions du milieu.

Les actions subies par les électrons normalement à la section du fil produisent un déplacement d'ensemble des électrons qui constitue le courant électrique ; ce mouvement se dissipe rapidement par suite des chocs des électrons contre les molécules et se transforme en agitation thermique. Le courant ne peut se maintenir que si les électrons subissent une action continue : la force électromotrice, cause du courant.

---

[102] B modification : "où".

[103] B modification : "sur les" remplacé par "suivant".

Les actions subies par les électrons normalement à la direction du fil [104] peuvent produire un courant électrique ; elles se transmettent à la masse du conducteur, et apparaissent sous forme d'effort mécanique exercé par le milieu sur le conducteur.

Les formules (2) peuvent donc être appliquées aux conducteurs, elles donneront l'action mécanique subie par le conducteur si e désigne sa charge électrique et

$$ev_x, \quad ev_y, \quad ev_z$$

les composantes du courant. Les actions subies par les électrons [p. 50] suivant la direction du fil entretiennent le courant. Elles constituent la force électromotrice du courant, proportionnelle à e_x, e_y, e_z. Elle est due à la différence de potentiel le long du fil et aux effets d'induction.

La loi d'induction s'exprime généralement de la manière suivante : la force électromotrice totale sur un circuit fermé est égale à la variation du flux du champ magnétique sur une surface limitée par le circuit. Cette loi est peu satisfaisante. Tout d'abord il est bien clair que la force électromotrice totale ne peut être que la somme des forces électromotrices en chaque point du conducteur ; il serait intéressant de les connaître. Mais surtout, l'action subie par les électrons du fil ne peut dépendre de la variation du champ magnétique, en dehors du fil. Cette loi doit pouvoir être considérée comme une déduction d'une loi qui ne fait intervenir que des grandeurs localisées en [105] point du fil.

On sait que le flux d'un vecteur sur une surface peut s'exprimer par une intégrale curvilique [*sic*] sur le contour de cette surface.

On a par le théorème de Stokes

$$\iint_S (h_x\,dydz + h_y\,dzdx + h_z\,dxdy)$$
$$= \int_C (\varphi_x\,dx + \varphi_y\,dy + \varphi_z\,dz)$$

ou[106]

$$(3) \qquad \begin{cases} h_x = \dfrac{\partial \varphi_y}{\partial z} - \dfrac{\partial \varphi_z}{\partial y} \\ h_y = \dfrac{\partial \varphi_z}{\partial x} - \dfrac{\partial \varphi_x}{\partial z} \\ h_z = \dfrac{\partial \varphi_x}{\partial y} - \dfrac{\partial \varphi_y}{\partial x} \end{cases}$$

[p. 51] Le vecteur φ_x, φ_y, φ_z est, au signe près, le potentiel vecteur. En désignant par φ le potentiel scalaire, la force électromotrice due à la différence de potentiel et à l'induction peut s'écrire

---

[104] B ajout : "ne".

[105] B ajout : "chaque".

[106] B modification : "où".

$$(4)\qquad \begin{cases} e_x = \dfrac{\partial \varphi_x}{\partial t} - \dfrac{\partial \varphi}{\partial x} \\ e_y = \dfrac{\partial \varphi_y}{\partial t} - \dfrac{\partial \varphi}{\partial y} \\ e_z = \dfrac{\partial \varphi_z}{\partial t} - \dfrac{\partial \varphi}{\partial z} \end{cases}$$

Ces formules expriment tout ce que nous apprend l'exploration des champs par le mouvement des particules et des conducteurs. Le mouvement des aiguilles aimantées ne peut nous apprendre rien de neuf ; car les propriétés des aimants peuvent être interprétées par l'hypothèse de courants particulaires [*sic*] ou de trajectoires fermées d'électrons.

Les formules que nous avons obtenues sont identiques aux [107] classiques de l'électromagnétisme lorsqu'on emploie un système d'unité tel que les unités électrostatiques et les unités électromagnétiques soient identiques. On sait qu'il suffit pour cela de choisir l'unité de temps par exemple de telle sorte que le rapport des unités des deux systèmes soit pris pour unité. Ce rapport a les dimensions d'une vitesse et est égal à la vitesse de la lumière dans le vide. Il suffit donc de prendre cette vitesse comme unité de vitesse.

Si on voulait employer des unités C.G.S., il faudrait introduire des facteurs numériques et préciser si on emploie des unités U.E.M. ou U.E.S.

Nous avons obtenu les équations de l'électromagnétisme [p. 52] pour un système particulier de coordonnées. Nous allons maintenant chercher des équations dont la forme algébrique ne dépende pas du choix des coordonnées et qui se réduisent aux équations que nous venons de trouver lorsqu'on emploie les coordonnées particulières convenables.

Ces équations peuvent être écrites sous une forme condensée en adoptant les notations suivantes. Nous désignerons les potentiels par

$$\varphi_1, \quad \varphi_2, \quad \varphi_3, \quad \varphi_4$$

au lieu de

$$\varphi_x, \quad \varphi_y, \quad \varphi_z, \quad \varphi$$

et nous poserons

$$(5)\qquad \begin{array}{ll} F_{14} = e_x & F_{23} = h_x \\ F_{24} = e_y & F_{31} = h_y \\ F_{34} = e_z & F_{12} = h_z \end{array}$$

[107] B ajout : "formules".

Les équations (3) et (4) s'écrivent alors

$$F_{\alpha\beta} = \frac{\partial \varphi_\alpha}{\partial x_\beta} - \frac{\partial \varphi_\beta}{\partial x_\alpha} \tag{6}$$

$$(\alpha, \beta = 1, 2, 3, 4)$$

Il n'y a que 6 équations distinctes car

$$F_{\alpha\beta} = -F_{\beta\alpha}$$

et $F_{\alpha\beta}$ est nul lorsque les deux indices sont égaux.

Les équations (2) peuvent être transformées d'une manière analogue, en remplaçant

$$ev_x, \quad ev_y, \quad ev_z, \quad e$$

respectivement par

$$\mathfrak{J}^1, \quad \mathfrak{J}^2, \quad \mathfrak{J}^3, \quad \mathfrak{J}^4.$$

[p. 53] et, en désignant les composantes

$$X, \quad Y, \quad Z$$

de la force par

$$-f_1, \quad -f_2, \quad -f_3$$

elles s'écrivent[108] $(\sigma = 1, 2, 3)$

$$f_\sigma + \sum_\alpha \mathfrak{J}^\alpha F_{\alpha\sigma} = 0 \tag{7}$$

Lorsque nous explorons les champs au moyen de particules non chargées, les trajectoires sont caractérisées par les équations aux variations

$$\delta \int ds = 0$$

ou[109] ds^2 est une forme quadratique des différentielles des coordonnées. Lorsque le champ s'explore au moyen d'une particule chargée de masse matérielle m et de masse électrique e cette variation ne sera plus nulle. L'hypothèse la plus simple que l'on puisse faire est que cette variation soit égale à la variation d'une forme linéaire de ces différentielles.

$$-\varphi_1\,dx_1 - \varphi_2\,dx_2 - \varphi_3\,dx_3 - \varphi_4\,dx_4$$

108 C correction (voir Annexe 2, p. 249) : "formule (7), au lieu de $F_{\alpha\sigma}$, lire $F_{\sigma\alpha}$. La démonstration qui suit doit être modifiée, en changeant le signe dans la formule (8)."

109 B modification : "où".

On écrira alors

$$\delta \int \left[m\, ds + e \sum_{\alpha} \varphi_\alpha\, dx_\alpha \right] = 0 \tag{8}$$

Cette loi garde la même forme algébrique lorsqu'on fait un changement quelconque de coordonnées. Elle est donc appliquable [*sic*] quel que soit le mode de repérage utilisé pour les divers événements.

Nous avons calculé (§ 5-7) le terme provenant de la variation du premier terme

$$f_\sigma = m \left\{ \sum_{\mu} g_{\mu\sigma} \frac{d^2 x_\mu}{ds^2} + \sum_{\mu} \sum_{\nu} \begin{bmatrix} \mu\,\nu \\ \sigma \end{bmatrix} \frac{dx_\mu}{ds} \frac{dx_\nu}{ds} \right\} \tag{9}$$

Ce terme doit donc être égal et de signe contraire à celui que [p. 54] l'on obtient en appliquant à la fonction

$$F = \sum_{\alpha} \varphi_\alpha u^\alpha \qquad \text{ou [sic]} \qquad u^\alpha = \frac{dx^\alpha}{ds}$$

la formule du calcul des variations (§ 5-5)

$$\frac{\partial F}{\partial x_\sigma} - \frac{d}{ds}\left(\frac{\partial F}{\partial u^\sigma} \right) = 0$$

Effectuons les dérivations

$$\frac{\partial F}{\partial x_\sigma} = \sum_{\alpha} \frac{\partial \varphi_\alpha}{\partial x_\sigma} u^\alpha$$

$$\frac{\partial F}{\partial u^\sigma} = \varphi_\sigma$$

$$\frac{d}{ds}\left(\frac{\partial F}{\partial u^\sigma} \right) = \sum_{\alpha} \frac{\partial \varphi_\sigma}{\partial x_\alpha} \frac{dx_\alpha}{ds} = \sum_{\alpha} \frac{\partial \varphi_\sigma}{\partial x_\alpha} u^\alpha$$

Il vient ainsi

$$f_\sigma + \sum_{\alpha} e u^\alpha \left(\frac{\partial \varphi_\alpha}{\partial x_\sigma} - \frac{\partial \varphi_\sigma}{\partial x_\alpha} \right) = 0$$

En posant

$$\jmath^\alpha = e \frac{dx_\alpha}{ds} \tag{10}$$

et

$$F_{\alpha\sigma} = \frac{\partial \varphi_\alpha}{\partial x_\sigma} - \frac{\partial \varphi_\sigma}{\partial x_\alpha} \tag{11}$$

on obtient

$$f_\sigma + \sum_\alpha \mathfrak{I}^\alpha F_{\alpha\sigma} = 0 \tag{12}$$

la[110] formule identique à l'équation (7).

Les équations (9) (10) (11) et (12) expriment, d'une manière conforme au principe de relativité, l'action des champs sur les particules matérielles électrisées.

Lorsqu'on emploie les coordonnées utilisées au début du paragraphe, l'équation (12) ($\sigma = 1, 2, 3$) se réduit en multipliant les deux membres par $\frac{ds}{dt}$ à l'équation (2) tandis que l'équation (9) diffère de (1) et ne peut lui être identifiée qu'approximativement.

[p. 55] Les équations (1) doivent être remplacées par

$$-m\frac{d^2x}{ds^2} + X\frac{ds}{dt} = 0$$

et les équations analogues.

On peut encore écrire[111]

$$\begin{aligned} m\frac{d}{dt}\frac{v_x}{\sqrt{1-v^2}} &= X \\ m\frac{d}{dt}\frac{v_y}{\sqrt{1-v^2}} &= Y \\ m\frac{d}{dt}\frac{v_z}{\sqrt{1-v^2}} &= Z \end{aligned} \tag{13}$$

Ce sont les équations de la dynamique de la relativité restreinte.

Elles peuvent être interprétées de la manière suivante. En les comparant aux équations classiques on pourra dire que l'on remplace la constante m par un facteur variable fonction de la vitesse. On parlera alors de masses variables. On peut, au contraire, transformer les équations de manière à les mettre sous la forme (1) en remplaçant X, Y, Z par une fonction de ces quantités et des vitesses. On parlera alors de forces variables avec la vitesse. Ce dernier point ne[112] paraît plus logique puisque X, Y, Z sont déjà fonction de la vitesse et qu'il s'agit simplement de modifier la forme de cette fonction.

1°) Masses variables :

Supposons tout d'abord la masse[113] perpendiculaire à la vitesse, la grandeur de la vitesse ne varie pas suivant la trajectoire, les équations (13) s'écrivent

$$\frac{m}{(1-v^2)^{\frac{1}{2}}}\frac{d^2x}{dt^2} = X$$

et les équations analogues.

---

[110] B suppression : "la".

[111] Le manuscrit ne numérote pas les équations qui suivent mais s'y réfère par le numéro (13).

[112] B modification : "de vue".

[113] B modification : "force". Cette correction, écrite au crayon et non à l'encre comme c'est le cas pour la toute grande majorité des autres corrections (sauf celles renseignées dans les notes de l'Annexe 1, voir p. 246), ne semble pas être de la main de G. Lemaître.

[p. 56] Supposons au contraire une force dirigée suivant la vitesse, et prenons l'axe des x dans leur direction commune, nous aurons

$$m\frac{d}{dt}\frac{v_x}{(1-v_x^2)^{\frac{1}{2}}}=\frac{m}{(1-v_x^2)^{\frac{3}{2}}}\frac{d^2x}{dt^2}=X$$

Nous pouvons donc conserver les équations fondamentales de la mécanique, en y remplaçant la masse constante par une masse variable dépendant des directions relative [*sic*] de la force et de la vitesse.

Nous aurons ainsi une masse transversale

$$\frac{m}{\left(1-\frac{v^2}{c^2}\right)^{\frac{1}{2}}}$$

Nous aurons[114] rétabli le facteur c qui s'introduit lorsque l'unité de temps est quelconque.

La masse longitudinale sera

$$\frac{m}{\left(1-\frac{v^2}{c^2}\right)^{\frac{3}{2}}}$$

2°) Forces variables :

On peut interpréter tout aussi simplement les équations (13) sans faire intervenir de masses variables.

On a

$$\frac{d}{dt}\frac{v_x}{\sqrt{1-v^2}}=\frac{1}{(1-v^2)^{\frac{1}{2}}}\frac{dv_x}{dt}+\frac{v_x}{(1-v^2)^{\frac{3}{2}}}\frac{1}{2}\frac{dv^2}{dt}$$

La première équation (13) s'écrit alors

$$X=\frac{m}{(1-v^2)^{\frac{1}{2}}}\frac{dv_x}{dt}+\frac{m}{(1-v^2)^{\frac{3}{2}}}\left(v_x\frac{dv_x}{dt}+v_y\frac{dv_y}{dt}+v_z\frac{dv_z}{dt}\right).$$

Multiplions cette équation et les équations analogues en y et [p. 57] en z respectivement par v_x, v_y, v_z, et additionnons membre à membre. Il vient

$$Xv_x+Yv_y+Zv_z=\left(\frac{m}{(1-v^2)^{\frac{1}{2}}}+\frac{mv^2}{(1-v^2)^{\frac{3}{2}}}\right)\left(v_x\frac{dv_x}{dt}+v_y\frac{dv_y}{dt}+v_z\frac{dv_z}{dt}\right)$$

ou

$$Xv_x+Yv_y+Zv_z=\frac{m}{(1-v^2)^{\frac{3}{2}}}\left(v_x\frac{dv_x}{dt}+v_y\frac{dv_y}{dt}+v_z\frac{dv_z}{dt}\right)$$

114 B modification : "avons".

Les équations du mouvement peuvent alors s'écrire

$$m\frac{d^2x}{dt^2} = [X - v_x(X v_x + Y v_y + Z v_z)]\sqrt{1-v^2}$$

ou, en prenant des unités quelconques,

$$m\frac{d^2x}{dt^2} = \left[X - \frac{v_x}{c^2}(X v_x + Y v_y + Z v_z)\right]\sqrt{1-\frac{v^2}{c^2}}$$

On peut donc dire que l'équation fondamentale de la mécanique est conservée, mais que l'expression de la force a changé. La notion de masse variable peut être utile dans certaines applications, il ne faut pourtant pas perdre de vue le caractère arbitraire de telles considérations. On pourrait toujours multiplier les deux membres des équations du mouvement par un facteur variable quelconque et considérer le produit de ce facteur par la masse comme une masse variable.

Chapitre III

Production du Champ par mouvement relatif

§ 7. Mouvement uniformément accéléré.

[p. 58] Nous avons vu qu'il est possible de choisir un système de coordonnées de telle sorte qu'au voisinage d'un point la géométrie soit euclidienne et le mouvement des points libres rectiligne et uniforme. Lorsque ces conditions sont réalisées dans un certain domaine le champ peut être représenté par la forme

$$ds^2 = -d\xi^2 - d\eta^2 - d\zeta^2 + c^2 d\tau^2$$

la géométrie de corps immobiles y est euclidienne. ξ, η, ζ sont les coordonnées cartésiennes par rapport à un système d'axes rectangulaires. τ est le temps marqué par des chronomètres immobiles, ces chronomètres sont interchangeables et leurs indications égales sont simultanées. Enfin la vitesse de la lumière est constante en tout point.

Nous appellerons un tel champ, un champ de Galilée.

Considérons un corps qui se déplace dans le champ de Galilée et supposons qu'un observateur explore le champ au moyen d'instruments liés à ce corps. Il mesure des longueurs et des temps, observe le mouvement des particules libres, ou les rayons lumineux. Comment vont se présenter à[115] cet observateur les propriétés du champ.

La solution de la mécanique classique est simple.

L'observateur entraîné rapporte ses mesures à trois axes rectangulaires en mouvement. Tout se passe comme si ces axes [p. 59] étaient en repos, mais les points libres en mouvement rectiligne et uniforme dans le système fixe, décrivent dans le système mobile des trajectoires dont l'accélération est déterminée par les forces d'inertie : réaction d'entraînement, force centrifuge et force centrifuge composée. La théorie du mouvement relatif calcule ces forces lorsqu'on donne les équations du mouvement du système mobile. Si ce mouvement est une translation rectiligne uniformément accélérée, les corps seront soumis à la seule réaction d'entraînement égale à l'accélération constante mais dirigée en sens contraire. Tout se passera comme si le système était immobile et s'il[116] y régnait un champ de gravitation constant. Cette accélération sera la même pour tous les mobiles. Les rayons lumineux la subiront aussi.

Dans la nouvelle mécanique, la théorie du mouvement relatif est moins immédiate. Nous savons qu'un corps en mouvement paraît contracté dans le sens du mouvement. Si donc, un corps est mis en mouvement il doit paraître se contracter davantage au fur et à mesure que sa vitesse augmente. Il faut pour cela que ses divers

[115] B modification : "pour".

[116] B modification : "qu'il".

points se rattrapent. Au contraire, si nous arrêtons brusquement un corps, nous [117] pouvons arrêter tous ses points simultanément, il faut que nous lui laissions perdre sa contraction apparente. Si nous arrêtions ses extrémités au même moment, nous le déformerions. En mesurant sa longueur directement après l'arrêt nous ne trouverions plus la même valeur qu'auparavant, nous constaterions une contraction réelle, précisément égale à la contraction apparente qui faussait les mesures indirectes de l'observateur fixe avant l'arrêt.

[p. 60] Un corps solide en mouvement est donc un corps qui paraît se contracter dans le sens de la vitesse proportionnellement à

$$\sqrt{1-\frac{v^2}{c^2}}$$

Un corps dont les divers points se déplaceraient dans le septième[118] fixe suivant les équations classiques du mouvement d'un solide, paraîtrait ne pas se déformer et par conséquent se déformerait en réalité. Un observateur qui mesurerait directement ses dimensions constaterait qu'elles varient.

Les équations du mouvement d'un solide ne peuvent être conservées que lorsque la vitesse est constante au même point (ξ, η, ζ), ce qui arrive pour une translation rectiligne uniforme, une rotation uniforme et la combinaison de ces deux mouvements (mouvement hélicoïdal).

En dehors de ces cas, le mouvement d'un solide est-il possible ? En particulier est-il possible de réaliser le mouvement d'un solide de manière à créer artificiellement un champ de gravitation constant ? Pouvons-nous trouver dans la [119] mécanique l'équivalent du mouvement d'entraînement uniformément accéléré ?

Il faut pour cela écrire dans le système fixe ξ, η, ζ, τ les équations du mouvement d'un corps dont les différents points décrivent des trajectoires parallèles (suivant l'axe des ξ par exemple) et qui subit suivant cette direction une contraction proportionnelle à

$$\sqrt{1-\frac{v^2}{c^2}}$$

Traçons dans ce corps des axes x, y, z qui coïncident à l'instant

$$\tau = 0$$

[p. 61] Avec[120] les axes ξ, η, ζ du système fixe.

Les coordonnées x, y, z de chaque point du mobile mesurées par mesures directes par rapport à ce système d'axes sont donc constantes.

---

[117] B ajout : "ne".

[118] B modification : "système".

[119] B ajout : "nouvelle". Cet ajout, au crayon, ne semble pas être de la main de G. Lemaître.

[120] B modification : "avec".

Les équations du mouvement d'un point quelconque (x,y,z) sont de la forme

$$(1) \qquad \begin{cases} \xi = f(\tau,x) \\ \eta = y \\ \zeta = z \end{cases}$$

puisque les mesures normales à la vitesse ne subissent pas la contraction.

La vitesse du point est

$$v = \frac{\partial \xi}{\partial \tau}$$

La contraction apparente est

$$\frac{\partial \xi}{\partial x} = \sqrt{1 - \frac{v^2}{c^2}}$$

Il faut donc que

$$(2) \qquad \left(\frac{\partial \xi}{\partial x}\right)^2 + \frac{1}{c^2}\left(\frac{\partial \xi}{\partial \tau}\right)^2 = 1$$

C'est une équation aux dérivées partielles du premier ordre. Nous allons l'intégrer en employant la méthode des caractéristiques.

Posons

$$p = \frac{\partial \xi}{\partial x} \quad , \quad q = \frac{\partial \xi}{\partial \tau}$$

Elle s'écrit

$$p^2 + \frac{1}{c^2} q^2 = 1$$

Les équations des caractéristiques sont alors

$$\frac{dx}{p} = \frac{d\tau}{\frac{1}{c^2} q} = \frac{-dp}{0} = \frac{-dq}{0} = \frac{d\xi}{p^2 + \frac{q^2}{c^2}}$$

[p. 62] Ces équations s'intègrent immédiatement : p et q doivent être constants. Les caractéristiques sont les lieux des points où la vitesse q est constante.
[121]

Nous pouvons la déterminer en nous donnant le mouvement d'un point M_0 du mobile.

$$\begin{cases} x_0 = 0 \\ \xi_0 = \varphi(\tau_0) \end{cases}$$

où φ est une fonction arbitraire de τ_0.

On aura alors

$$q_0 = \frac{d\varphi}{d\tau_0} \quad , \quad p_0 = \sqrt{1 - \frac{1}{c^2}\left(\frac{d\varphi}{d\tau_0}\right)^2}$$

[121] B ajout : "L'intégrale est un lieu de caractéristiques".

Une caractéristique passant par M_0 a pour équations

$$\frac{x}{p_0} = \frac{\tau - \tau_0}{\frac{1}{c^2} q_0} = \frac{\xi - \xi_0}{1} \quad , \quad p = p_0 \quad , \quad q = q_0$$

ou

$$(3) \qquad \begin{cases} \xi = \varphi(\tau_0) + \dfrac{x}{\sqrt{1 - \frac{1}{c^2}\left(\frac{d\varphi}{d\tau_0}\right)^2}} \\[2ex] \tau = \tau_0 + \dfrac{1}{c^2} \dfrac{x}{\sqrt{1 - \frac{1}{c^2}\left(\frac{d\varphi}{d\tau_0}\right)^2}} \dfrac{d\varphi}{d\tau_0} \end{cases}$$

Nous obtenons ainsi une représentation paramétrique de l'intégrale. On obtiendrait la fonction $\xi = f(\tau, x)$ en éliminant τ_0 entre ces deux équations.

Voyons maintenant comment apparaissent les propriétés du champ à un observateur qui se sert d'instruments de mesure liés au solide.

Nous savons que ces propriétés dépendent de [p. 63] l'invariant ds^2, dont l'expression au moyen des coordonnées du système fixe est

$$ds^2 = -d\xi^2 - d\eta^2 - d\zeta^2 + c^2 d\tau^2$$

Exprimons ces différentielles au moyen des coordonnées du système mobile

$$d\xi = \frac{dx}{\sqrt{1 - \frac{1}{c^2}\left(\frac{d\varphi}{d\tau_0}\right)^2}} + \left[1 + \frac{1}{c^2} \frac{x}{\left[1 - \frac{1}{c^2}\left(\frac{d\varphi}{d\tau_0}\right)^2\right]^{\frac{3}{2}}} \frac{d^2\varphi}{d\tau_0^2}\right] \frac{d\varphi}{d\tau_0} d\tau_0$$

$$d\eta = dy$$

$$d\zeta = dz$$

$$d\tau = \frac{1}{c^2} \frac{d\varphi}{d\tau_0} \frac{dx}{\sqrt{1 - \frac{1}{c^2}\left(\frac{d\varphi}{d\tau_0}\right)^2}} + \left[1 + \frac{1}{c^2} \frac{x}{\left[1 - \frac{1}{c^2}\left(\frac{d\varphi}{d\tau_0}\right)^2\right]^{\frac{3}{2}}} \frac{d^2\varphi}{d\tau_0^2}\right] d\tau_0$$

ds^2 peut donc s'écrire

$$\begin{aligned} ds^2 &= -dx^2 - dy^2 - dz^2 \\ &\quad + c^2 \left[1 - \frac{1}{c^2}\left(\frac{d\varphi}{d\tau_0}\right)^2\right] \left[1 + \frac{1}{c^2} \frac{x}{\left[1 - \frac{1}{c^2}\left(\frac{d\varphi}{d\tau_0}\right)^2\right]^{\frac{3}{2}}} \frac{d^2\varphi}{d\tau_0^2}\right]^2 d\tau_0^2 \end{aligned}$$

Cette expression définit complètement les propriétés du champ.

Quelle forme faut-il donner à la fonction φ pour que l'accélération que prend un corps abandonné à lui-même ne dépende pas du temps mesuré par un chronomètre entraîné avec le corps ? Nous avons calculé (§ 5-14) les valeurs des symboles de Christoffel dans le cas où g_{44} est le seul potentiel qui ne soit pas constant. Il en résulte que l'accélération est

$$\frac{d^2x}{ds^2} = -\left\{ \begin{matrix} 44 \\ i \end{matrix} \right\} = -\frac{1}{2}\frac{\partial g_{44}}{\partial x}$$

[p. 64] Le temps (t)[122] marqué par un chronomètre entraîné est au point

$$dt = \sqrt{1 - \frac{1}{c^2}\left(\frac{d\varphi}{d\tau_0}\right)^2}\, d\tau_0$$

Prenons comme coordonnée au lieu de τ_0 le temps (t)[123] défini par l'équation

$$t = \int_0^{\tau_0} \sqrt{1 - \frac{1}{c^2}\left(\frac{d\varphi}{d\tau_0}\right)^2}\, d\tau_0 \tag{4}$$

L'invariant s'écrit alors

$$ds^2 = -dx^2 - dy^2 - dz^2 + c^2 \left[1 + \frac{1}{c^2} \frac{x}{\left[1 - \frac{1}{c^2}\left(\frac{d\varphi}{d\tau_0}\right)^2\right]^{\frac{3}{2}}} \frac{d^2\varphi}{d\tau_0^2} \right]^2 dt^2 \tag{5}$$

Pour que l'accélération soit indépendante de (t)[124] au point

$$x = 0$$

il faut que le coefficient de t[125] soit égal à une constante

$$\frac{\frac{d^2\varphi}{d\tau_0^2}}{\left[1 - \frac{1}{c^2}\left(\frac{d\varphi}{d\tau_0}\right)^2\right]^{\frac{3}{2}}} = g \tag{6}$$

L'invariant est alors

$$ds^2 = -dx^2 - dy^2 - dz^2 + c^2\left(1 + \frac{gx}{c^2}\right)^2 dt^2 \tag{7}$$

[122] B suppression : parenthèses supprimées.

[123] B suppression : parenthèses supprimées.

[124] B suppression : parenthèses supprimées.

[125] Il faut lire bien sûr : le coefficient de x.

L'équation du mouvement d'un point libre est

$$\frac{1}{c^2}\frac{d^2x}{ds^2} = -g\left(1+\frac{gx}{c^2}\right)$$

L'observateur entraîné n'aura plus aucun moyen de s'apercevoir d'une variation quelconque du champ. Il attribuera l'accélération des mobiles à la présence d'un champ de gravitation constant.

En faisant le changement de variable

$$\left(1+\frac{gx}{c^2}\right)^2 = 1+\frac{2gX}{c^2} \tag{8}$$

on peut mettre ds^2 sous la forme

$$ds^2 = -\frac{dX^2}{1+\frac{2gX}{c^2}} - dy^2 - dz^2 + c^2\left(1+\frac{2gX}{c^2}\right)dt^2 \tag{9}$$

[p. 65] Voyons ce que deviennent les équations générales du mouvement rectiligne d'un solide dans le cas particulier que nous venons d'examiner. Nous obtiendrons ainsi les équations correspondant au mouvement uniformément accéléré de l'ancienne mécanique. La fonction $\varphi(\tau_0)$ vérifie alors la relation (6).

Cette équation s'intègre immédiatement.

On obtient, en choisissant la constante d'intégration de telle sorte que $\frac{d\varphi}{d\tau_0}$ s'annule avec τ_0,

$$\frac{\frac{d\varphi}{d\tau_0}}{\left[1-\frac{1}{c^2}\left(\frac{d\varphi}{d\tau_0}\right)^2\right]^{\frac{1}{2}}} = g\tau_0;$$

résolvons cette équation par rapport à $\frac{d\varphi}{d\tau_0}$

$$\frac{d\varphi}{d\tau_0} = \frac{g\tau_0}{\sqrt{1+\left(\frac{g\tau_0}{c}\right)^2}}$$

Intégrons

$$\varphi = \int_0^{\tau_0}\frac{g\tau_0\,d\tau_0}{\sqrt{1+\left(\frac{g\tau_0}{c}\right)^2}} = \frac{c^2}{g}\left[\sqrt{1+\left(\frac{g\tau_0}{c}\right)^2}-1\right]$$

Portons les valeurs de φ et de $\frac{d\varphi}{d\tau_0}$ dans les équations (3) ; nous obtenons les équations du mouvement sous forme paramétrique.

$$\begin{cases} \xi = \sqrt{1+\left(\frac{g\tau_0}{c}\right)^2}\left(\frac{c^2}{g}+x\right)-\frac{c^2}{g} \\ \tau = \tau_0\left(1+\frac{gx}{c^2}\right) \end{cases} \tag{10}$$

L'élimination de τ_0 entre ces deux équations se fait maintenant sans difficulté. On obtient

$$\text{(11)} \qquad \xi = \frac{c^2}{g}\left[-1+\sqrt{\left(1+\frac{gx}{c^2}\right)^2+\left(\frac{g\tau}{c}\right)^2}\right]$$

Cherchons enfin les équations par lesquelles on passe des coordonnées x, t du système mobile à celles ξ, τ du système fixe. Le temps t est défini par l'équation (4) qui s'écrit en tenant [p. 66] compte de la valeur de $\frac{d\varphi}{d\tau_0}$.

$$t = \int_0^{\tau_0} \frac{d\tau_0}{\sqrt{1+\left(\frac{g\tau_0}{c}\right)^2}} = \frac{c}{g}\operatorname{arc\,sh}\frac{g\tau_0}{c}$$

d'où

$$\tau_0 = \frac{c}{g}\operatorname{sh}\frac{gt}{c}$$

Les équations définissant la transformation demeurent[126] en exprimant τ_0 au moyen de t

$$\text{(12)} \qquad \begin{cases} \xi = \dfrac{c^2}{g}\left[-1+\left(1+\dfrac{gx}{c^2}\right)\operatorname{ch}\dfrac{gt}{c}\right] \\ \tau = \dfrac{c}{g}\left(1+\dfrac{gx}{c^2}\right)\operatorname{sh}\dfrac{gt}{c} \end{cases}$$

En les résolvant par rapport à x et t, on trouve

$$x = \frac{c^2}{g}\left[-1+\sqrt{\left(1+\frac{g\xi}{c^2}\right)^2-\left(\frac{g\tau}{c}\right)^2}\right]$$

$$t = \frac{1}{2}\frac{c}{g}\lg\frac{1+\frac{g\xi}{c^2}+\frac{g\tau}{c}}{1+\frac{g\xi}{c^2}-\frac{g\tau}{c}}$$

On obtiendrait immédiatement les équations analogues aux précédentes lorsqu'on emploie au lieu de x la coordonnée X définie par l'équation (8).

[126] B modification : "deviennent".

§ **8** **Rotation uniforme.**

[p. 67] Le champ de Galilée peut se représenter en coordonnées semi-polaires r, θ, z.

$$x = r\cos\theta$$
$$y = r\sin\theta$$

par l'expression

$$ds^2 = -dr^2 - r^2 d\theta^2 - dz^2 + c^2 dt^2$$

Si nous imaginons un système tournant autour de l'axe des z du système de Galilée avec une vitesse angulaire constante ω, dans le nouveau système ds^2 s'exprimera par

$$ds^2 = -dr^2 - r^2(d\theta - \omega\, dt)^2 - dz^2 + c^2 dt^2 \tag{1}$$

où θ est maintenant mesuré par rapport à ces axes entraînés dans le mouvement. Nous allons étudier les propriétés géométriques des corps immobiles dans ce système. Nous posons[127]

$$\delta r = \delta\theta = \delta z = 0$$

dans l'équation de simultanéité (§ 3-1).

On obtient

$$-r^2(d\theta - \omega\, dt)(-\omega\,\delta t) + c\, dt\,\delta t = 0$$

ou[128]

$$\omega r^2(d\theta - \omega\, dt) + c^2 dt = 0 \tag{2}$$

Nous avons vu quelles particularités présente dans ce cas la simultanéité sur un contour fermé, et comment l'expérience de Monsieur Sagnac les met en évidence.

Éliminons dt entre l'équation fondamentale (1) et [p. 68] l'équation de simultanéité, nous devons obtenir l'élément de distance. Nous aurons en changeant les signes

$$d\sigma^2 = dr^2 + \frac{r^2\, d\theta^2}{1 - \frac{\omega^2 r^2}{c^2}} + dz^2 \tag{3}$$

Nous voyons que la longueur d'un rayon

$$d\theta = dz = 0$$

est égale à

$$\int_0^r dr = r$$

[127] B modification : "poserons".

[128] Le manuscrit ne numérote pas l'équation qui suit mais s'y réfère par le numéro (2).

tandis que celle de la circonférence

$$dr = 0 \qquad dz = 0$$

est

$$\int_0^{2\pi} \frac{r\,d\theta}{\sqrt{1-\frac{\omega^2 r^2}{c^2}}} = \frac{2\pi r}{\sqrt{1-\frac{\omega^2 r^2}{c^2}}}$$

Le rapport de la circonférence au diamètre est plus grand que π.

On obtiendrait immédiatement ce résultat en appliquant la conection[129] de Lorentz aux mesures indirectes lues sur des instruments immobiles dans le champ de Galilée.

Un observateur qui photographierait d'un point de l'axe, un disque tournant autour de cet axe, obtiendrait une image du disque, sur laquelle le rapport de la circonférence au diamètre est égal à π puisque la géométrie des corps immobiles dans le champ de Galilée est euclidienne.

La conection[130] de Lorentz est nulle pour le diamètre qui se déplace normalement à sa direction, la circonférence seule est plus grande qu'elle ne paraît, dans le rapport

$$\frac{1}{\sqrt{1-\frac{v^2}{c^2}}}$$

[p. 69] Des observateurs effectuant des mesures géométriques de haute précision avec des règles identiques trouvent[131] que la géométrie sur le disque n'est pas euclidienne.

Ils pourront naturellement toujours rejeter cette conclusion s'ils préfèrent admettre que leurs règles se déforment par suite de la présence du champ centrifuge. Ils pourront faire diverses hypothèses sur ces déformations. Le plus simple est de supposer que le mètre se raccourcit dans le rapport $\sqrt{1-\frac{\omega^2}{c^2}r^2}$ lorsqu'on le dirige normalement au rayon pour mesurer la circonférence. Si on fait le changement de variable

$$\frac{r^2}{1-\frac{\omega^2}{c^2}r^2} = \rho^2 \tag{4}$$

d'où on tire

$$dr = \frac{d\rho}{\left(1+\frac{\omega^2}{c^2}\rho^2\right)^{\frac{3}{2}}}$$

[129] B modification : "correction".

[130] B modification : "correction".

[131] B modification : "trouveront".

où ρ représente alors la longueur de la circonférence divisée par 2π, l'élément de distance prend la forme[132]

$$(5) \qquad d\sigma^2 = \frac{d\rho^2}{\left(1+\frac{\omega^2}{c^2}\rho^2\right)^3} + \rho^2 d\theta^2 + dz^2$$

On peut donc supposer que la géométrie reste euclidienne mais que les règles dirigées suivant la direction de la force centrifuge sont allongées dans le rapport[133]

$$\left(1+\frac{\omega^2}{c^2}\rho^2\right)^{\frac{3}{2}} = \frac{1}{\left(1-\frac{\omega^2}{c^2}r^2\right)^{\frac{3}{2}}}$$

On voit que la géométrie est euclidienne sur un cylindre circulaire ayant pour axe l'axe de rotation.

[p. 70] Enfin on peut chercher quelle dilatation ou quelle contraction indépendante de la direction il faut attribuer aux règles pour pouvoir prétendre que la géométrie euclidienne plane[134].

Il faut pour cela mettre l'élément de distance, sur le disque, sous la forme

$$dr^2 + \frac{r^2 d\theta^2}{1-\frac{\omega^2}{c^2}r^2} = \frac{dy^2 + y^2 d\theta^2}{x^2}$$

où x est la dilatation supposée des mètres, et (y)[135] la longueur du rayon corrigée de cette dilatation.

On doit intégrer les deux équations

$$dr = \frac{dy}{x}$$

et

$$\frac{r}{\sqrt{1-\frac{\omega^2}{c^2}r^2}} = \frac{y}{x}$$

Divisons membre à membre

$$\frac{dr}{r}\sqrt{1-\frac{\omega^2}{c^2}r^2} = \frac{dy}{y}$$

[132] B modification : dans l'équation, le 3 en exposant est barré à l'encre rouge.

[133] B modification : l'exposant $3/2$ qui apparaît deux fois dans cette équation est barré à l'encre rouge.

[134] B modification : "euclidienne plane" remplacé par "sur un disque tournant est la géométrie euclidienne plane".

[135] B suppression : parenthèses supprimées.

et intégrons.

$$y = \frac{2r}{1+\sqrt{1-\frac{\omega^2}{c^2}r^2}}\, e^{\sqrt{1-\frac{\omega^2}{c^2}r^2}-1}$$

et

$$x = \frac{2\sqrt{1-\frac{\omega^2}{c^2}r^2}}{1+\sqrt{1-\frac{\omega^2}{c^2}r^2}}\, e^{\sqrt{1-\frac{\omega^2}{c^2}r^2}-1}$$

(x)[136] tend vers zéro lorsque (r)[137] tend vers $(\frac{c}{\omega})$[138]. Pour $(\frac{\omega}{c}r)$[139] suffisamment petit, on a

$$x = 1 - \frac{3}{4}\frac{\omega^2}{c^2}r^2 - \frac{19}{132}\frac{\omega^6}{c^6}r^6 + \cdots$$

Cherchons quelle forme doit avoir une surface de révolution pour que la géométrie y soit euclidienne lorsque la surface est animée d'une rotation (ω)[140] autour de son axe.

[p. 71] D'après (5), le profil de la surface est déterminé par l'équation

$$\frac{d\rho^2}{\left(1+\frac{\omega^2}{c^2}\rho^2\right)^3} + dz^2 = d\rho^2$$

On aura alors sur la surface

$$d\sigma^2 = d\rho^2 + \rho^2 d\theta^2$$

qui montre que la géométrie y est euclidienne.

Si nous employons la variable (r)[141] liée à ρ par (4)

$$dr^2 + dz^2 = \frac{dr^2}{\left(1-\frac{\omega^2}{c^2}r^2\right)^3}$$

Le profil de la génératrice de la surface de révolution cherché [*sic*] est donné par

$$z = \int \sqrt{\left(\frac{1}{1-\frac{\omega^2}{c^2}r^2}\right)^3 - 1}\, dr$$

(z)[142] tend vers l'infini lorsque (r)[143] tend vers $\frac{c}{\omega}$.

---

136 B suppression : parenthèses supprimées.

137 B suppression : parenthèses supprimées.

138 B suppression : parenthèses supprimées.

139 B suppression : parenthèses supprimées.

140 B suppression : parenthèses supprimées.

141 B suppression : parenthèses supprimées.

142 B suppression : parenthèses supprimées.

143 B suppression : parenthèses supprimées.

Pour r petit on aura en développant en série

$$z = \frac{\sqrt{3}}{2}\frac{\omega}{c}r^2\left(1 + \frac{1}{2}\frac{\omega^2}{c^2}r^2 + \cdots\right)$$

Cette surface est donc osculatrice à un paraboloïde, à l'origine et asymptotique au cylindre $(r = \frac{c}{\omega})$. Sur une surface animée de la rotation ω la géométrie est la même que sur un plan euclidien. Supposons qu'on arrête le mouvement de rotation. Un observateur qui arpenterait la surface constaterait, pendant l'arrêt, les mêmes déformations de la surface que celles qu'on constaterait, au repos, en observant les déformations sur[144] une surface plane qu'on appliquerait sur la surface de révolution.

[p. 72] Le mouvement des mobiles par rapport au système tournant ne présente rien de particulier. La théorie classique du mouvement relatif est appliquable [*sic*] puisque les équations de transformation entre les coordonnées des deux systèmes peuvent être conservées.

Remarquons enfin que si la vitesse de rotation vient à changer le mobile doit nécessairement se déformer puisque le rapport

$$\frac{\pi}{\sqrt{1 - \frac{\omega^2}{c^2}r^2}}$$

de la circonférence au diamètre doit changer. Il faut donc bien que la circonférence ou le diamètre change de longueur ou que tous deux se déforment.

144 B modification : "d' ".

§ 9 Équation générale des champs d'inertie.

[p. 73] Lorsque les potentiels sont constants le mouvement d'un point libre est rectiligne et uniforme ; nous avons affaire à ce que nous avons appelé un champ de Galilée. Si nous faisons un changement des[145] coordonnées, les potentiels ne sont plus constants, les dérivées secondes des coordonnées ne s'annulent plus suivant la trajectoire des points libres ; nous obtenons ce que nous appellerons un champ d'inertie.

Le champ d'inertie est donc caractérisé par le fait qu'il est possible par un changement de coordonnées

$$\xi_\sigma = \xi_\sigma(x_1, x_2, x_3, x_4) \quad (\sigma = 1, 2, 3, 4) \tag{1}$$

de transformer la forme fondamentale ds^2 en une forme quadratique à coefficients constants

$$ds^2 = \sum_\sigma \sum_\tau \gamma_{\sigma\tau} d\xi_\sigma d\xi_\tau \tag{2}$$

Nous avons vu que dans ce cas les dérivées des équations de transformation vérifient les relations $(\alpha, \beta, \tau = 1, 2, 3, 4)$ (§ 5-19)

$$\frac{\partial^2 \xi_\tau}{\partial x_\alpha \partial x_\beta} = \sum_\sigma \frac{\partial \xi_\tau}{\partial x_\sigma} \left\{ \begin{matrix} \alpha\beta \\ \sigma \end{matrix} \right\} \tag{3}$$

où le symbole de Christoffel de deuxième espèce est formé avec les potentiels $g_{\mu\nu}$ du système de coordonnées x_1, x_2, x_3, x_4.

Nous nous proposons d'éliminer de ces équations les fonctions ξ_σ. Prenons la dérivée de (3) par rapport à x_γ.

$$\frac{\partial^3 \xi_\tau}{\partial x_\alpha \partial x_\beta \partial x_\gamma} = \sum_\sigma \frac{\partial^2 \xi_\tau}{\partial x_\sigma \partial x_\gamma} \left\{ \begin{matrix} \alpha\beta \\ \sigma \end{matrix} \right\} + \sum_\sigma \frac{\partial \xi_\tau}{\partial x_\sigma} \frac{\partial}{\partial x_\gamma} \left\{ \begin{matrix} \alpha\beta \\ \sigma \end{matrix} \right\}$$

[p. 74] Transformons la dérivée seconde du second membre en appliquant l'équation (3), nous aurons, en remplaçant l'indice sommatoire σ par ρ,[146]

$$\frac{\partial^2 \xi_\tau}{\partial x_\rho \partial x_\gamma} = \frac{\partial \xi_\tau}{\partial x_\sigma} \left\{ \begin{matrix} \rho\gamma \\ \sigma \end{matrix} \right\}$$

Il vient

$$\frac{\partial^3 \xi_\tau}{\partial x_\alpha \partial x_\beta \partial x_\gamma} = \sum_\sigma \frac{\partial \xi_\tau}{\partial x_\sigma} \left[\sum_\rho \left\{ \begin{matrix} \rho\gamma \\ \sigma \end{matrix} \right\} \left\{ \begin{matrix} \alpha\beta \\ \rho \end{matrix} \right\} + \frac{\partial}{\partial x_\gamma} \left\{ \begin{matrix} \alpha\beta \\ \sigma \end{matrix} \right\} \right]$$

---

[145] B modification : "de".

[146] Dans l'équation qui suit, il faut lire bien sûr : $\frac{\partial^2 \xi_\tau}{\partial x_\rho \partial x_\gamma} = \sum_\sigma \frac{\partial \xi_\tau}{\partial x_\sigma} \left\{ \begin{matrix} \rho\gamma \\ \sigma \end{matrix} \right\}$.

Cette équation doit être vérifiée quels que soient

$$(\alpha, \beta, \gamma, \tau = 1, 2, 3, 4)$$

Elle est donc encore vérifiée si on permute les indices α et γ. Le premier membre n'est pas modifié. En soustrayant membre à membre les deux expressions, nous obtenons

$$\sum_\sigma \frac{\partial \xi_\tau}{\partial x_\sigma} R^\sigma_{\alpha\beta\gamma} = 0 \tag{4}$$

où nous avons posé[147,148]

$$\begin{aligned} R^\sigma_{\alpha\beta\gamma} = & -\frac{\partial}{\partial x_\alpha}\left\{ \begin{matrix} \beta\gamma \\ \sigma \end{matrix} \right\} + \frac{\partial}{\partial x_\gamma}\left\{ \begin{matrix} \alpha\beta \\ \sigma \end{matrix} \right\} \\ & -\sum_\rho \left\{ \begin{matrix} \rho\alpha \\ \sigma \end{matrix} \right\}\left\{ \begin{matrix} \beta\gamma \\ \rho \end{matrix} \right\} + \sum_\rho \left\{ \begin{matrix} \rho\gamma \\ \sigma \end{matrix} \right\}\left\{ \begin{matrix} \alpha\beta \\ \rho \end{matrix} \right\} \end{aligned} \tag{5}$$

L'équation (4) pour $(\tau = 1, 2, 3, 4)$ peut être considérée comme un système de quatre équations linéaires et homogènes à quatre inconnues

$$R^\sigma_{\alpha\beta\gamma} \qquad (\sigma = 1, 2, 3, 4)$$

Le déterminant de ce système est le déterminant fonctionnel des [p. 75] équations de transformation (1). Nous le supposerons naturellement différent de zéro. Les équations (4) sont donc équivalentes à

$$\begin{gathered} R^\sigma_{\alpha\beta\gamma} = 0 \\ (\alpha, \beta, \gamma, \sigma = 1, 2, 3, 4) \end{gathered} \tag{6}$$

Ce sont les équations générales des champs d'inertie.

En géométrie ces équations sont les équations d'un espace euclidien.

À deux variables elles expriment la condition pour qu'une surface soit appliquable [*sic*] sur un plan euclidien.

Lorsque la forme fondamentale peut être réduite par un changement de coordonnées à une forme à coefficients constants, les expressions (5) ne s'annulent plus. Elles jouissent de propriétés importantes sur lesquelles Einstein a basé l'étude des champs de gravitation dans le cas général.

---

[147] Le manuscrit ne numérote pas l'équation qui suit mais s'y réfère par le numéro (5).

[148] C correction (voir Annexe 2, p. 249) : "Le tenseur de Riemann est défini par une expression de signe contraire à l'expression habituelle. Cette définition est conservée dans tout le mémoire, sauf : § 12 [p 102] (voir p. 213) et [p 109] (voir p. 221). Dans ce dernier cas il en résulte une erreur dans le calcul de ρ ! En rectifiant, on obtient ([p. 112], 1ere formule – voir p. 223) $\rho' = \left(1 + \frac{2V}{c^2}\right)\rho + \frac{4}{\kappa c^2} g^2$."

Pour établir ces propriétés nous utiliserons un système de coordonnées particulier, pour lequel les calculs se simplifient beaucoup.

Nous avons vu qu'on pouvait toujours disposer de la valeur en un point (M)[149], des dérivées secondes des fonctions de transformation (1), de telle sorte qu'en ce point les dérivées des potentiels s'annulent dans le système ξ_σ. Il suffit pour cela de leur donner en (M)[150] la valeur (3). Nous pouvons de plus disposer des dérivées premières

$$\frac{\partial \xi_\tau}{\partial x_\sigma}$$

de telle sorte que les potentiels $\gamma_{\sigma\tau}$ se réduisent en (M)[151] [p. 76] aux valeurs

$$\gamma_{\sigma\tau} = g_\sigma^\tau = \begin{cases} 1 & \text{si} \quad \sigma = \tau \\ 0 & \text{si} \quad \sigma \neq \tau \end{cases}$$

Certaines coordonnées seront imaginaires. Cela n'a aucun inconvénient pour les calculs que nous faisons ici. Il est d'ailleurs facile de repasser aux variables réelles.

En portant dans l'expression (2) de (ds^2)[152] les valeurs des différentielles tirées de (1), on doit obtenir

$$ds^2 = \sum_\alpha \sum_\beta g_{\alpha\beta}\, dx_\alpha dx_\beta$$

il en résulte que

$$g_{\alpha\beta} = \sum_\sigma \sum_\tau \frac{\partial \xi_\sigma}{\partial x_\alpha} \frac{\partial \xi_\tau}{\partial x_\beta} \gamma_{\sigma\tau} \tag{7}$$

Les symboles de Christoffel qui figurent dans (5) ont pour définition

$$\left\{ \begin{matrix} \alpha\beta \\ \sigma \end{matrix} \right\} = \frac{1}{2} \sum_\tau g^{\sigma\tau} \left(\frac{\partial g_{\alpha\tau}}{\partial x_\beta} + \frac{\partial g_{\beta\tau}}{\partial x_\alpha} - \frac{\partial g_{\alpha\beta}}{\partial x_\tau} \right) \tag{8}$$

outre les potentiels $g_{\sigma\tau}$ ces expressions contiennent les quantités $g^{\sigma\tau}$ associées à ces potentiels. Nous les avons définies par les équations.

$$\sum_\rho g_{\alpha\rho}\, g^{\beta\rho} = g_\alpha^\beta \tag{9}$$

Si nous posons de même

$$\sum_\rho \gamma_{\alpha\rho}\, \gamma^{\beta\rho} = g_\alpha^\beta \tag{10}$$

---

[149] B suppression : parenthèses supprimées.

[150] B suppression : parenthèses supprimées.

[151] B suppression : parenthèses supprimées.

[152] B suppression : parenthèses supprimées.

les $g^{\alpha\beta}$ s'exprimeront en fonction des $\gamma^{\sigma\tau}$ par l'équation

$$(11) \qquad g^{\alpha\beta} = \sum_{\sigma}\sum_{\tau} \frac{\partial x_\alpha}{\partial \xi_\sigma}\frac{\partial x_\beta}{\partial \xi_\tau}\gamma^{\sigma\tau}$$

Il suffit de montrer que cette expression vérifie la définition (9). On a en effet par (7) et (11)[153]

$$\sum_{\rho} g_{\alpha\rho}\, g^{\beta\rho} = \sum_{\rho}\sum_{\sigma}\sum_{\tau}\sum_{\sigma_1}\sum_{\tau_1} \frac{\partial \xi_\sigma}{\partial x_\alpha}\frac{\partial \xi_\tau}{\partial x_\rho}\frac{\partial \xi_\rho}{\partial x_{\tau_1}}\frac{\partial x_\beta}{\partial \xi_{\sigma_1}}\gamma_{\sigma\tau}\gamma^{\sigma_1\tau_1}$$

[p. 77] Effectuons la sommation en ρ

$$\sum_{\rho} \frac{\partial \xi_\tau}{\partial x_\rho}\frac{\partial x_\rho}{\partial \xi_{\tau_1}} = \frac{d\xi_\tau}{d\xi_{\tau_1}} = g^{\tau}_{\tau_1}$$

puis celle en τ et τ_1, en vertu de (10),

$$\sum_{\tau}\sum_{\tau_1} g^{\tau}_{\tau_1}\gamma_{\sigma\tau}\gamma^{\sigma_1\tau_1} = \sum_{\tau}\gamma_{\sigma\tau}\gamma^{\sigma_1\tau} = g^{\sigma_1}_{\sigma}$$

Il reste ainsi

$$\sum_{\sigma}\sum_{\sigma_1} \frac{\partial \xi_\sigma}{\partial x_\alpha}\frac{\partial x_\beta}{\partial \xi_{\sigma_1}} g^{\sigma_1}_{\sigma} = \sum_{\sigma} \frac{\partial \xi_\sigma}{\partial x_\alpha}\frac{\partial x_\beta}{\partial \xi_\sigma} = \frac{dx_\beta}{dx_\alpha} = g^{\beta}_{\alpha}$$

c.q.f.d.

De l'équation (10) il résulte qu'en (M)[154] les équations

$$(12) \qquad \gamma_{\alpha\beta} = g^{\beta}_{\alpha} \quad \text{et} \quad \frac{\partial \gamma_{\alpha\beta}}{\partial \xi_\gamma} = 0$$

entraînent[155]

$$(13) \qquad \gamma^{\alpha\beta} = g^{\beta}_{\alpha} \quad \text{et} \quad \frac{\partial \gamma^{\alpha\beta}}{\partial \xi_\gamma} = 0$$

Voyons ce que devient en M l'expression (5) lorsqu'on exprime les symboles de Christoffel au moyen des fonctions ξ et des γ.

Les termes de cette expression contiennent les potentiels $\gamma_{\alpha\beta}$ et $\gamma^{\alpha\beta}$ et les dérivées premières et secondes de ces quantités.

---

153 Dans l'équation qui suit, il faut lire bien sûr :

$$\sum_{\rho} g_{\alpha\rho}\, g^{\beta\rho} = \sum_{\rho}\sum_{\sigma}\sum_{\tau}\sum_{\sigma_1}\sum_{\tau_1} \frac{\partial \xi_\sigma}{\partial x_\alpha}\frac{\partial \xi_\tau}{\partial x_\rho}\frac{\partial x_\rho}{\partial \xi_{\tau_1}}\frac{\partial x_\beta}{\partial \xi_{\sigma_1}}\gamma_{\sigma\tau}\gamma^{\sigma_1\tau_1}.$$

154 B suppression : parenthèses supprimées.

155 Le manuscrit ne numérote pas les relations qui suivent mais s'y réfère par le numéro (13).

Les termes qui ne contiennent aucune dérivée sont ceux que l'on obtient en annulant les dérivées, c'est à dire [*sic*] en supposant que les $\gamma_{\alpha\beta}$ sont des constantes ; nous avons vérifié que ces termes s'annulent (6).

Les termes qui contiennent des dérivées premières s'annulent en (M)[156] en vertu de (12) et (13). Il nous reste donc à calculer ceux qui contiennent des dérivées secondes.

Ces dernières ne se trouvent que dans les deux premiers termes de (5).

$$-\frac{\partial}{\partial x_\alpha}\left\{\begin{matrix}\beta\gamma\\ \sigma\end{matrix}\right\}+\frac{\partial}{\partial x_\gamma}\left\{\begin{matrix}\alpha\beta\\ \sigma\end{matrix}\right\}$$

[p. 78] Calculons le second en ne tenant compte que des termes contant[157] des dérivées secondes.

Nous obtiendrons ensuite le premier en permutant les indices α et γ.

Nous aurons

$$\frac{\partial}{\partial x_\gamma}\left\{\begin{matrix}\alpha\beta\\ \sigma\end{matrix}\right\}=\frac{1}{2}\sum_\tau g^{\sigma\tau}\left(\frac{\partial^2 g_{\alpha\tau}}{\partial x_\beta\partial x_\gamma}+\frac{\partial^2 g_{\beta\tau}}{\partial x_\alpha\partial x_\gamma}-\frac{\partial^2 g_{\alpha\beta}}{\partial x_\tau\partial x_\gamma}\right)+\cdots$$

Les dérivées secondes sont égales d'après (7) à

$$\frac{\partial^2 g_{\alpha\tau}}{\partial x_\beta\partial x_\gamma}=\sum_\lambda\sum_\rho\frac{\partial\xi_\lambda}{\partial x_\alpha}\frac{\partial\xi_\rho}{\partial x_\tau}\frac{\partial^2\gamma_{\lambda\rho}}{\partial x_\beta\partial x_\gamma}+\cdots$$

mais

$$\frac{\partial^2\gamma_{\lambda\rho}}{\partial x_\beta\partial x_\gamma}=\sum_\mu\sum_\nu\frac{\partial^2\gamma_{\lambda\rho}}{\partial\xi_\mu\partial\xi_\nu}\frac{\partial\xi_\mu}{\partial x_\beta}\frac{\partial\xi_\nu}{\partial x_\gamma}+\cdots$$

les termes non écrits ne contenant pas de dérivées secondes.

Il vient donc

$$\frac{\partial^2 g_{\alpha\tau}}{\partial x_\beta\partial x_\gamma}=\sum_\lambda\sum_\mu\sum_\nu\sum_\rho\frac{\partial\xi_\lambda}{\partial x_\alpha}\frac{\partial\xi_\mu}{\partial x_\beta}\frac{\partial\xi_\nu}{\partial x_\gamma}\frac{\partial\xi_\rho}{\partial x_\tau}\frac{\partial^2\gamma_{\lambda\rho}}{\partial\xi_\mu\partial\xi_\nu}\tag{14}$$

Les autres dérivées s'obtiennent en permutant les indices α,β,γ,τ ; cela revient à permuter les indices correspondants λ,μ,ν,ρ.

D'autre part, (11) et (13)

$$g^{\sigma\tau}=\sum_\eta\frac{\partial x_\sigma}{\partial\xi_\eta}\frac{\partial x_\tau}{\partial\xi_\eta}$$

[156] B suppression : parenthèses supprimées.

[157] B modification : "contenant".

Il vient ainsi

$$\frac{\partial}{\partial x_\gamma}\left\{\begin{matrix}\alpha\beta\\ \sigma\end{matrix}\right\} =$$

$$\frac{1}{2}\sum_\lambda\sum_\mu\sum_\nu\sum_\rho\sum_\eta\sum_\tau \frac{\partial\xi_\lambda}{\partial x_\alpha}\frac{\partial\xi_\mu}{\partial x_\beta}\frac{\partial\xi_\nu}{\partial x_\gamma}\frac{\partial\xi_\rho}{\partial x_\tau}\frac{\partial x_\tau}{\partial\xi_\eta}\frac{\partial x_\sigma}{\partial\xi_\eta}$$
$$\times\left(\frac{\partial^2\gamma_{\lambda\rho}}{\partial\xi_\mu\partial\xi_\nu}+\frac{\partial^2\gamma_{\mu\rho}}{\partial\xi_\lambda\partial\xi_\nu}-\frac{\partial^2\gamma_{\lambda\mu}}{\partial\xi_\rho\partial\xi_\nu}\right)+\cdots$$

[p. 79] Effectuons la sommation en τ

$$\sum_\tau \frac{\partial\xi_\rho}{\partial x_\tau}\frac{\partial x_\tau}{\partial\xi_\eta} = g_\rho^\eta$$

et celle en η

$$\sum_\eta g_\rho^\eta \frac{\partial x_\sigma}{\partial\xi_\eta} = \frac{\partial x_\sigma}{\partial\xi_\rho}$$

Il vient ainsi

$$\frac{\partial}{\partial x_\gamma}\left\{\begin{matrix}\alpha\beta\\ \sigma\end{matrix}\right\} =$$

$$\frac{1}{2}\sum_\lambda\sum_\mu\sum_\nu\sum_\rho \frac{\partial\xi_\lambda}{\partial x_\alpha}\frac{\partial\xi_\mu}{\partial x_\beta}\frac{\partial\xi_\nu}{\partial x_\gamma}\frac{\partial x_\sigma}{\partial\xi_\rho}\left(\frac{\partial^2\gamma_{\lambda\rho}}{\partial\xi_\mu\partial\xi_\nu}+\frac{\partial^2\gamma_{\mu\rho}}{\partial\xi_\lambda\partial\xi_\nu}-\frac{\partial^2\gamma_{\lambda\mu}}{\partial\xi_\rho\partial\xi_\nu}\right)+\cdots$$

Il nous reste à permuter les indices α et γ et à soustraire l'expression ainsi obtenue. Tous les termes non écrits se détruiront. Au lieu de permuter les indices α et γ, on peut permuter les indices sommatoires correspondants λ et μ.
Il vient ainsi

(15) $$R_{\alpha\beta\gamma}^\sigma = \sum_\lambda\sum_\mu\sum_\nu\sum_\rho \frac{\partial\xi_\lambda}{\partial x_\alpha}\frac{\partial\xi_\mu}{\partial x_\beta}\frac{\partial\xi_\nu}{\partial x_\gamma}\frac{\partial x_\sigma}{\partial\xi_\rho}P_{\lambda\mu\nu}^\rho$$

où

(16) $$P_{\lambda\mu\nu}^\rho = \frac{1}{2}\left(\frac{\partial^2\gamma_{\lambda\rho}}{\partial\xi_\mu\partial\xi_\nu}-\frac{\partial^2\gamma_{\mu\lambda}}{\partial\xi_\nu\partial\xi_\rho}+\frac{\partial^2\gamma_{\nu\mu}}{\partial\xi_\rho\partial\xi_\lambda}-\frac{\partial^2\gamma_{\rho\nu}}{\partial\xi_\lambda\partial\xi_\mu}\right)$$

Nous obtenons ainsi une nouvelle expression de $R_{\alpha\beta\gamma}^\sigma$.

Rappelons la signification des fonctions qui y figurent.
ξ_1, ξ_2, ξ_3, ξ_4 sont des fonctions des coordonnées définissant un système de coordonnées (1) jouissant de propriétés spéciales (12 et 13) au point (M)[158] où on

[158] B suppression : parenthèses supprimées.

applique la formule (15). Les dérivées figurant en (16) sont les valeurs en (M)[159] des dérivées des potentiels dans le système $(\xi_1, \xi_2, \xi_3, \xi_4)$.

[p. 80] Ces équations sont appliquables [*sic*] en un point quelconque (M)[160]. Mais il ne faut pas perdre de vue que les ξ sont des fonctions différentes en chaque point. Il serait, par exemple, inadmissible de dériver les expressions (15).

Supposons que l'on veuille utiliser au lieu du système de coordonnées (x_1, x_2, x_3, x_4) de nouvelles coordonnées (x'_1, x'_2, x'_3, x'_4) liées aux premières par les équations.

$$x_\sigma = x_\sigma(x'_1, x'_2, x'_3, x'_4) \qquad x'_\sigma = x'_\sigma(x_1, x_2, x_3, x_4)$$
$$(\sigma = 1,2,3,4) \qquad (\sigma = 1,2,3,4)$$

En remplaçant les coordonnées x_σ par leurs valeurs en fonction de nouvelles coordonnées dans les fonctions ξ_σ on obtient des fonctions des x'_σ que nous désignerons encore par ξ_σ.

En désignant par un indice les expressions calculées dans le nouveau système on aura

$$R'^{\sigma}_{\alpha\beta\gamma} = \sum_\lambda \sum_\mu \sum_\nu \sum_\rho \frac{\partial \xi_\lambda}{\partial x'_\alpha} \frac{\partial \xi_\mu}{\partial x'_\beta} \frac{\partial \xi_\nu}{\partial x'_\gamma} \frac{\partial x'_\sigma}{\partial \xi_\rho} P^{\rho}_{\lambda\mu\nu}$$

Nous aurons par la règle de dérivation des fractions de fonction.

$$\frac{\partial \xi_\lambda}{\partial x'_\alpha} = \sum \frac{\partial \xi_\lambda}{\partial x_{\lambda_1}} \frac{\partial x_{\lambda_1}}{\partial x'_\alpha}$$
$$\frac{\partial \xi_\mu}{\partial x'_\beta} = \sum \frac{\partial \xi_\mu}{\partial x_{\mu_1}} \frac{\partial x_{\mu_1}}{\partial x'_\beta}$$
$$\frac{\partial \xi_\nu}{\partial x'_\gamma} = \sum \frac{\partial \xi_\nu}{\partial x_{\nu_1}} \frac{\partial x_{\nu_1}}{\partial x'_\gamma}$$
$$\frac{\partial x'_\sigma}{\partial \xi_\rho} = \sum \frac{\partial x'_\sigma}{\partial x_{\rho_1}} \frac{\partial x_{\rho_1}}{\partial \xi_\rho}$$

En posant d'après (15)

$$R^{\rho_1}_{\lambda_1,\mu_1,\nu_1} = \sum_\lambda \sum_\mu \sum_\nu \sum_\rho \frac{\partial \xi_\lambda}{\partial x_{\lambda_1}} \frac{\partial \xi_\mu}{\partial x_{\mu_1}} \frac{\partial \xi_\nu}{\partial x_{\nu_1}} \frac{\partial x_{\rho_1}}{\partial \xi_\rho} P^{\rho}_{\lambda\mu\nu}$$

[p. 81] il vient finalement

$$(17) \qquad R'^{\sigma}_{\alpha\beta\gamma} = \sum_{\lambda_1} \sum_{\mu_1} \sum_{\nu_1} \sum_{\rho_1} \frac{\partial x_{\lambda_1}}{\partial x'_\alpha} \frac{\partial x_{\mu_1}}{\partial x'_\beta} \frac{\partial x_{\nu_1}}{\partial x'_\gamma} \frac{\partial x'_\sigma}{\partial x_{\rho_1}} R^{\rho_1}_{\lambda_1\mu_1\nu_1}$$

---

[159] B suppression : parenthèses supprimées.

[160] B suppression : parenthèses supprimées.

C'est l'équation suivant laquelle se transforme $R^{\sigma}_{\alpha\beta\gamma}$ lorsqu'on change de système de coordonnées. En particulier lorsqu'on emploie le système ξ_σ d'un point (M)[161] les équations (15) se réduisent en (M)[162] à

$$R^{\sigma}_{\alpha\beta\gamma} = P^{\sigma}_{\alpha\beta\gamma}$$

L'expression (16) est l'expression réduite de $R^{\sigma}_{\alpha\beta\gamma}$. C'est à cela que se réduit $R^{\sigma}_{\alpha\beta\gamma}$ en un point (M)[163] lorsqu'on emploie un système de coordonnées particulier. Cette expression rigoureuse en (M)[164] pourra être considérée comme approchée dans un domaine suffisamment petit autour de (M)[165], car les conditions (12) sur lesquelles elle est basée et qui sont vérifiées exactement en (M)[166] sont vérifiées avec une approximation donnee dans un domaine suffisamment petit.

Un groupe de quantités qui se transforment lors d'un changement de coordonnées par les équations (17) ou par les équations de la forme générale[167]

$$(18)\qquad T'^{\beta_1\beta_2\cdots}_{\alpha_1\alpha_2\cdots} = \sum_{\sigma_1}\sum_{\sigma_2}\cdots\sum_{\tau_1}\sum_{\tau_2}\cdots\frac{\partial x_{\sigma_1}}{\partial x'_{\alpha_1}}\frac{\partial x_{\sigma_2}}{\partial x'_{\alpha_2}}\cdots\frac{\partial x'_{\beta_1}}{\partial x_{\tau_1}}\frac{\partial x'_{\beta_2}}{\partial x_{\tau_1}}\cdots T^{\tau_1\tau_2\cdots}_{\sigma_1\sigma_2\cdots}$$

porte le nom de tenseur.

Chaque $T^{\beta_1\beta_2\cdots}_{\alpha_1\alpha_2\cdots}$ est une composante du tenseur.

Le raisonnement que nous venons de faire pour le tenseur de Riemann $R^{\sigma}_{\alpha\beta\gamma}$ s'étend facilement à un tenseur quelconque $T^{\beta_1\beta_2\cdots}_{\alpha_1\alpha_2\cdots}$. Il en résulte le théorème suivant :

Théorème

S'il est possible de faire correspondre à un point [p. 82] quelconque un système de coordonnées $\xi_1, \xi_2, \xi_3, \xi_4$ et des nombres $\Theta^{\tau_1\tau_2\cdots}_{\sigma_1\sigma_2\cdots}$ de telle sorte que les quantités $T^{\beta_1\beta_2\cdots}_{\alpha_1\alpha_2\cdots}$ soient définies dans un système quelconque x_1, x_2, x_3, x_4 par les équations

$$(19)\qquad T^{\beta_1\beta_2\cdots}_{\alpha_1\alpha_2\cdots} = \sum_{\sigma_1}\sum_{\sigma_2}\cdots\sum_{\tau_1}\sum_{\tau_2}\cdots\frac{\partial \xi_{\sigma_1}}{\partial x_{\alpha_1}}\frac{\partial \xi_{\sigma_2}}{\partial x_{\alpha_2}}\cdots\frac{\partial x_{\beta_1}}{\partial \xi_{\tau_1}}\frac{\partial x_{\beta_2}}{\partial \xi_{\tau_2}}\cdots \Theta^{\tau_1\tau_2\cdots}_{\sigma_1\sigma_2\cdots}$$

$T^{\beta_1\beta_2\cdots}_{\alpha_1\alpha_2\cdots}$ est un tenseur, c'est à dire [*sic*] se transforme lors d'un changement de coordonnées par les équations (18).

$\Theta^{\tau_1\tau_2\cdots}_{\sigma_1\sigma_2\cdots}$ est la valeur des composantes dans le système $(\xi_1, \xi_2, \xi_3, \xi_4)$.

[161] B suppression : parenthèses supprimées.

[162] B suppression : parenthèses supprimées.

[163] B suppression : parenthèses supprimées.

[164] B suppression : parenthèses supprimées.

[165] B suppression : parenthèses supprimées.

[166] B suppression : parenthèses supprimées.

[167] Dans l'équation qui suit, il faut lire bien sûr :

$$T'^{\beta_1\beta_2\cdots}_{\alpha_1\alpha_2\cdots} = \sum_{\sigma_1}\sum_{\sigma_2}\cdots\sum_{\tau_1}\sum_{\tau_2}\cdots\frac{\partial x_{\sigma_1}}{\partial x'_{\alpha_1}}\frac{\partial x_{\sigma_2}}{\partial x'_{\alpha_2}}\cdots\frac{\partial x'_{\beta_1}}{\partial x_{\tau_1}}\frac{\partial x'_{\beta_2}}{\partial x_{\tau_2}}\cdots T^{\tau_1\tau_2\cdots}_{\sigma_1\sigma_2\cdots}.$$

Naturellement il n'est pas nécessaire qu'il y ait des deux sortes d'indices, supérieurs et inférieurs.

Nous avons rencontré des tenseurs n'ayant que des indices inférieurs, ils sont dits covariants. Ce sont les potentiels $g_{\mu\nu}$. Leur caractère tensoriel résulte immédiatement de (7).

Comme tenseur contrevariant, c'est à dire [*sic*] à indices supérieurs, il faut citer tout d'abord la vitesse propre

$$u^\sigma \equiv \frac{dx_\sigma}{ds} = \sum_\tau \frac{\partial x_\sigma}{\partial \xi_\tau} \left(\frac{d\xi_\tau}{ds} \right)$$

et les potentiels associés $g^{\mu\nu}$ dont le caractère tensoriel résulte de (11).

Il suit immédiatement de la définition (18) du caractère tensoriel que les composantes de deux tenseurs de même espèce peuvent être combinées deux à deux par addition. Les quantités obtenues sont des composantes d'un tenseur de même espèce que les deux premiers.

$$A^{\beta_1\beta_2\cdots}_{\alpha_1\alpha_2\cdots} + B^{\beta_1\beta_2\cdots}_{\alpha_1\alpha_2\cdots} = C^{\beta_1\beta_2\cdots}_{\alpha_1\alpha_2\cdots}$$

En combinant[168] deux à deux les composantes de deux tenseurs, on obtient les composantes d'un tenseur qu'on [p. 83] représente en réunissant en[169] une même lettre les indices des tenseurs composants.

$$A^{\beta_1\beta_2\cdots}_{\alpha_1\alpha_2\cdots} \cdot B^{\delta_1\delta_2\cdots}_{\gamma_1\gamma_2\cdots} = C^{\beta_1\beta_2\ \cdots\ \delta_1\delta_2\cdots}_{\alpha_1\alpha_2\ \cdots\ \gamma_1\gamma_2\cdots}$$

Enfin on peut combiner entre elles les composantes d'un même tenseur en ajoutant entre elles les quatre composantes pour lesquelles un indice inférieur et un indice supérieur prennent ensemble les valeurs 1, 2, 3, 4, les autres indices gardant même valeur.

On formera ainsi le tenseur contracté. Ainsi en contractant $T^{\beta_1\beta_2\cdots}_{\alpha_1\alpha 2\ldots}$ par rapport aux indices α_1 et β_1 on obtient :

$$\sum_\sigma T^{\sigma\beta_2\cdots}_{\sigma\alpha_2\cdots} = T^{\beta_2\cdots}_{\alpha_2\cdots}$$

Le caractère tensoriel du nouveau tenseur s'établit facilement. On aurait en changeant les coordonnées (18)

$$\begin{aligned} T'^{\beta_2\cdots}_{\alpha_2\cdots} &= \sum_\sigma T'^{\sigma\beta_2\cdots}_{\sigma\alpha_2\cdots} \\ &= \sum_\sigma \sum_{\sigma_1} \sum_{\sigma_2} \cdots \sum_{\tau_1} \sum_{\tau_2} \cdots \frac{\partial x_{\sigma_1}}{\partial x'_\sigma} \frac{\partial x_{\sigma_2}}{\partial x'_{\alpha_2}} \cdots \frac{\partial x'_\sigma}{\partial x_{\tau_1}} \frac{\partial x'_{\beta_2}}{\partial x_{\tau_2}} \cdots T^{\tau_1\tau_2\cdots}_{\sigma_1\sigma_2\cdots} \end{aligned}$$

[168] B modification : "multipliant".

[169] B modification : "sur".

Effectuons la sommation en σ

$$\sum_{\sigma} \frac{\partial x_{\sigma_1}}{\partial x'_{\sigma}} \frac{\partial x'_{\sigma}}{\partial x_{\tau_1}} = \frac{dx_{\sigma_1}}{dx_{\tau_1}} = g_{\tau_1}^{\sigma_1}$$

puis celles en σ_1 et τ_1

$$\sum_{\sigma_1} \sum_{\tau_1} g_{\tau_1}^{\sigma_1} T_{\sigma_1 \sigma_2 \cdots}^{\tau_1 \tau_2 \cdots} = \sum_{\sigma_1} T_{\sigma_1 \sigma_2 \cdots}^{\sigma_1 \tau_2 \cdots} = T_{\sigma_2 \cdots}^{\tau_2 \cdots}$$

Il reste finalement

$$T'^{\beta_2 \cdots}_{\alpha_2 \cdots} = \sum_{\sigma_2} \cdots \sum_{\tau_2} \cdots \frac{\partial x_{\sigma_2}}{\partial x'_{\alpha_2}} \cdots \frac{\partial x'_{\beta_2}}{\partial x_{\tau_2}} \cdots T_{\sigma_2 \cdots}^{\tau_2 \cdots}$$

qui est la définition même (18) du caractère tensoriel du tenseur contracté.

Les deux dernières opérations, multiplication et contraction, s'effectuent souvent simultanément et s'appellent alors produit intérieur. Le tenseur g_{α}^{β} est ainsi le produit intérieur de $g_{\alpha\beta}$ et $g^{\alpha\beta}$.
[p. 84] On a en effet

$$g_{\alpha}^{\beta} = \sum_{\sigma} g_{\alpha\sigma} g^{\sigma\beta}$$

c'est l'équation par laquelle nous avons défini les $g^{\alpha\beta}$ [170].

Les tenseurs qu'on obtient en formant le produit intérieur d'un tenseur par l'un ou l'autre des tenseurs fondamentaux $g_{\alpha\beta}$ et $g^{\alpha\beta}$ sont désignés par la même lettre. Il suffit de les distinguer par le nombre et la position des indices. Ces tenseurs sont dits associés. En contractant le tenseur de Riemann, on obtient le tenseur de Rieman [*sic*] contracté.

(20) $$R_{\alpha\beta} = \sum_{\sigma} R_{\alpha\beta\sigma}^{\sigma}$$

Les tenseurs associés seront

(20_1) $$R_{\alpha}^{\beta} = \sum_{\sigma} g^{\beta\sigma} R_{\alpha\sigma}$$

et

(20_2) $$R^{\alpha\beta} = \sum_{\sigma} g^{\alpha\sigma} R_{\sigma}^{\beta}$$

[170] B ajout : ce membre de phrase est mis entre parenthèses.

En formant le produit intérieur de ces tenseurs par g_{α_β} on n'obtient pas de nouveaux tenseurs, on aura

$$(20_3) \qquad \sum_\sigma g_{\alpha\sigma} R^{\beta\sigma} = \sum_\sigma \sum_\tau g_{\alpha\sigma} g^{\sigma\tau} R^\beta_\tau = R^\beta_\alpha$$

car

$$\sum_\sigma g_{\alpha\sigma} g^{\sigma\tau} = g^\tau_\alpha$$

De même

$$(20_4) \qquad \sum_\sigma g_{\beta\sigma} R^\sigma_\alpha = R_{\alpha\beta}$$

En contractant R^β_α on obtient l'invariant

$$(20_5) \qquad R = \sum_\sigma R^\sigma_\sigma$$

La définition du caractère tensoriel se réduit en effet dans ce cas à

$$R' = R$$

On peut former au moyen de ce tenseur de nouveaux tenseurs[171]

$$(20_6) \qquad R g_{\alpha\beta} \quad , \quad R g^{\alpha\beta}$$

[p. 85] On ne peut les désigner par la même lettre que $R_{\alpha\beta}$, $R^{\alpha\beta}$.

Le calcul tensoriel nous permet ainsi de former des expressions qui se transforment simplement lors d'un changement de coordonnées.

Les nouvelles composantes sont des fonctions linéaires et homogènes des anciennes composantes. Il en résulte que si toutes les composantes d'un tenseur s'annulent dans un système elles s'annulent dans tout autre système.

Si donc une loi physique s'exprime par l'annulation de toutes les composantes d'un tenseur, elle sera valable dans tout système de coordonnées. Nous aurons obtenu conformément au principe de relativité une loi dont l'expression algébrique est indépendante du mode de repérage adopté pour les divers événements.

La méthode introduite par Einstein est donc la suivante : trouver une loi tensorielle[172] qui, lorsqu'on adopte le mode de référence utilisé par les observateurs se confondent avec les lois empiriques qui résument leurs observations avec l'approximation dont leurs mesures est susceptible.

---

[171] Le manuscrit ne numérote pas les tenseurs qui suivent mais s'y réfère par le numéro (20_6).

[172] B modification : "des lois tensorielles".

L'expression réduite du tenseur de Riemann contracté est d'après (15)

$$P_{\lambda\mu} = \sum_\rho P^\rho_{\lambda\mu\rho} = -\frac{1}{2}\sum_\rho \frac{\partial^2 \gamma_{\lambda\mu}}{\partial \xi_\rho^2}$$
$$+\frac{1}{2}\frac{\partial}{\partial \xi_\mu}\sum_\rho \left(\frac{\partial \gamma_{\lambda\rho}}{\partial \xi_\rho} - \frac{1}{2}\frac{\partial \gamma_{\rho\rho}}{\partial \xi_\lambda}\right)$$
$$+\frac{1}{2}\frac{\partial}{\partial \xi_\lambda}\sum_\rho \left(\frac{\partial \gamma_{\mu\rho}}{\partial \xi_\rho} - \frac{1}{2}\frac{\partial \gamma_{\rho\rho}}{\partial \xi_\mu}\right)$$

[p. 86] Il en résulte que le tenseur contracté est symétrique

$$P_{\lambda\mu} = P_{\mu\lambda}$$

et, (20) et (21)[173],

$$R_{\alpha\beta} = R_{\beta\alpha}$$

Le système ξ_σ a été déterminé en s'imposant en un point M les conditions (12). Ces conditions déterminent en (M) [*sic*] la valeur des dérivées premières et secondes $\frac{\partial \xi_\sigma}{\partial x_\alpha}$ et $\frac{\partial^2 \xi_\sigma}{\partial x_\alpha \partial x_\beta}$ des fonctions ξ_σ de la transformation (1).

Ne pourrions-nous pas disposer de la valeur en (M) [*sic*] des dérivées troisièmes de ces fonctions de manière à simplifier l'expression réduite de $R_{\alpha\beta}$? Si nous pouvions[174] choisir ces dérivées de manière que, en (M) [*sic*]

(21)
$$\frac{\partial}{\partial \xi_\mu}\sum_\rho \left(\frac{\partial \gamma_{\lambda\rho}}{\partial \xi_\rho} - \frac{1}{2}\frac{\partial \gamma_{\rho\rho}}{\partial \xi_\lambda}\right) = 0$$
$$(\lambda, \mu = 1,2,3,4)$$

l'expression réduite de $R_{\alpha\beta}$ se réduit[175] à son premier terme

(22)
$$P_{\lambda\mu} = -\sum_\rho \frac{\partial^2 \gamma_{\lambda\mu}}{\partial \xi_\rho^2}$$

Désignons par $A(\alpha\beta)$ la valeur en (M) [*sic*] de

$$\frac{\partial}{\partial x_\beta}\sum_\rho \left(\frac{\partial g_{\alpha\beta}}{\partial x_\rho} - \frac{1}{2}\frac{\partial g_{\rho\rho}}{\partial x_\alpha}\right) = A(\alpha\beta)$$

[176] ce que devient cette expression lorsqu'on l'exprime au moyen des $\gamma_{\mu\nu}$ et des ξ_σ en s'imposant, en plus des conditions (12) et (13), les relations (21).

---

[173] Il faut lire bien sûr : (20) et (20_2).

[174] B modification : "pouvons".

[175] B modification : "réduira".

[176] B ajout : "et calculons".

Les termes contenant les dérivées secondes des $\gamma_{\mu\nu}$ s'annulent en vertu de (21). Cela résulte immédiatement du calcul des dérivées secondes que nous avons effectué plus haut (14).

[p. 87] Les termes contenant les dérivées premières s'annulent puisque ces dérivées s'annulent (12) et (13).

Il reste à calculer les termes ne contenant pas de dérivées, il faut pour cela calculer $A(\alpha\beta)$ en supposant les $\gamma_{\sigma\tau}$ constants.

Nous aurons en dérivant $g_{\alpha\beta}$ (7), où nous posons d'après (12) $\gamma_{\sigma\tau} = g_{\sigma}^{\tau}$,

$$\frac{\partial g_{\alpha\beta}}{\partial x_\gamma} = \sum_\sigma \left[\frac{\partial \xi_\sigma}{\partial x_\alpha}\frac{\partial^2 \xi_\sigma}{\partial x_\beta \partial x_\gamma} + \frac{\partial^2 \xi_\sigma}{\partial x_\alpha \partial x_\gamma}\frac{\partial \xi_\sigma}{\partial x_\beta}\right]$$

portant cette valeur dans $A(\alpha\beta)$,

$$A(\alpha\beta) = \frac{\partial}{\partial x_\beta}\sum_\sigma\sum_\rho \left[\begin{array}{c} \dfrac{\partial \xi_\sigma}{\partial x_\alpha}\dfrac{\partial^2 \xi_\sigma}{\partial x_\rho^2} + \dfrac{\partial^2 \xi_\sigma}{\partial x_\alpha \partial x_\rho}\dfrac{\partial \xi_\sigma}{\partial x_\rho} \\ -\dfrac{1}{2}\dfrac{\partial \xi_\sigma}{\partial x_\rho}\dfrac{\partial^2 \xi_\sigma}{\partial x_\alpha \partial x_\rho} - \dfrac{1}{2}\dfrac{\partial^2 \xi_\sigma}{\partial x_\alpha \partial x_\rho}\dfrac{\partial \xi_\sigma}{\partial x_\rho} \end{array}\right.$$

et en réduisant les termes semblables

$$A(\alpha\beta) = \frac{\partial}{\partial x_\beta}\sum_\sigma\sum_\rho \frac{\partial \xi_\sigma}{\partial x_\alpha}\frac{\partial^2 \xi_\sigma}{\partial x_\rho^2}$$

Effectuons la dérivation.

$$A(\alpha\beta) = \sum_\sigma \left[\frac{\partial \xi_\sigma}{\partial x_\alpha}\frac{\partial}{\partial x_\beta}\sum_\rho \frac{\partial^2 \xi_\sigma}{\partial x_\rho^2} + \frac{\partial^2 \xi_\sigma}{\partial x_\alpha \partial x_\beta}\sum_\rho \frac{\partial^2 \xi_\sigma}{\partial x_\rho^2}\right]$$

Est-il possible quelles que soient les valeurs de $A(\alpha\beta)$ et des dérivées $\frac{\partial \xi_\sigma}{\partial x_\alpha}$ et $\frac{\partial^2 \xi_\sigma}{\partial x_\alpha \partial x_\beta}$, de déterminer les dérivées troisièmes de manière à satisfaire à cette relation pour $(\alpha, \beta = 1, 2, 3, 4)$?

Pour $(\alpha = 1, 2, 3, 4)$ elle peut être considérée comme un système linéaire en

$$\frac{\partial}{\partial x_\beta}\sum_\rho \frac{\partial^2 \xi_\sigma}{\partial x_\rho^2}$$

[p. 88] Le déterminant de ce système est différent de zéro. C'est le déterminant fonctionnel des équations de transformation (1). Le système peut donc être résolu. On obtient des équations de la forme

$$\frac{\partial}{\partial x_\beta}\sum_\rho \frac{\partial^2 \xi_\sigma}{\partial x_\rho^2} = a(\beta\sigma) \tag{23}$$

Où[177] $a(\beta\sigma)$ représentent des fonctions des $A(\alpha\beta)$, $\frac{\partial \xi_\sigma}{\partial x_\alpha}$, $\frac{\partial^2 \xi_\sigma}{\partial x_\alpha \partial x_\beta}$.

---

177 B modification : "où".

Les relations (21) pourront toujours être vérifiées si nous pouvons, quels que soient les $a(\beta\sigma)$, déterminer les valeurs en (M)[178] des dérivées troisièmes de manière à vérifier les équations (23).
On y satisfait en posant

$$\frac{\partial^3 \xi_\sigma}{\partial x_\alpha \partial x_\beta \partial x_\gamma} = \frac{a(\alpha\sigma)\, a(\beta\sigma)\, a(\gamma\sigma)}{\sum_\rho a(\rho\sigma)^2}$$

Il est donc toujours possible de déterminer le système ξ de telle sorte que l'expression réduite du tenseur Riemann contracté [*sic*] soit donnée par l'expression (22)

$$P_{\lambda\mu} = -\frac{1}{2}\left(\frac{\partial^2 \gamma_{\lambda\mu}}{\partial \xi_1^2} + \frac{\partial^2 \gamma_{\lambda\mu}}{\partial \xi_2^2} + \frac{\partial^2 \gamma_{\lambda\mu}}{\partial \xi_3^2} + \frac{\partial^2 \gamma_{\lambda\mu}}{\partial \xi_4^2}\right) \tag{24}$$

Nous avons employé un système ξ pour lequel la forme fondamentale se réduit du[179] M à

$$ds^2 = d\xi_1^2 + d\xi_2^2 + d\xi_3^2 + d\xi_4^2$$

Les coordonnées ξ_σ seront imaginaires.
Nous pouvons les écrire

$$\begin{aligned} \xi_1 &= x\sqrt{-1} \\ \xi_2 &= y\sqrt{-1} \\ \xi_3 &= z\sqrt{-1} \\ \xi_4 &= t \end{aligned}$$

[p. 89] ds^2 s'écrira alors

$$ds^2 = -dx^2 - dy^2 - dz^2 + dt^2$$

Les $\gamma_{\lambda\mu}$ se transforment en $g_{\lambda\mu}$ comme les $P_{\lambda\mu}$ en $R_{\lambda\mu}$. Cette transformation ne fait qu'introduire une ou plusieurs fois le facteur constant $\sqrt{-1}$. Ces facteurs se détruisent dans les deux membres de (24).

La forme réduite du tenseur contracté sera donc dans le système (x, y, z, t)

$$R_{\lambda\mu} = \frac{1}{2}\left(\frac{\partial^2 g_{\lambda\mu}}{\partial x^2} + \frac{\partial^2 g_{\lambda\mu}}{\partial y^2} + \frac{\partial^2 g_{\lambda\mu}}{\partial z^2} - \frac{\partial^2 g_{\lambda\mu}}{\partial t^2}\right) \tag{25}$$

[178] B suppression : parenthèses supprimées.

[179] B modification : "en".

Chapitre IV. La gravitation.

§ 10. Potentiel Newtonien et potentiel retardé.

Les observations astronomiques sont résumées avec précision par la loi de la gravitation universelle de Newton.

Pour obtenir cette loi, on emploie un mode de repérage particulier : des coordonnées cartésiennes par rapport à trois axes rectangulaires immobiles ou animés d'une translation rectiligne et uniforme, par rapport à l'ensemble des étoiles ; le temps est mesuré par la rotation de la terre par rapport aux étoiles.

L'accélération d'un corps libre peut être attribuée à l'action des divers astres. Elle dépend d'un potentiel, V. C'est-à-dire que les composantes de l'accélération peuvent s'écrire

$$(1) \qquad \begin{cases} \dfrac{d^2x}{dt^2} = -\dfrac{\partial V}{\partial x} \\ \dfrac{d^2y}{dt^2} = -\dfrac{\partial V}{\partial y} \\ \dfrac{d^2z}{dt^2} = -\dfrac{\partial V}{\partial z} \end{cases}$$

[p. 90] Le potentiel V se calcule alors par l'équation

$$(2) \qquad V = -K \sum \frac{m}{r}$$

où m et r désignent respectivement la masse des divers astres et leur distance au point où se calcule le potentiel. K est la constante de l'attraction ; en unités C.G.S.

$$(2') \qquad K = 6{,}7 \cdot 10^{-8}$$

Ses dimensions sont $L^3 T^{-2} M^{-1}$.

L'équation (2) exprime l'action des masses sur le champ, l'équation (1) traduit la réaction du champ sur les points libres qui y circulent.

Les équations (2) peuvent être écrites sous une autre forme.

Si nous désignons par ρ la densité de la matière en un point ξ, η, ζ, la masse contenue dans un élément de volume $d\xi\, d\eta\, d\zeta$ situé en ce point sera $\rho\, d\xi\, d\eta\, d\zeta$. La somme de l'équation (2) pourra s'écrire sous forme d'une intégrale triple étendue à tout l'espace. Le potentiel est[180] un point x, y, z à l'instant (t) [*sic*] sera

$$(3) \qquad V(x,y,z,t) = -K \int \frac{\rho(\xi,\eta,\zeta,t)}{r}\, d\xi\, d\eta\, d\zeta$$

[180] B modification : "en".

où (r)[181] est défini par la relation

$$r^2 = (x-\xi)^2 + (y-\eta)^2 + (z-\zeta)^2$$

Inversement on peut exprimer la densité ρ en fonction du potentiel. On obtient alors l'équation bien connue de Poisson.

$$\Delta V = \frac{\partial^2 V}{\partial x^2} + \frac{\partial^2 V}{\partial y^2} + \frac{\partial^2 V}{\partial z^2} = 4\pi K\rho \tag{4}$$

En un point où il n'y a pas de matière cette équation se réduit à

$$\Delta V = 0 \tag{5}$$

[p. 91] et porte le nom de l'équation de Laplace. ΔV s'appelle le laplacien de V.

Le potentiel Newtonien n'est qu'une solution particulière de l'équation de Poisson. On l'obtient en imposant à V des conditions aux limites : lorsqu'on s'éloigne de[182] l'infini dans une direction quelconque, V doit tendre vers une limite indépendante de cette direction. L'accélération d'un point infiniment éloigné de toute masse est alors nulle. Naturellement cette condition ne peut être réalisée que lorsqu'on emploie un système d'axes particulier ou aussi tout système dont le mouvement par rapport à celui-ci est rectiligne et uniforme.

La théorie Newtonienne de la gravitation suppose une action immédiate des masses attirantes. Si leur action se propage de proche en proche avec une certaine vitesse (c)[183], l'influence de masses, situées à la distance (r)[184] d'un point (x,y,z), ne s'y fera sentir qu'après un temps $\frac{r}{c}$. En tenant compte de ce retard, on devra écrire l'équation du potentiel sous la forme

$$V(x,y,z,t) = -K\int \frac{\rho(\xi,\eta,\zeta,t-\frac{r}{c})}{r}\,d\xi\,d\eta\,d\zeta \tag{6}$$

c'est l'équation du potentiel retardé.

Si on pose

$$\Box V = \frac{\partial^2 V}{\partial x^2} + \frac{\partial^2 V}{\partial y^2} + \frac{\partial^2 V}{\partial z^2} - \frac{1}{c^2}\frac{\partial^2 V}{\partial t^2} \tag{7}$$

l'équation de Poisson devient

$$\Box V = 4\pi K\rho \tag{8}$$

$\Box V$ s'appelle le dalembertien [*sic*] de V.

---

[181] B suppression : parenthèses supprimées.

[182] B modification : "à".

[183] B suppression : parenthèses supprimées.

[184] B suppression : parenthèses supprimées.

[p. 92] L'équation de Laplace s'écrit maintenant

$$\square V = 0 \tag{9}$$

Ces formules rendent aussi bien compte de l'expérience que celles de la théorie primitive lorsque la vitesse de propagation (c)[185] est suffisamment grande ; par exemple, de l'ordre de celle de la lumière.

Lorsque nous choisissons l'unité de temps de telle sorte que la vitesse de propagation soit prise pour unité, le dalembertien [*sic*] s'écrit

$$\square V = \frac{\partial^2 V}{\partial x^2} + \frac{\partial^2 V}{\partial y^2} + \frac{\partial^2 V}{\partial z^2} - \frac{\partial^2 V}{\partial t^2} \tag{10}$$

La constante de Gauss K doit alors être divisée par le carré (c^2)[186] de la vitesse de propagation dans le système C.G.S.

Au moyen des symboles opératoires définis ci-dessus les expressions réduites du tenseur contracté obtenues à la fin du § 9 peuvent s'écrire

$$\square g_{\mu\nu} = 2R_{\mu\nu} \tag{11}$$

Elles sont valables au voisinage d'un point (M)[187] où les dérivées premières des $g_{\mu\nu}$ sont nulles et ou [*sic*] (ds^2)[188] se réduit à

$$ds^2 = -dx^2 - dy^2 - dz^2 + dt^2$$

On déduira alors comme on déduit (6) de (8)

$$g_{\mu\nu}(x,y,z,t) = -\frac{1}{2\pi}\int \frac{R_{\mu\nu}(\xi,\eta,\zeta,t-r)}{r}\,d\xi\,d\eta\,d\zeta \tag{12}$$

Les potentiels $g_{\mu\nu}$ jouent dans la théorie d'Einstein le rôle des potentiels newtoniens[189] V, ainsi que nous nous en sommes rendus compte du[190] § 1. Nous voyons maintenant que $R_{\mu\nu}$ doit représenter les masses matérielles. Il nous faut donc égaler $R_{\mu\nu}$ à [p. 93] un tenseur de même nature dépendant de la répartition des masses. Outre $R_{\mu\nu}$, les tenseurs $g_{\mu\nu}$ et $Rg_{\mu\nu}$ (§ 9-20_6) peuvent encore figurer dans la relation cherchée. Si $T_{\mu\nu}$ est un tenseur représentant la matière, nous pourrons essayer de formuler les lois de la gravitation par l'équation générale.

$$T_{\mu\nu} = c_1 R_{\mu\nu} + c_2 R g_{\mu\nu} + c_3 g_{\mu\nu} \tag{13}$$

où c_1, c_2, c_3 sont des constantes indéterminées.

---

[185] B suppression : parenthèses supprimées.

[186] B suppression : parenthèses supprimées.

[187] B suppression : parenthèses supprimées.

[188] B suppression : parenthèses supprimées.

[189] B modification : "du potentiel newtonnien" [*sic*].

[190] B modification : "au".

Nous chercherons ensuite quelles valeurs il faut donner à ces constantes pour que (13) satisfasse aux conditions qui doivent[191] vérifier la loi de la gravitation.

Au lieu d'une équation entre tenseurs covariants on peut considérer l'équation équivalente entre les tenseurs associés à ceux-ci.

$$T^{\mu\nu} = c_1 R^{\mu\nu} + c_2 R g^{\mu\nu} + c_3 g^{\mu\nu} \tag{14}$$

§ **11. Le tenseur d'énergie matérielle.**

L'état de la matière est caractérisé par sa densité et sa vitesse. Ces grandeurs physiques sont-elles des tenseurs ? Nous savons que la vitesse propre

$$u^{\sigma} = \frac{dx_{\sigma}}{ds} \tag{1}$$

est un tenseur. La densité, ρ, varie lorsqu'on change de système de coordonnées.

Lorsqu'on se borne à changer les coordonnées d'espace on raisonne de la manière suivante : la masse contenue dans un certain volume doit être indépendante de la manière dont on repère les divers points de ce volume. Comme la densité est le quotient de la masse par le volume qu'elle occupe, le produit [p. 94] de la densité par l'élément de volume doit être un invariant.

Nous raisonnons[192] de même dans le cas général. La masse contenue dans un domaine d'univers doit être indépendante de la manière dont on en repère les divers points ; nous appellerons densité, la masse contenue dans l'unité de volume d'univers.

$$\rho \, dx_1 \, dx_2 \, dx_3 \, dx_4 \tag{2}$$

sera donc un invariant.

Lorsqu'on emploie des coordonnées propres de la matière (§ 3) une des coordonnées est l'invariant (ds)[193] mesuré en suivant la matière dans son mouvement éventuel. Le produit de ρ par les trois autres différentielles des coordonnées sera donc encore un invariant. Les deux définitions de la densité que nous venons d'envisager sont donc des équivalentes.

Nous avons vu en étudiant la géométrie générale comment se transforme l'élément de volume (§ 1-9).

$$\sqrt{\gamma} \, dx_1 \, dx_2 \, dx_3$$

où γ est le déterminant des potentiels est un invariant.

---

[191] B modification : "que doit".

[192] B modification : "raisonnerons".

[193] B suppression : parenthèses supprimées.

Cette propriété se généralise immédiatement à quatre coordonnées. En tenant compte de ce que le déterminant g des $g_{\mu\nu}$ est négatif, nous obtenons que

$$\sqrt{-g}\, dx_1\, dx_2\, dx_3\, dx_4 \tag{3}$$

est un invariant.

Le quotient des deux invariants (2) et (3)

$$\frac{\rho}{\sqrt{-g}}$$

est donc un invariant.

Nous pouvons combiner cet invariant avec le [p. 95] contrevariant (1) nous formerons ainsi par multiplication un contrevariant de second ordre.[194]

$$T^{\mu\nu} = \frac{\rho}{\sqrt{-g}} \frac{dx_\mu}{ds} \frac{dx_\nu}{dt} \tag{4}$$

que nous appellerons le tenseur d'énergie matérielle. C'est le tenseur que nous introduirons dans les équations (§ 10-14).

On formera sans difficulté le covariant associé à $T^{\mu\nu}$

$$T_{\mu\nu} = \sum_\sigma \sum_\tau g_{\mu\sigma} g_{\nu\tau} T^{\sigma\tau} \tag{5}$$

l'invariant[195] obtenu par contraction [196]

$$T = \sum_\mu \sum_\nu g_{\mu\nu} T^{\mu\nu} = \frac{\rho}{\sqrt{-g}} \sum_\mu \sum_\nu g_{\mu\nu} \frac{dx_\mu}{ds} \frac{dx_\nu}{ds}$$

D'après la définition de ds^2 cet invariant se réduit à

$$T = \frac{\rho}{\sqrt{-g}} \tag{6}$$

[194] Dans l'équation qui suit, il faut lire bien sûr : $T^{\mu\nu} = \frac{\rho}{\sqrt{-g}} \frac{dx_\mu}{ds} \frac{dx_\nu}{ds}$.

[195] B modification : "L'invariant".

[196] B ajout : "est".

[p. 96] Nous désignerons sous le nom de densité tensionnelles[197] le produit d'un tenseur par $\sqrt{-g}$; nous écrivons[198] les densités tensionnelles[199] comme les tenseurs correspondants, mais en employant des lettres italiques.

Nous aurons ainsi

$$\mathcal{T}^{\mu\nu} = \rho \frac{dx_\mu}{ds} \frac{dx_\nu}{ds} \tag{7}$$

Il est possible d'écrire les équations du mouvement d'un point libre en fonction du tenseur d'énergie matérielle.

La solution de ce problème sera une première étape dans la recherche des équations générales de la gravitation, car celles-ci doivent rendre compte du mouvement particules libres [*sic*] sous l'influence du champ en même temps que de la production du champ par les masses.

Nous avons vu qu'on pouvait toujours choisir un système de coordonnées de telle sorte qu'en un point *M* le mouvement des points libres soit rectiligne et uniforme. Il faut pour cela que les dérivées des potentiels s'annulent en ce point. Nous chercherons donc tout d'abord à exprimer au moyen du tenseur matériel à quelle condition le mouvement des points est rectiligne et uniforme, nous obtiendrons ainsi une certaine équation valable dans un système particulier de coordonnées. Il nous suffira, ensuite de trouver une équation tensionelle[200] qui se réduise à cette équation lorsque les dérivées des potentiels sont constantes[201]. Cette nouvelle équation sera l'équation du mouvement des points libres dans le cas général.

Considérons une petite masse matérielle en mouvement uniforme [202] la vitesse de tous ses points peut être supposée la même et la densité est constante.

Formons l'expression :[203]

$$\begin{aligned} f^\mu &= \sum_\sigma \frac{\partial \mathcal{T}^{\mu\sigma}}{\partial x_\sigma} \\ &= \rho \sum_\sigma \frac{\partial \left(\frac{dx_\mu}{ds}\right)}{\partial x_\sigma} + \frac{dx_\mu}{ds} \sum_\sigma \frac{\partial}{\partial x_\sigma}\left(\rho \frac{dx_\sigma}{ds}\right) \end{aligned} \tag{8}$$

[p. 97] On a

$$\sum_\sigma \frac{\partial \left(\frac{dx_\mu}{ds}\right)}{\partial x_\sigma} \frac{dx_\sigma}{ds} = \frac{d^2 x_\mu}{ds^2}$$

---

[197] B modification : "tensorielles".

[198] B modification : "écrirons".

[199] B modification : "tensorielles".

[200] B modification : "tensorielle".

[201] B modification : "nulles".

[202] Il convient d'ajouter un point-virgule.

[203] Dans les relations qui suivent, il faut lire bien sûr :

$$f^\mu = \sum_\sigma \frac{\partial \mathcal{T}^{\mu\sigma}}{\partial x_\sigma} = \rho \sum_\sigma \frac{\partial \left(\frac{dx_\mu}{ds}\right)}{\partial x_\sigma} \frac{dx_\sigma}{ds} + \frac{dx_\mu}{ds} \sum_\sigma \frac{\partial}{\partial x_\sigma}\left(\rho \frac{dx_\sigma}{ds}\right).$$

Pour les divers points du mobile on a donc

$$f^{\mu} = 0$$

puisque l'accélération est nulle et que la densité et la vitesse sont constantes. Réciproquement si nous posons

$$f^{\mu} = 0$$

Nous[204] pouvons en déduire

$$\begin{aligned} 0 &= \sum_{\mu} \sum_{\nu} g_{\mu\nu} f^{\mu} \frac{dx_{\nu}}{ds} \\ &= \rho \sum_{\mu} \sum_{\nu} g_{\mu\nu} \frac{d^2 x_{\mu}}{ds^2} \frac{dx_{\nu}}{ds} + \sum_{\sigma} \frac{\partial}{\partial x_{\sigma}} \left(\rho \frac{dx_{\sigma}}{ds} \right) \end{aligned}$$

car

$$\sum_{\mu} \sum_{\nu} g_{\mu\nu} \frac{dx_{\mu}}{ds} \frac{dx_{\nu}}{ds} = 1$$

En dérivant cette dernière expression on trouve puisque les dérivées des $g_{\mu\nu}$ sont nulles au point M,[205]

$$\sum_{\mu} \sum_{\nu} \frac{d^2 x_{\mu}}{ds^2} \frac{dx_{\nu}}{ds} = 0$$

Les équations

$$f^{\mu} = 0 \qquad (\mu = 1, 2, 3, 4)$$

ont donc pour conséquence

$$\sum_{\sigma} \frac{\partial}{\partial x_{\sigma}} \left(\rho \frac{dx_{\sigma}}{ds} \right) = 0 \tag{9}$$

et

$$\frac{d^2 x_{\mu}}{ds^2} = 0 \qquad (\mu = 1, 2, 3, 4)$$

Les dernières équations expriment que le mouvement d'un point est rectiligne et uniforme. C'est bien l'équation du mouvement des points libres, au point M, pour le système de coordonnées que nous avons employé.

La première est l'équation de continuité. Elle exprime la conservation de la masse. Pour nous en rendre compte employons les coordonnées propres du mobile x, y, z, t (§ 3). ds sera alors égal à cdt. Formons au moyen du premier membre de l'équation une intégrale de volume étendue à l'espace occupé par le mobile. Cette intégrale sera nulle et nous aurons

204 B modification : "nous".

205 Dans l'équation qui suit, il faut lire bien sûr : $\sum_{\mu} \sum_{\nu} g_{\mu\nu} \frac{d^2 x_{\mu}}{ds^2} \frac{dx_{\nu}}{ds} = 0$.

$$\frac{1}{c}\iiint\left[\frac{\partial}{\partial x}\left(\rho\frac{dx}{dt}\right)+\frac{\partial}{\partial y}\left(\rho\frac{dy}{dt}\right)+\frac{\partial}{\partial z}\left(\rho\frac{dz}{dt}\right)\right]dxdydz$$
$$+\frac{1}{c}\frac{d}{dt}\iiint\rho\,dxdydz=0$$

[p. 98] Transformons la première intégrale par la formule de Green. En désignant par dS l'élément de la surface limitant le corps et par ℓ, m, n les cosinus directeur [*sic*] de la normale extérieure à cette surface, nous obtenons.

$$\iint\rho\left(\ell\frac{dx}{dt}+m\frac{dy}{dt}+n\frac{dz}{dt}\right)dS$$

Cette intégrale représente la quantité de matière qui sort du volume. Elle est nulle puisque nous intégrons sur la surface même du mobile.

Il reste donc

$$\frac{d}{dt}\iiint\rho\,dxdydz=0$$

qui exprime que la masse totale du corps reste constante, nous obtiendrons donc les équations tensionnelles [*sic*] qui expriment la conservation de la matière et le mouvement des particules libres en annulant un tenseur f^{μ} qui se réduise [*sic*] à

$$\sum_{\sigma}\frac{\partial\mathcal{T}^{\mu\sigma}}{\partial x_{\sigma}}$$

Lorsque les dérivées des potentiels sont constantes[206].

* * *

Pour former ce tenseur nous utiliserons le théorème qui nous a déjà servi au § 9 [207].

Le tenseur d'énergie matérielle se transforme de la manière suivante

$$(10)\qquad T^{\alpha\beta}=\sum_{\mu}\sum_{\nu}\frac{\partial x_{\alpha}}{\partial\xi_{\mu}}\frac{\partial x_{\beta}}{\partial\xi_{\nu}}\Theta^{\mu\nu}$$

où $\Theta^{\mu\nu}$ désigne les composantes de ce tenseur dans le système ξ_{σ}.
Inversement on aura

$$(11)\qquad \Theta^{\alpha\beta}=\sum_{\mu}\sum_{\nu}\frac{\partial\xi_{\alpha}}{\partial x_{\mu}}\frac{\partial\xi_{\beta}}{\partial x_{\nu}}T^{\mu\nu}$$

En résolvant ces équations ($\mu,\nu=1,2,3,4$) par rapport aux $T^{\mu\nu}$ on doit retrouver les équations précédentes. Nous utiliserons ce fait dans [p. 99] un instant.

---

[206] B modification : "nulles".

[207] B ajout : "pour établir le caractère tensoriel des divers tenseurs. Nous représenterons de nouveau par ξ_1, ξ_2, ξ_3 ξ_4 le système de coordonnées utilisé au § 9."

Dérivons $\Theta^{\alpha\beta}$ par rapport à x_γ nous aurons

$$\frac{\partial\Theta^{\alpha\beta}}{\partial x_\gamma} = \sum_\rho \frac{\partial\xi_\rho}{\partial x_\gamma}\frac{\partial\Theta^{\alpha\beta}}{\partial\xi_\rho}$$
$$= \sum_\mu\sum_\nu \frac{\partial^2\xi_\alpha}{\partial x_\mu\partial x_\gamma}\frac{\partial\xi_\beta}{\partial x_\nu}T^{\mu\nu}$$
$$+ \sum_\mu\sum_\nu \frac{\partial\xi_\alpha}{\partial x_\mu}\frac{\partial^2\xi_\beta}{\partial x_\nu\partial x_\gamma}T^{\mu\nu}$$
$$+ \sum_\mu\sum_\nu \frac{\partial\xi_\alpha}{\partial x_\mu}\frac{\partial\xi_\beta}{\partial x_\nu}\frac{\partial T^{\mu\nu}}{\partial x_\gamma}$$

Au point M nous pouvons calculer les dérivées secondes par la formule (§ 9-9)[208]

$$\frac{\partial^2\xi_\alpha}{\partial x_\mu\partial x_\gamma} = \sum_\tau \frac{\partial\xi_\alpha}{\partial x_\tau}\left\{\begin{matrix}\gamma\mu\\ \tau\end{matrix}\right\}$$

Le premier des trois termes de l'équation précédente s'écrit ainsi

$$\sum_\mu\sum_\nu\sum_\tau \frac{\partial\xi_\alpha}{\partial x_\tau}\frac{\partial\xi_\beta}{\partial x_\nu}\left\{\begin{matrix}\gamma\mu\\ \tau\end{matrix}\right\}T^{\mu\nu}$$

ou en échangeant les indices sommatoires μ et τ

$$\sum_\mu\sum_\nu\sum_\tau \frac{\partial\xi_\alpha}{\partial x_\mu}\frac{\partial\xi_\beta}{\partial x_\nu}\left\{\begin{matrix}\gamma\tau\\ \mu\end{matrix}\right\}T^{\tau\nu}$$

le second terme s'écrit de même

$$\sum_\mu\sum_\nu\sum_\tau \frac{\partial\xi_\alpha}{\partial x_\mu}\frac{\partial\xi_\beta}{\partial x_\nu}\left\{\begin{matrix}\gamma\tau\\ \nu\end{matrix}\right\}T^{\mu\tau}$$

On aura ainsi

$$\sum_\rho \frac{\partial\xi_\rho}{\partial x_\gamma}\frac{\partial\Theta^{\alpha\beta}}{\partial\xi_\rho} = \sum_\mu\sum_\nu \frac{\partial\xi_\alpha}{\partial x_\mu}\frac{\partial\xi_\beta}{\partial x_\nu}\left(\frac{\partial T^{\mu\nu}}{\partial x_\gamma} + \sum_\tau\left\{\begin{matrix}\gamma\tau\\ \mu\end{matrix}\right\}T^{\tau\nu} + \sum_\tau\left\{\begin{matrix}\gamma\tau\\ \nu\end{matrix}\right\}T^{\mu\tau}\right)$$

Résolvons cette équation par rapport à la quantité [209] parenthèse dans le second membre. Le calcul est le même que celui qui permet de passer de l'équation (11) à l'équation (10). Il vient[210]

$$\frac{\partial T^{\mu\nu}}{\partial x_\gamma} + \sum_\tau\left\{\begin{matrix}\gamma\tau\\ \mu\end{matrix}\right\}T^{\tau\nu} + \sum_\tau\left\{\begin{matrix}\gamma\tau\\ \nu\end{matrix}\right\} = \sum_\mu\sum_\nu\sum_\rho \frac{\partial x_\alpha}{\partial\xi_\mu}\frac{\partial x_\beta}{\partial\xi_\nu}\frac{\partial\xi_\rho}{\partial x_\gamma}\frac{\partial\Theta^{\alpha\beta}}{\partial\xi_\rho}$$

---

[208] Il faut lire bien sûr : (§ 9-3).

[209] B ajout : "entre".

[210] Dans l'équation qui suit, il faut lire bien sûr :

$$\frac{\partial T^{\mu\nu}}{\partial x_\gamma} + \sum_\tau\left\{\begin{matrix}\gamma\tau\\ \mu\end{matrix}\right\}T^{\tau\nu} + \sum_\tau\left\{\begin{matrix}\gamma\tau\\ \nu\end{matrix}\right\}T^{\mu\tau} = \sum_\mu\sum_\nu\sum_\rho \frac{\partial x_\alpha}{\partial\xi_\mu}\frac{\partial x_\beta}{\partial\xi_\nu}\frac{\partial\xi_\rho}{\partial x_\gamma}\frac{\partial\Theta^{\alpha\beta}}{\partial\xi_\rho}.$$

En appliquant le théorème du § 9, nous voyons que le premier membre de cette équation est un tenseur.
Le tenseur contracté correspondant

$$f^{\mu}=\sum_{\sigma}\frac{\partial T^{\mu\sigma}}{\partial x_{\sigma}}+\sum_{\sigma}\sum_{\tau}\left\{ {\sigma\tau \atop \mu} \right\}T^{\tau\sigma}+\sum_{\sigma}\sum_{\tau}\left\{ {\sigma\tau \atop \sigma} \right\}T^{\mu\tau}$$

peut se simplifier si on remarque que (§ 5-12)

$$\sum_{\sigma}\left\{ {\sigma\tau \atop \sigma} \right\}=\frac{1}{2}\frac{1}{g}\frac{\partial g}{\partial x_{\tau}}=\frac{1}{\sqrt{-g}}\frac{\partial\sqrt{-g}}{\partial x_{\tau}}$$

il vient ainsi [p. 100]

$$f^{\mu}=\frac{1}{\sqrt{-g}}\sum_{\sigma}\frac{\partial}{\partial x_{\sigma}}\left(T^{\mu\sigma}\sqrt{-g}\right)+\sum_{\sigma}\sum_{\tau}\left\{ {\sigma\tau \atop \mu} \right\}T^{\tau\sigma}$$

La densité tensionelle [*sic*] correspondante sera[211]

(11) $$\mathfrak{f}^{\mu}=\sum_{\sigma}\frac{\partial\mathfrak{T}^{\mu\sigma}}{\partial x_{\sigma}}+\sum_{\sigma}\sum_{\tau}\left\{ {\sigma\tau \atop \mu} \right\}\mathfrak{T}^{\tau\sigma}$$

Le second memebre [*sic*] se réduit à son premier terme lorsque les dérivées des potentiels sont nulles, car les symboles de Christoffel s'annulent dans ce cas.

Nous pouvons donc conclure que les équations tensionelles[212]

(12) $$f^{\mu}=0 \qquad (\mu=1,2,3,4)$$

expriment en fonction du tenseur d'énergie matérielle, les équations du mouvement des points libres et le principe de conservation de la matière.

REMARQUE. – Dans la démonstration du caractère tensionel[213] de f^{μ} nous n'avons pas fait usage de la symétrie de $\mathfrak{T}^{\mu\nu}$.

Cette démonstration subsiste encore pour une densité tensionelle[214] pour laquelle

$$\mathfrak{T}^{\sigma\tau}=-\mathfrak{T}^{\tau\sigma}$$

le tenseur f^{μ} se réduit dans ce cas à son premier terme

(13) $$f^{\mu}=\sum_{\sigma}\frac{\partial\mathfrak{T}^{\mu\sigma}}{\partial x_{\sigma}}$$

---

[211] Le manuscrit attribue le numéro (11) à l'équation qui suit alors qu'une autre équation de ce paragraphe le porte déjà ; il sera fait référence à l'équation qui suit par le numéro (11').

[212] B modification : "tensorielles".

[213] B modification : "tensoriel".

[214] B modification : "tensorielle".

§ 12. ÉQUATIONS GÉNÉRALES DE LA MÉCANIQUE [215] DE [216] GRAVITATION.

[p. 101] Nous avons vu à la fin du § 10 que les équations du champ de gravitation peuvent être mises sous la forme générale (§ 10-12)[217]

$$(1) \qquad T_{\mu\nu} = c_1 R_{\mu\nu} + c_2 R g_{\mu\nu} + c_3 g_{\mu\nu}$$

ou sous la forme contrevariante équivalente (§ 10-12)[218].
Nous [219] proposons maintenant de déterminer les constantes encore arbitraires c_1, c_2, c_3 de telle sorte que les équations

$$(2) \qquad f^\alpha = 0$$

qui expriment en fonction de $\mathcal{T}^{\mu\nu}$ les équations de la dynamique des points libres soient des conséquences de l'équation (1).

Puisque les équations (1) & (1) [*sic*] sont des équations tensorielles il suffira de faire cette démonstration en un point quelconque M en employant un système de coordonnées particulier au moyen duquel la démonstration soit plus facile.

Nous emploierons le système que nous avons utilisé du [*sic*] § 9, et que nous avons désigné par ξ_1, ξ_2, ξ_3, ξ_4. Il sera plus commode ici de le désigner par x_1, x_2, x_3, x_4. Avec les notations actuelles les conditions (§ 9-12 et 13) s'écrivent

$$(3) \qquad g_{\alpha\beta} = g_\alpha^\beta \quad , \quad \frac{\partial g_{\alpha\beta}}{\partial x_\gamma} = 0$$

et

$$(4) \qquad g^{\alpha\beta} = g_\alpha^\beta \quad , \quad \frac{\partial g^{\alpha\beta}}{\partial x_\gamma} = 0$$

Nous avons montré qu'il était toujours possible de [220] les imposer en un point M.
On en déduit, que en M,

$$\sqrt{-g} = 1$$

et que les symboles de Christoffel s'annulent.

[p. 102] Le tenseur f^α se réduit à

$$f^\alpha = \sum_\tau \frac{\partial T^{\alpha\tau}}{\partial x_\tau}$$

---

[215] B ajout : "ET".
[216] B ajout : "la".
[217] Il faut lire bien sûr : (§ 10-13).
[218] Il faut lire bien sûr : (§ 10-14).
[219] B ajout : "nous".
[220] B ajout : "se".

ou encore en employant le tenseur covariant associé à $T^{\alpha\beta}$ (§ 11-5)

$$f^\alpha = \sum_\tau \frac{\partial T_{\alpha\tau}}{\partial x_\tau}$$

à cause de (3).

En remplaçant $T_{\alpha\tau}$ par sa valeur (1) nous devons avoir

$$c_1 \sum_\tau \frac{\partial R_{\alpha\tau}}{\partial x_\tau} + c_2 \sum_\tau \frac{\partial (R g_{\alpha\tau})}{\partial x_\tau} + c_3 \sum_\tau \frac{\partial g_{\alpha\tau}}{\partial x_\tau} = 0$$

Le dernier terme s'annule (3). Le second se transforme de la manière suivante

$$\sum_\tau \frac{\partial (R g_{\alpha\tau})}{\partial x_\tau} = \sum_\tau g_{\alpha\tau} \frac{\partial R}{\partial x_\tau} = \frac{\partial R}{\partial x_\alpha}$$

et peut encore s'écrire par suite de la définition de l'invariant

$$R = \sum_\sigma \sum_\tau g^{\sigma\tau} R_{\sigma\tau} ,$$

$$\frac{\partial R}{\partial x_\alpha} = \sum_\sigma \sum_\tau \frac{\partial}{\partial x_\alpha} (g^{\sigma\tau} R_{\sigma\tau}) = \sum_\tau \frac{\partial R_{\tau\tau}}{\partial x_\alpha}$$

Il s'agit donc de choisir les constantes c_1 et c_2 de telle sorte que

$$c_1 \sum_\tau \frac{\partial R_{\alpha\tau}}{\partial x_\tau} + c_2 \sum_\tau \frac{\partial R_{\tau\tau}}{\partial x_\alpha} = 0 \tag{6}$$

Rappelons la définition [221] de Riemann [222]

$$\begin{aligned} R_{\lambda\mu} = \sum_\sigma \left[-\frac{\partial}{\partial x_\sigma} \begin{Bmatrix} \lambda\,\mu \\ \sigma \end{Bmatrix} + \frac{\partial}{\partial x_\mu} \begin{Bmatrix} \lambda\,\sigma \\ \sigma \end{Bmatrix} \right] \\ + \sum_\sigma \sum_\tau \left[\begin{Bmatrix} \lambda\,\sigma \\ \tau \end{Bmatrix} \begin{Bmatrix} \mu\,\tau \\ \sigma \end{Bmatrix} - \begin{Bmatrix} \lambda\,\mu \\ \sigma \end{Bmatrix} \begin{Bmatrix} \sigma\,\tau \\ \tau \end{Bmatrix} \right] \end{aligned} \tag{7}$$

et celle du symbole de Christoffel

$$\begin{Bmatrix} \lambda\,\mu \\ \nu \end{Bmatrix} = \frac{1}{2} \sum_\sigma g^{\sigma\nu} \left(\frac{\partial g_{\lambda\sigma}}{\partial x_\mu} + \frac{\partial g_{\mu\sigma}}{\partial x_\lambda} - \frac{\partial g_{\lambda\mu}}{\partial x_\sigma} \right) \tag{8}$$

Calculons à quoi se réduit en M l'expression

$$\frac{\partial R_{\lambda\mu}}{\partial x_\nu}$$

lorsqu'on emploie les coordonnées particulières définies plus haut.

---

[221] B ajout : "du tenseur".

[222] B ajout : "contracté (§ 9-20 et 5)".

Les termes provenant de la dérivation des produits de symboles de Christoffel (2° ligne de 7) s'annulent avec ces symboles.

Les autres sont formés au moyen des dérivées secondes des symboles de Christoffel.

Ces dérivées se réduisent (8) et (3 et 4) à

$$\frac{\partial^2}{\partial x_\alpha \partial x_\beta}\left\{\begin{matrix}\lambda\,\mu\\ \nu\end{matrix}\right\} = \frac{1}{2}\left(\frac{\partial^3 g_{\lambda\nu}}{\partial x_\mu \partial x_\alpha \partial x_\beta} + \frac{\partial^3 g_{\mu\nu}}{\partial x_\lambda \partial x_\alpha \partial x_\beta} - \frac{\partial^3 g_{\lambda\mu}}{\partial x_\nu \partial x_\alpha \partial x_\beta}\right)$$

[p. 103] Dans chaque dérivée figurent les cinq indices $\alpha\,\beta\,\lambda\,\mu\,\nu$. Nous pouvons caractériser chaque dérivée par les deux indices du potentiel et écrire en abrégé

$$\frac{1}{2}\frac{\partial^3 g_{\lambda\nu}}{\partial x_\mu \partial x_\alpha \partial x_\beta} = (\lambda\nu)$$

Nous aurons ainsi

$$\frac{\partial^2}{\partial x_\alpha \partial x_\beta}\left\{\begin{matrix}\lambda\,\mu\\ \nu\end{matrix}\right\} = (\lambda\nu) + (\mu\nu) - (\lambda\mu)$$

les cinq indices étant $(\alpha\,\beta\,\lambda\,\mu\,\nu)$.

Avec cette notation

$$\frac{\partial R_{\lambda\mu}}{\partial x_\nu} = \sum_\sigma \left[-\frac{\partial^2}{\partial x_\nu \partial x_\sigma}\left\{\begin{matrix}\lambda\,\mu\\ \sigma\end{matrix}\right\} + \frac{\partial^2}{\partial x_\nu \partial x_\mu}\left\{\begin{matrix}\lambda\,\sigma\\ \sigma\end{matrix}\right\}\right]$$

s'écrit

$$\begin{aligned}\frac{\partial R_{\lambda\mu}}{\partial x_\nu} &= -(\lambda\sigma) - (\mu\sigma) + (\lambda\mu)\\ &\quad + (\lambda\sigma) + (\sigma\sigma) - (\lambda\sigma)\end{aligned}$$

ou[223]

(9) $$\frac{\partial R_{\lambda\mu}}{\partial x_\nu} = -(\lambda\sigma) - (\mu\sigma) + (\lambda\mu) + (\sigma\sigma)$$

Les cinq indices sont

$$\lambda \quad \mu \quad \nu \quad \sigma \quad \sigma$$

Il faut sommer par rapport à σ.

Nous avons à calculer les deux expressions qui figurent dans (6)

$$\sum_\tau \frac{\partial R_{\alpha\tau}}{\partial x_\tau} \quad , \quad \sum_\tau \frac{\partial R_{\tau\tau}}{\partial x_\alpha}$$

Nous obtiendrons la première en posant dans (9)

$$\lambda = \alpha \quad , \quad \mu = \nu = \tau$$

[223] Le manuscrit ne numérote pas l'équation qui suit mais s'y réfère par le numéro (9).

et en sommant par rapport à τ.

$$\sum_\tau \frac{\partial R_{\alpha\tau}}{\partial x_\tau} = -(\alpha\sigma) - (\tau\sigma) + (\alpha\tau) + (\sigma\sigma)$$

Les cinq indices sont $\alpha\,\sigma\,\sigma\,\tau\,\tau$. On peut échanger les deux indices sommatoires σ et τ.

$$(\alpha\sigma) = (\alpha\tau)$$

En réduisant les termes semblables, il reste donc

$$\sum_\tau \frac{\partial R_{\alpha\tau}}{\partial x_\tau} = -(\tau\sigma) + (\sigma\sigma) \tag{10}$$

La seconde expression se calcule en posant dans (9).

$$\nu = \alpha \qquad \lambda = \mu = \tau$$

et en sommant par rapport à τ.

$$\sum_\tau \frac{\partial R_{\tau\tau}}{\partial x_\alpha} = -(\tau\sigma) - (\tau\sigma) + (\tau\tau) + (\sigma\sigma)$$

[p. 104] Les cinq indices sont encore $\alpha\,\sigma\,\sigma\,\tau\,\tau$. On a

$$(\sigma\sigma) = (\tau\tau)$$

et par conséquent

$$\sum_\tau \frac{\partial R_{\tau\tau}}{\partial x_\alpha} = -2(\tau\sigma) + 2(\sigma\sigma) \tag{11}$$

L'équation s'écrit donc (10 et 11)

$$c_1 \sum_\tau \frac{\partial R_{\alpha\tau}}{\partial x_\tau} + c_2 \sum_\tau \frac{\partial R_{\tau\tau}}{\partial x_\alpha} = (c_1 + 2c_2)\,[-(\tau\sigma) + (\sigma\sigma)] = 0$$

On y satisfait identiquement en posant

$$c_1 + 2c_2 = 0 \tag{12}$$

Moyennant cette condition les équations (2) sont une conséquence de l'équation (1) lorsqu'on emploie le système particulier que nous avons utilisé dans la démonstration. Il en sera encore de même lorsqu'on emploie un système de coordonnées quelconque. Car lorsque toutes les composantes d'un tenseur s'annulent en un point dans un système de coordonnées les composantes du tenseur s'annulent encore en ce point lorsqu'on emploie un système quelconque de coordonnées.

Les équations de la gravitation doivent donc être de la forme

$$T_{\mu\nu} = c_1 \left(R_{\mu\nu} - \frac{1}{2} g_{\mu\nu} R \right) + c_3 g_{\mu\nu}$$

On peut résoudre ces équations par rapport à $R_{\mu\nu}$.
On à [*sic*] en effet en observant qui [*sic*]

$$\sum_\mu \sum_\nu g_{\mu\nu} g^{\mu\nu} = \sum_\mu g_\mu^\mu = 4 \quad ,$$

$$T = \sum_\mu \sum_\nu g^{\mu\nu} T_{\mu\nu} = c_1(R - 2R) + 4c_3 = -c_1 R + 4c_3$$

On entre qui [*sic*]

(13) $$R_{\mu\nu} = -\kappa \left(T_{\mu\nu} - \frac{1}{2} g_{\mu\nu} T \right) + \lambda g_{\mu\nu} \, ,$$

en posant

$$\kappa = -\frac{1}{c_1} \qquad \text{et} \qquad \lambda = \frac{c_3}{c_1}$$

C'est sous cette forme que l'on écrit habituellement les équations générales de la gravitation. On en déduit comme nous l'avons vu les équations générales de la dynamique.

(14) $$f^\mu \equiv \sum_\sigma \frac{\partial \mathfrak{T}^{\mu\sigma}}{\partial x_\sigma} + \sum_\sigma \sum_\tau \left\{ \begin{matrix} \sigma\tau \\ \mu \end{matrix} \right\} \mathfrak{T}^{\sigma\tau} = 0$$

[p. 105] Nous allons étudier dans le paragraphe 13 les conséquences des équations de la gravitation lorsqu'on y pose.

$$\lambda = 0$$

Nous verrons ensuite quelles modifications s'introduisent lorsque cette constante n'est pas nulle.

§ 13. APPLICATIONS ASTRONOMIQUES ;

[p. 106] Il est toujours possible de choisir un système de coordonnées tel qu'en un point M, ds^2 se réduise à

$$ds^2 = -dx^2 - dy^2 - dz^2 + dt^2 \tag{1}$$

les dérivées des $g_{\mu\nu}$ soient nulles, et les dalembertiens [*sic*] des $g_{\mu\nu}$ se réduisent à

$$\Box g_{\mu\nu} = 2R_{\mu\nu} \tag{2}$$

Ces conditions peuvent toujours être réalisées, avec une certaine approximation dans un domaine.

Lorsqu'on les suppose réalisées dans l'ensemble du système solaire on obtient une approximation qui correspond à celle de la mécanique classique.

En recherchant une seconde approximation on parvient à expliquer sans hypothèse particulière le mouvement du périhelie [*sic*] de Mercure dont la théorie classique ne rendait pas compte.

L'équation (1) exprime que la géométrie est euclidienne et que les coordonnées x, y, z peuvent être considérées comme des coordonnées cartésiennes par rapport à trois axes rectangulaires. Le temps t est le temps marqué par un chronomètre immobile, il sera mesuré par exemple par la rotation de la terre par rapport aux étoiles. L'unité de temps est choisie de telle sorte que l'unité de même[224] soit la vitesse de propagation de la lumière dans le vide. C'est le temps nécessaire à la lumière pour parcourir l'unité de longueur.

Les dérivées des $g_{\mu\nu}$ sont nulles approximativement. Plus elles sont petites plus grand est le domaine ou [*sic*] l'équation (1) et les conséquences qui s'en déduisent sont acceptables.

Nous admettrons que l'équation (1) ainsi que l'annulation des dérivées des $g_{\mu\nu}$ sont vérifiées exactement, à la limite, lorsqu'on s'éloigne indéfiniment de toute masse.

[p. 107] Les équations (2) se calculent par la formule (§ 10 et[225] 12)

$$g_{\mu\nu}(x,y,z,t) = -\frac{1}{2\pi}\int \frac{R_{\mu\nu}(\xi,\eta,\zeta,t-r)}{r}\, d\xi\, d\eta\, d\zeta$$

ou en tenant compte des équations de la gravitation

$$R_{\mu\nu} = -\kappa\left(T_{\mu\nu} - \frac{1}{2}g_{\mu\nu}T\right) \tag{3}$$

$$g_{\mu\nu} = \frac{\kappa}{2\pi}\int \frac{T_{\mu\nu} - \frac{1}{2}g_{\mu\nu}T}{r}\, d\xi\, d\eta\, d\zeta \tag{4}$$

224 B modification : "vitesse".

225 B modification : "–".

On pourra généralement négliger le retard dans le calcul des potentiels, la vitesse des astres étant petite par rapport à la vitesse unité (vitesse de la lumière).

Rappelons la définition de $T_{\mu\nu}$[226]

$$T_{\mu\nu} = \sum_\sigma \sum_\tau g_{\mu\sigma} g_{\nu\tau} T^{\sigma\tau}$$
$$= \frac{\rho}{\sqrt{-g}} \sum_\sigma \sum_\tau g_{\mu\sigma} g_{\nu\tau} \frac{\partial x_\sigma}{ds} \frac{dx_\tau}{ds}$$

Les vitesses des astres étant petites, nous pourrons poser

$$\frac{dx_1}{ds} = v_x, \quad \frac{dx_2}{ds} = v_y, \quad \frac{dx_3}{ds} = v_z, \quad \frac{dx_4}{ds} = \sqrt{1 - v^2}$$

En particulier lorsque les vitesses v_x, v_y, v_z des astres sont nulles ou négligeables, $T_{\mu\nu}$ se réduit à

$$T_{\mu\nu} = \frac{\rho}{\sqrt{-g}} g_{\mu 4} g_{\nu 4}$$

Si κ est une quantité petite, et nous verrons de suite qu'il en est bien ainsi, nous pouvons dans la résolution des équations[227] (4) remplacer sous le signe intégrale $g_{\mu\nu}$ par les valeurs approchées tirées de (1) que nous appelons $\delta_{\mu\nu}$.

$$\delta_{\mu\nu} = \begin{matrix} -1 & 0 & 0 & 0 \\ 0 & -1 & 0 & 0 \\ 0 & 0 & -1 & 0 \\ 0 & 0 & 0 & 1 \end{matrix}$$

Tous les $T_{\mu\nu}$ sont alors nuls sauf

$$T_{44} = \rho$$

L'invariant T est égal à

$$T = \frac{\rho}{\sqrt{-g}} \cong \rho$$

En posant

(5) $$\varpi = \frac{\kappa}{4\pi} \int \frac{\rho}{r} d\xi \, d\eta \, d\zeta$$

les équations (4) ont pour solution

(6) $$g_{\mu\nu} = \delta_{\mu\nu} + g_\mu^\nu \varpi$$

---

226 Dans les équations qui suivent, il faut lire bien sûr :
$$T_{\mu\nu} = \sum_\sigma \sum_\tau g_{\mu\sigma} g_{\nu\tau} T^{\sigma\tau} = \frac{\rho}{\sqrt{-g}} \sum_\sigma \sum_\tau g_{\mu\sigma} g_{\nu\tau} \frac{dx_\sigma}{ds} \frac{dx_\tau}{ds}.$$

227 Remarquons que le tapuscrit qui portait, comme c'est souvent le cas, "équatations" a, en cet endroit, été à nouveau corrigé au crayon et non à l'encre, sans que l'on puisse affirmer, cette fois, qu'il s'agisse d'une autre main que celle de G. Lemaître.

où g_μ^ν désigne toujours

$$g_\mu^\nu = \begin{cases} 1 & \text{si} \quad \mu = \nu \\ 0 & \text{si} \quad \mu \neq \nu \end{cases}$$

Le déterminant des $g_{\mu\nu}$ s'écrira aussi[228]

$$g = \begin{vmatrix} -1+\varpi & 0 & 0 & 0 \\ 0 & -1+\varpi & 0 & 0 \\ 0 & 0 & -1+\varpi & 0 \\ 0 & 0 & 0 & 1+\varpi \end{vmatrix} \cong -1+2\varpi$$

[p. 108] Les $g_{\mu\nu}$ sont égaux aux mineurs des éléments correspondants de ce déterminant, divisés par la valeur du déterminant. Ils sont donnés par le tableau ($\frac{1}{g} \cong -1-2\varpi$).

$$\begin{matrix} g^{\mu\nu} = -1-\varpi & 0 & 0 & 0 \\ 0 & -1-\varpi & 0 & 0 \\ 0 & 0 & -1-\varpi & 0 \\ 0 & 0 & 0 & 1-\varpi \end{matrix}$$

c'est à dire [*sic*]

(7) $$g^{\mu\nu} = \delta_{\mu\nu} - g_\mu^\nu \varpi$$

Enfin

(8) $$\sqrt{-g} = 1 - \varpi$$

Les équations du mouvement d'un point libre se calculeront par l'équation (§ 5-15), ($i = 1, 2, 3$)

(9) $$\frac{d^2 x_i}{dx_4^2} + \sum_\mu \sum_\nu \left[\left\{ \begin{matrix} \mu\,\nu \\ i \end{matrix} \right\} - \left\{ \begin{matrix} \mu\,\nu \\ 4 \end{matrix} \right\} \frac{dx_i}{dx_4} \right] \frac{dx_\mu}{dx_4} \frac{dx_\nu}{dx_4} = 0$$

où il faudra remplacer les coordonnées x_1, x_2, x_3, x_4 respectivement par x, y, z, t.

Les symboles de Christoffel se calculent par les formules (§ 5-13)

(10) $$\begin{aligned} \left\{ \begin{matrix} \sigma\,\sigma \\ \tau \end{matrix} \right\} &= -\frac{1}{2} \frac{1}{g_{\tau\tau}} \frac{\partial g_{\sigma\sigma}}{\partial x_\tau} \cong -\frac{1}{2} \delta_{\tau\tau} \frac{\partial \varpi}{\partial x_\tau} \qquad (\sigma \neq \tau) \\ \left\{ \begin{matrix} \sigma\,\tau \\ \sigma \end{matrix} \right\} &= \frac{1}{2} \frac{1}{g_{\sigma\sigma}} \frac{\partial g_{\sigma\sigma}}{\partial x_\tau} \cong \frac{1}{2} \delta_{\sigma\sigma} \frac{\partial \varpi}{\partial x_\tau} \end{aligned}$$

228 B modification : "ainsi".

Les équations du mouvement sera [*sic*] ainsi[229]

$$(11) \qquad \frac{d^2x}{dt^2} + \frac{1}{2}\left(1 - 3v_x^2 + v_y^2 + v_z^2\right)\frac{\partial\varpi}{\partial x} - v_x v_y \frac{\partial\varpi}{\partial y} - v_x v_z \frac{\partial\varpi}{\partial z} - \frac{1}{2} v_x \frac{\partial\varpi}{\partial t} = 0$$

et les équations analogues.

Telles sont les équations approchées du mouvement, lorsque les masses produisant le champ sont immobiles et qu'on adopte l'approximation indiquée au début de ce paragraphe.

L'accélération est fonction de la vitesse du mobile. Cette vitesse est écrite en prenant la vitesse de la lumière comme unité, elle est très petite. Le terme principal est donc bien le terme indépendant des vitesses.

$$\frac{d^2x}{dt^2} + \frac{1}{2}\frac{\partial\varpi}{\partial x} = 0$$

[p. 109] ϖ est proportionnel au potentiel newtonnien [*sic*]. Cette équation est équivalente à l'équation (§ 10-1). ϖ doit être le double du potentiel newtonien. Il faut pour cela choisir la constante κ de telle sorte que

$$\frac{\kappa}{4\pi} = 2\frac{K}{c^2}$$

[230] étant la valeur de la constante de l'attraction (§ 10 et 2) lorsqu'on choisit l'unité de temps comme nous l'avons fait ici.

Nous aurons ainsi

$$(12) \qquad \kappa = \frac{8\pi K}{c^2} = 1{,}87 \cdot 10^{-27}$$

qui est bien une quantité petite.

Pour trouver les équations du mouvement en unités C.G.S. il suffit de multiplier les équations du mouvement par c^2.

La formule (11) est une formule approchée. Pour obtenir une meilleure approximation, nous pouvons nous contenter de l'approximation déjà obtenue pour tous les termes de (9) contenant les quantités petites v_x, v_y, v_z, le seul dont nous devons obtenir une valeur plus approchée est donc

$$\left\{ \begin{matrix} 44 \\ i \end{matrix} \right\} = \sum_\sigma g^{i\sigma} \left[\begin{matrix} 44 \\ \sigma \end{matrix} \right] \cong (1+\varpi)\left(-\frac{\partial g_{4i}}{\partial x_4} + \frac{1}{2}\frac{\partial g_{44}}{\partial x_i}\right)$$

$\frac{\partial g_{4i}}{\partial x_4}$ est nulle en première approximation puisque g_{4i} s'annule en première approximation. Les variations des potentiels avec le temps doivent être petites puisque

[229] C correction (voir Annexe 2, p. 249) : "formule (11), lire
$$\frac{d^2x}{dt^2} + \frac{1}{2}(1 - 3v_x^2 + v_y^2 + v_z^2)\frac{\partial\varpi}{\partial x} - 2v_x v_y \frac{\partial\varpi}{\partial y} - 2v_x v_z \frac{\partial\varpi}{\partial z} - \frac{1}{2}v_x(3 - v^2)\frac{\partial\varpi}{\partial t} = 0."$$

[230] Un espace destiné à être complété à la main semble être resté vide.

les vitesses des astres peuvent être supposées petites. Si nous envisageons le cas ordinaire [231] quasi stationnaire, nous pourrions dans le calcul de l'approximation actuelle négliger les dérivées des potentiels par rapport au temps. Il nous reste donc à trouver une meilleure approximation de $\frac{\partial g_{44}}{\partial x_i}$ en négligeant les dérivées par rapport à x_4. Rappelons la définition de R_{44}.

$$R_{44} = \sum_\sigma \left[-\frac{\partial}{\partial x_\sigma} \left\{ \begin{matrix} 44 \\ \sigma \end{matrix} \right\} + \frac{\partial}{\partial x_4} \left\{ \begin{matrix} 4\sigma \\ \sigma \end{matrix} \right\} \right]$$
$$+ \sum_\sigma \sum_\tau \left[\left\{ \begin{matrix} 4\sigma \\ \tau \end{matrix} \right\} \left\{ \begin{matrix} 4\tau \\ \sigma \end{matrix} \right\} - \left\{ \begin{matrix} \sigma\tau \\ \sigma \end{matrix} \right\} \left\{ \begin{matrix} 44 \\ \tau \end{matrix} \right\} \right]$$

[p. 110] Le second terme est nul si nous supposons que les dérivées par rapport à x_4 s'annulent.

Calculons les autres en considérant ϖ comme une quantité petite du premier ordre et en négligeant les termes d'ordre supérieur au second.

Nous aurons

$$\frac{\partial}{\partial x_4} = 0$$
$$\left\{ \begin{matrix} 44 \\ \sigma \end{matrix} \right\} = -\frac{1}{2} \sum_\tau g^{\sigma\tau} \frac{\partial g_{44}}{\partial x_\tau}$$

et, par conséquent[232]

$$-\sum_\sigma \frac{\partial}{\partial x_\sigma} \left\{ \begin{matrix} 44 \\ \sigma \end{matrix} \right\} = \frac{1}{2} \sum_\sigma \sum_\tau g^{\sigma\tau} \frac{\partial^2 g_{44}}{\partial x_\sigma \partial x_\tau} + \frac{1}{2} \sum_\sigma \sum_\tau \frac{\partial g^{\sigma\tau}}{\partial x_\sigma} \frac{\partial g_{44}}{\partial x_\sigma}$$

Nous aurons en première approximation

$$g^{\sigma\tau} = -g_\sigma^\tau (1+\varpi) \qquad (\sigma, \tau \neq 4)$$
$$g^{44} = 1 + \varpi$$

Nous pouvons donc écrire

$$-\sum_\sigma \frac{\partial}{\partial x_\sigma} \left\{ \begin{matrix} 44 \\ \sigma \end{matrix} \right\} = -\frac{1}{2} (1+\varpi) \sum_\sigma \frac{\partial^2 g_{44}}{\partial x_\sigma^2} - \frac{1}{2} \sum_\sigma \left(\frac{\partial \varpi}{\partial x_\sigma} \right)^2$$

Pour calculer les termes de la seconde ligne du développement de R_{44} nous avons à remplacer les ~~symboles de Christoffel par leur valeur en première approximation.~~
~~Remarquons que les — sont nuls sauf~~
~~et~~
~~Il vient alors~~

---

[231] B ajout : "d'un champ".

[232] Dans l'équation qui suit, il faut lire bien sûr :

$$-\sum_\sigma \frac{\partial}{\partial x_\sigma} \left\{ \begin{matrix} 44 \\ \sigma \end{matrix} \right\} = \frac{1}{2} \sum_\sigma \sum_\tau g^{\sigma\tau} \frac{\partial^2 g_{44}}{\partial x_\sigma \partial x_\tau} + \frac{1}{2} \sum_\sigma \sum_\tau \frac{\partial g^{\sigma\tau}}{\partial x_\sigma} \frac{\partial g_{44}}{\partial x_\tau}.$$

~~d'autre part (   )~~
~~il vient donc~~
~~On obtient donc finalement~~
~~Résolvons cette équation par rapport à~~
~~Nous aurons à l'approximation actuelle~~
~~Et en remplaçant   par sa valeur~~[233]
[p. 111] symboles de Christoffel par leurs valeurs en première approximation (10).
Remarquons que les $\left\{ \begin{matrix} 4\sigma \\ \tau \end{matrix} \right\}$ sont nuls sauf

$$\left\{ \begin{matrix} 44 \\ \tau \end{matrix} \right\} = \frac{1}{2} \frac{\partial \varpi}{\partial x_\tau} \qquad (\tau \neq 4)$$

et

$$\left\{ \begin{matrix} 4\sigma \\ 4 \end{matrix} \right\} = \frac{1}{2} \frac{\partial \varpi}{\partial x_\sigma} \qquad (\sigma \neq 4)$$

Il vient alors

$$\sum_\sigma \sum_\tau \left\{ \begin{matrix} 4\sigma \\ \tau \end{matrix} \right\} \left\{ \begin{matrix} 4\tau \\ \sigma \end{matrix} \right\} = \frac{1}{2} \sum_\sigma \left(\frac{\partial \varpi}{\partial x_\sigma} \right)^2$$

D'autre part (§ 5-12)

$$\sum_\sigma \left\{ \begin{matrix} \sigma\tau \\ \sigma \end{matrix} \right\} = \frac{1}{\sqrt{-g}} \frac{\partial \sqrt{-g}}{\partial x_\tau} = -\frac{\partial \varpi}{\partial x_\tau} \,;$$

il vient donc

$$-\sum_\sigma \sum_\tau \left\{ \begin{matrix} \sigma\tau \\ \sigma \end{matrix} \right\} \left\{ \begin{matrix} 44 \\ \tau \end{matrix} \right\} = -\frac{1}{2} \sum_\sigma \left(\frac{\partial \varpi}{\partial x_\sigma} \right)^2$$

On obtient donc finalement

$$R_{44} = -\frac{1}{2}(1+\varpi) \sum_\sigma \frac{\partial^2 g_{44}}{\partial x_\sigma^2} - \frac{1}{2} \sum_\sigma \left(\frac{\partial \varpi}{\partial x_\sigma} \right)^2$$

Résolvons cette équation par rapport à

$$\sum_\sigma \frac{\partial^2 g_{44}}{\partial x_\sigma^2} = \Box g_{44}$$

Nous aurons à l'approximation actuelle

$$\Box g_{44} = -2(1+\varpi) R_{44} + \sum_\sigma \left(\frac{\partial \varpi}{\partial x_\sigma} \right)^2$$

[233] B suppression : texte rayé.

Et, en remplaçant R_{44} par sa valeur

$$R_{44} = \kappa\left(T_{44} - \frac{1}{2}T\right) = -\frac{1}{2}\kappa\rho$$

on trouve finalement

$$\Box g_{44} = -\kappa(1-\varpi)\rho + \sum_{\sigma}\left(\frac{\partial\varpi}{\partial x_{\sigma}}\right)^2$$

On obtiendra donc g_{44} sous la forme

$$g_{44} = 1 + \varpi$$

en remplaçant dans le calcul de ϖ (5) la densité ρ par une densité fictive

$$\rho' = (1-\varpi)\rho - \frac{1}{\kappa}\sum_{\sigma}\left(\frac{\partial\varpi}{\partial x_{\sigma}}\right)^2$$

En désignant par V le potentiel newtonien en unités C.G.S. et par g l'accélération correspondante

$$g^2 = \left(\frac{\partial V}{\partial x}\right)^2 + \left(\frac{\partial V}{\partial y}\right)^2 + \left(\frac{\partial V}{\partial z}\right)^2,$$

[p. 112] cette expression peut s'écrire

$$\rho' = \left(1 - \frac{2V}{c^2}\right)\rho^2 - \frac{1}{\kappa c^4}g^2$$

La constante

$$\frac{1}{\kappa c^4} = 0{,}66 \cdot 10^{-15}$$

Le terme en g^2 est le seul terme que nous ayons rencontré jusqu'à présent où apparaisse le caractère non homogène des équations de la gravitation. Le fait que ces équations ne sont pas homogènes a pour conséquence que l'accélération d'un mobile dans le champ de plusieurs masses n'est pas exactement la résultante des accélérations que lui communiquerait chacune des masses si elle agissait seule. La différence est très faible mais elle pourrait se faire sentir dans certains phénomènes, tels le mouvement du périhélie des planètes.

Nous avons ainsi trouvé les équations du manquement[234] des astres avec une approximation qui dépasse[235] amplement suffisante pour les applications astronomiques. Ce sont l'équation (11) et les équations analogues où on doit calculer ϖ en employant la densité fictive ρ' au lieu de la densité réelle ρ.

* * *

[234] B modification : "mouvement".

[235] B suppression : "qui dépasse".

Nous avons effectué ce calcul en supposant négligeable l'influence de la vitesse et de la rotation des masses attirantes. Il est facile de tenir compte de cette influence. [236, 237]

$$\begin{array}{llll} T_{\mu\nu} = \rho\, v_x^2 & , \rho\, v_x v_y & , \rho\, v_x v_z & , \rho\, v_x \\ \rho\, v_y v_x & , \rho\, v_y^2 & , \rho\, v_y v_z & , \rho\, v_y \\ \rho\, v_z v_x & , \rho\, v_z v_y & , \rho\, v_z^2 & , \rho\, v_z \\ \rho\, v_x & , \rho\, v_y & , \rho\, v_z & , \rho\,(1-v^2) \end{array}$$

On peut calculer séparément l'intégrale (4) pour chaque astre.
Si nous désignons par v_{0x}, v_{0y}, v_{0z} la vitesse du centre de gravité de l'astre et par ω_x, ω_y, ω_z sa rotation nous pourrons écrire [p. 113]

$$v_x = v_{0x} + z\,\omega_y - y\,\omega_z$$
$$v_y = v_{0y} + x\,\omega_z - z\,\omega_x$$
$$v_z = v_{0z} + y\,\omega_x - x\,\omega_y$$

où x, y, z sont les coordonnées des points par rapport à des axes passant par le centre de gravité.

Désignons par $d\Pi$ l'élément de volume. On aura

$$\int \rho\, v_x^2\, d\Pi = M\, v_{0x}^2 + \frac{I}{2}\left(\omega_y^2 + \omega_z^2\right)$$

où M est la masse de l'astre et I son mouvement[238] d'inertie par rapport à un axe passant par le centre de gravité. (Nous supposons que l'astre a une symétrie sphérique).

Nous aurons de même

$$\int \rho\, v_x v_y\, d\Pi = M\, v_{0x} v_{0y} - \frac{I}{2}\,\omega_x \omega_y$$

Les valeurs de

$$\int T_{\mu\nu}\, d\Pi$$

sont alors données pour chaque astre par le tableau suivant (nous avons supprimé les indices zéro des vitesses)

[236] B ajout : "Le tenseur $T_{\mu\nu}$ peut s'écrire".

[237] C correction (voir Annexe 2, p. 249) : "dernière formule, lire

$$\begin{array}{lllll} T_{\mu\nu} = & \rho v_x^2 & , \cdots & , \cdots & , -\rho v_x \\ & \cdots & , \cdots & , \cdots & , -\rho v_y \\ & \cdots & , \cdots & , \cdots & , -\rho v_z \\ & -\rho v_x & , -\rho v_y & , -\rho v_z & , \rho(1+v^2) \end{array}"$$

[238] B modification : "moment".

$$\begin{array}{llll}
Mv_x^2 + \frac{I}{2}(\omega_y^2 + \omega_z^2) & , Mv_y v_x - \frac{I}{2}\omega_y\omega_x & , Mv_z v_x - \frac{I}{2}\omega_z\omega_x & , Mv_x \\
Mv_x v_y - \frac{I}{2}\omega_x\omega_y & , Mv_y^2 + \frac{I}{2}(\omega_x^2 + \omega_z^2) & , Mv_z v_y - \frac{I}{2}\omega_z\omega_y & , Mv_y \\
Mv_x v_z - \frac{I}{2}\omega_x\omega_z & , Mv_y v_z - \frac{I}{2}\omega_y\omega_z & , Mv_z^2 + \frac{I}{2}(\omega_x^2 + \omega_y^2) & , Mv_z \\
Mv_x & , Mv_y & , Mv_z & , M(1 - v^2) - I\omega^2
\end{array}$$

On aura ainsi par exemple

$$g_{11} = -1 + \varpi + \frac{\kappa}{2\pi}\sum \frac{Mv_x^2 + \frac{I}{2}(\omega_y^2 + \omega_z^2)}{r}$$

$$g_{12} = \frac{\kappa}{2\pi}\sum \frac{Mv_x v_y - \frac{I}{2}\omega_x\omega_y}{r}$$

$$g_{14} = \frac{\kappa}{2\pi}\sum \frac{Mv_x}{r}$$

$$g_{14} = 1 + \varpi - \frac{\kappa}{2\pi}\sum \frac{Mv^2 + I\omega^2}{r}$$

On formerait les autres expressions analogues par permutation des indices x, y, z.

[p. 114] **Conséquences géométriques.**

Si nous nous en tenons au cas où nous pouvons négliger l'influence de la vitesse des masses attirantes, la forme fondamentale s'écrit

$$ds^2 = (-1 + \varpi)(dx^2 + dy^2 + dz^2) + (1 + \varpi)dt^2$$

La géométrie correspondante est caractérisée par l'élément de distance

$$d\sigma^2 = (1 - \varpi)(dx^2 + dy^2 + dz^2)$$

On voit que la géométrie n'est pas euclidienne.

On peut faire la carte de l'espace dans un espace euclidien. En considérant x, y, z comme les coordonnées cartésiennes de l'espace euclidien dans lequel on fait la carte, on voit que l'échelle est indépendante de la direction des longueurs et égale à

$$\frac{1}{\sqrt{1 - \varpi}} \cong 1 + \frac{1}{2}\varpi$$

ou, d'après la signification de $\varpi = \frac{2V}{c^2}$,

$$1 + \frac{V}{c^2}$$

où V est le potentiel newtonien en unités C.G.S.

On peut interprèter [*sic*] ce résultat, en disant que la géométrie est euclidienne mais que les mètres subissent une déformation proportionnelle à

$$\sqrt{1-\varpi} \cong 1-\frac{V}{c^2}$$

Naturellement, on pourrait faire d'autres interprétations en se servant, non d'une carte conforme comme nous l'avons fait ici, mais de tout autre carte.

Ainsi dans le cas où il n'y a qu'une seule masse attirante,

$$\varpi = -2K\frac{m}{r}$$

[p. 115] on aura en utilisant des coordonnées polaires

$$d\sigma^2 = \left(1+\frac{2Km}{r}\right)\left[dr^2 + r^2\left(\sin^2\theta\, d\varphi^2 + d\theta^2\right)\right]$$

Faisons le changement de coordonnées

$$\rho^2 = r^2\left(1+\frac{2Km}{r}\right)$$

d'où on tire

$$\rho\, d\rho = r\left(1+\frac{Km}{r}\right)dr$$

et

$$\left(1+\frac{2Km}{r}\right)dr^2 \cong \frac{\rho^2}{r^2}d\rho^2 \cong \frac{d\rho^2}{1-\frac{2Km}{\rho}}$$

il veut[239]

$$d\sigma^2 = \frac{d\rho^2}{1-\frac{2Km}{\rho}} + \rho^2\left(\sin^2\theta\, d\varphi^2 + d\theta^2\right)$$

On voit que ρ est alors l'élément de longueur dans une direction normale au rayon vecteur.

On peut dire que la géométrie est euclidienne mais que les mètres subissent une dilatation suivant le rayon vecteur proportionnellement à

$$\frac{1}{\sqrt{1-\frac{2Km}{\rho}}}$$

[239] B modification : "il vient".

Dans le cas d'une masse unique au repos, Schwarzschild a trouvé la valeur exacte de ds^2

$$ds^2 = -\frac{dr^2}{1+\varpi} - r^2\left(\sin^2\theta\, d\varphi^2 + d\theta^2\right) + (1+\varpi)\,dt^2$$

$$\varpi = -2\frac{K}{c^2}\frac{m}{r}$$

Si nous supposons que la masse s'éloigne indéfiniment suivant l'axe des x négatifs, nous obtenons à la limite, r et m tendant vers l'infini,

$$ds^2 = -\frac{dx^2}{1+\varpi} - dy^2 - dz^2 + (1+\varpi)\,dt^2$$

$$\varpi = \frac{2gx}{c^2} \quad , \quad g = \lim \frac{Km}{r^2}$$

C'est la forme que nous avons trouvée (§ 7-9) en étudiant le champ de gravitation artificiel obtenu par un mouvement uniformément accéléré.
Nous voyons que ce champ est rigoureusement équivalent au champ produit par une masse infinie, infiment [*sic*] éloignée.

[p. 116] **<u>Pesanteur des rayons lumineux</u>**
Les rayons lumineux sont des géodésiques qui annulent ds.
On voit que leur vitesse, indépendante de la direction, varie d'un point à l'autre et est égale à

$$\frac{d\sigma}{dt} = \sqrt{1+\varpi} \cong 1 + \frac{K}{c^2}V$$

Il en résulte que les rayons lumineux ne se propagent pas en ligne droite. Comparons l'accélération subie suivant l'axe des x par un rayon lumineux de vitesse

$$v_x = 0 \quad , \quad v_y = 1 \quad , \quad v_z = 0$$

à celle qui [*sic*] subit au même point une masse matérielle au repos.
La formule (11) donne pour le rayon

$$\frac{d^2x}{dt^2} + \frac{\partial\varpi}{\partial x} = 0$$

et pour la masse

$$\frac{d^2x}{dt^2} + \frac{1}{2}\frac{\partial\varpi}{\partial x} = 0$$

L'accélération du rayon lumineux normalement à sa direction de propagation est donc le double de celle qui[240] subirait une masse en repos au même point. On sait que les observations faites durant l'éclipse du 29 Mai 1919 confirment ces conclusions.

---

[240] B modification : "que".

Déplacement vers le rouge des raies du spectie [*sic*] solaire
Le temps propre sur les différents astres est égal à

$$ds = \sqrt{1+\varpi} \cdot dt \cong \left(1 + \frac{V}{c^2}\right) dt$$

Des chronomètres identiques marquent le même intervalle [241] entre le commencement et la fin de chacune des périodes semblables par lesquelles ils mesurent le temps ; l'intervalle *dt* correspondant sera donc différent en des endroits où le potentiel est différent.
Si on admet que les éléments chimiques ont une constitution identique sur le soleil et la terre, nous devons admettre que la période des longueurs d'ondes émises par ces éléments sur le soleil sont plus lentes dans le rapport[242]

$$1 + \frac{V_T - V_S}{c^2}$$

[p. 117] La vérification expérimentale de cette conclusion est rendue très difficile par suite de certaines anomalies qui se présentent dans l'observation des raies solaires et surtout par suite de la difficulté de déterminer quelle influence a la pression sur les déplacements observés. Quoi qu'il en soit[243] le résultat des recherches, sans être peut-être définitif, est nettement favorable [244] conclusions de la nouvelle théorie.

Rayé[245]
Rayé[246]

[241] Un espace destiné à être complété semble être resté vide.

[242] En raison du nombre de modifications apportées par B, donnons ce que devient cette phrase en imprimant en caractères italiques les différences : "Si on admet que les éléments chimiques ont une constitution identique sur le soleil et *sur* la terre, *on doit conclure* que *les* longueurs d'onde *caractéristiques de* ces éléments *sont plus grandes* sur le soleil dans le rapport".

[243] B modification : "Quoi qu'il en soit" remplacé par "Cependant".

[244] B ajout : "aux".

[245] Nous croyons pouvoir lire : "Louvain, le 30 mai 1922,"

[246] Nous croyons pouvoir lire : "G. Lemaître"

§ 14. LES ÉTOILES FIXES.

[p. 118] Lorsqu'on veut appliquer la théorie de la gravitation de Newton à l'ensemble des étoiles, on rencontre [247] difficultés auxquelles la théorie nouvelle donne une solution satisfaisante.

On peut supposer tout d'abord que les étoiles sont en nombre fini. La densité de l'univers stellaire tend alors vers zéro lorsqu'on s'éloigne indéfiniment. Cette conception donne lieu à plusieurs difficultés. La lumière émise par les étoiles s'éloigne sans retour. L'énergie des étoiles se dissipe peu à peu vers l'infini. De plus il paraît bien difficile de se rendre compte de la stabilité de l'univers. On peut comparer la multitude des étoiles aux molécules d'un gaz. Les forces centrales qui agissent entre les étoiles sont analogues à celles que suppose entre les molécules la théorie cinétique des gaz. Il est impossible d'admettre la stabilité d'une masse gazeuse qui n'est pas contenue par une paroi et dont la densité tend vers zéro, lorsqu'on s'éloigne à l'infini ; il n'est pas plus possible d'admettre une solution semblable pour l'ensemble des étoiles. On pourrait renoncer à la loi de Newton et supposer que le potentiel croit sans limite à l'infini. Les étoiles qui tiendraient[248] à s'éloigner définitivement seraient ainsi ramenées définitivement[249] vers l'amas des autres. On devrait alors constater que la vitesse[250] des étoiles éloignées sont beaucoup plus grande[251] que celles des étoiles rapprochées. L'observation ne décèle rien de semblable.

Si au contraire on admet qu'en s'éloignant définitivement[252] on rencontre une densité moyenne d'étoiles égale à celle des étoiles que nous observons dans le domaine qui est accessible à nos [p. 119] observations, on suppose nécessairement que les étoiles sont en nombre infini ; la loi de Newton ne peut s'appliquer, on devait[253] constater une force infinie dirigée vers le centre de gravité de l'ensemble des étoiles. On ne voit d'ailleurs pas très bien ce que pourrait être le centre de gravité d'une masse homogène indéfinie, [254] on ne peut naturellement songer à accepter ces forces infinies.

On peut éviter ces dernières difficultés en modifiant la loi de Newton, il suffit de remplacer le potentiel Newtonien V défini par l'équation de Poisson.

$$\square V = 4\pi K \rho \tag{1}$$

où K est la constante de l'attraction et ρ la densité de la matière par un potentiel V défini par l'équation

[247] B ajout : "des".

[248] B modification : "tendraient".

[249] B modification : "nécessairement".

[250] B modification : "les vitesses".

[251] B modification : "grandes".

[252] B modification : "indéfiniment".

[253] B modification : "devrait".

[254] B ajout : "et l' ".

(2) $$\Box V - \lambda V = 4\pi K \rho$$

où λ désigne une constante universelle.

Lorsqu'on considère un domaine contenant un très grand nombre d'étoiles, la densité peut être considérée comme une constante ρ_0 et l'équation modifiée de Poisson admet la solution constante.

$$V = -\frac{4\pi K}{\lambda}\rho_0$$

On peut donc concevoir un univers indéfini où la densité moyenne est constante. Il reste la difficulté qu'il faut supposer un nombre infini d'étoiles.

Quelles modifications devons-nous apporter aux équations tensionelles[255] de la gravitation pour qu'elles se réduisent à la théorie que nous venons de développer lorsqu'on emploie les coordonnées usuelles ?

Les équations :

$$R_{\mu\nu} = -\kappa\left(T_{\mu\nu} - \frac{1}{2}g_{\mu\nu}T\right)$$

se réduisent dans le cas où la matière est en repos à

$$R_{\mu\nu} = -\kappa g_{\mu}^{\nu}\frac{\rho}{\sqrt{-g}}$$

et en employant des coordonnées particulières, on peut écrire

$$\Box g_{\mu\nu} = \frac{R_{\mu\nu}}{2} = -\frac{\kappa}{2}g_{\mu}^{\nu}\rho$$

[p. 120] Ces équations sont équivalentes à l'équation de Poisson sous sa forme habituelle (1). On obtient en première approximation

(3) $$g_{\mu\nu} = \delta_{\mu\nu} + 2V$$

Nous avons vu que les équations de la gravitation peuvent être mises sous une forme plus générale (§ 12-13)

(4) $$R_{\mu\nu} - \lambda g_{\mu\nu} = -\kappa\left(T_{\mu\nu} - \frac{1}{2}g_{\mu\nu}T\right)$$

On voit que la solution approchée de $g_{\mu\nu}$ (3) réduira ces équations à la forme nouvelle de l'équation de Poisson

$$\Box V - \lambda V = 4\pi K \rho$$

Intégrons les nouvelles équations.

[255] B modification : "tensorielles".

Dans un univers de densité constante

$$\rho = \rho_0$$

Désignons toujours par $\delta_{\mu\nu}$ les coefficients de la forme

$$ds^2 = -dx^2 - dy^2 - dz^2 + dt^2$$

Nous allons démontrer que les équations (4) admettent la solution

$$(5) \qquad \begin{cases} g_{44} = 1 \qquad g_{i4} = 0 \\ g_{ij} = -\gamma_{ij} \qquad\qquad (i, j = 1, 2, 3) \end{cases}$$

où les γ_{ij} représentent les potentiels métriques d'un espace sphérique de courbure constante R.

Ces potentiels s'obtiennent en éliminant x_4 entre les équations

$$d\sigma^2 = dx_1^2 + dx_2^2 + dx_3^2 + dx_4^2$$

et

$$R^2 = x_1^2 + x_2^2 + x_3^2 + x_4^2$$

En posant

$$(6) \qquad r^2 = x_1^2 + x_2^2 + x_3^2$$

on trouve

$$(7) \qquad \begin{aligned} d\sigma^2 &= \sum_{\mu}^{3}\sum_{\nu}^{3} \gamma_{\mu\nu}\, dx_\mu dx_\nu \\ \gamma_{\mu\nu} &= g_\mu^\nu + \frac{x_\mu x_\nu}{R^2 - r^2} \end{aligned}$$

Dans le problème actuel tous les points doivent être équivalents. La solution (5) satisfait à cette condition. Celà [*sic*] résulte de la manière dont on calcule les $\gamma_{\mu\nu}$.

Il suffit donc de faire la démonstration pour un point quelconque. Nous choisirons

$$x_1 = x_2 = x_3 = x_4 = 0$$

[p. 121] Les dérivées premières des $g_{\mu\nu}$ s'annulent en ce point. Les dérivées secondes y sont toutes nulles sauf

$$\frac{\partial^2 g_{ij}}{\partial x_i \partial x_j} = -\frac{1}{R^2} \qquad (i, j = 1, 2, 3)$$

Les $R_{\mu\nu}$ se réduisent alors à

$$R_{\mu\nu} = \sum_{\sigma}^{3} \frac{\partial}{\partial x_\sigma} \begin{bmatrix} \mu\nu \\ \sigma \end{bmatrix} - \sum_{\sigma}^{3} \frac{\partial}{\partial x_\nu} \begin{bmatrix} \mu\sigma \\ \sigma \end{bmatrix}$$

On trouve pour $(\mu = \nu = 1,2,3)$

$$R_{\mu\mu} = \frac{1}{2}\frac{1}{R^2}$$

tandis que $R_{\mu\nu}$ s'annule pour $(\mu = \nu = 4)$ et pour $(\mu \neq \nu)$.

Les équations (4) deviennent alors pour $(\mu = \nu = 1,2,3)$

$$\frac{2}{R^2} + \lambda = -\frac{\kappa\rho}{2}$$

et pour $(\mu = \nu = 4)$

$$-\lambda = -\frac{\kappa\rho}{2}$$

Ces équations sont vérifiées si nous posons

$$\lambda = \frac{\kappa\rho}{2} = \frac{1}{R^2} \tag{8}$$

La géométrie des corps immobiles dans le champ est caractérisée par l'élément de distance (7).

Les $\gamma_{\mu\nu}$ sont les potentiels métriques de la géométrie sphérique (géométrie de Riemann au sens étroit). L'espace, quoique sans borne, a pourtant un volume fini. Nous ne sommes donc plus amenés comme précédemment à supprimer[256] infini le nombre des étoiles.

Calculons le volume total de l'espace.

Nous savons qu'il est égal à l'intégrale étendue à tout l'espace (§ 1-8).

$$\int \sqrt{\gamma}\, dx_1 dx_2 dx_3$$

[p. 122] où γ est le déterminant des potentiels $(\gamma_{\mu\nu})$[257].

$$\gamma = \begin{vmatrix} 1 + \frac{x_1^2}{R^2 - r^2} & \frac{x_2 x_1}{R^2 - r^2} & \frac{x_3 x_1}{R^2 - r^2} \\ \frac{x_1 x_2}{R^2 - r^2} & 1 + \frac{x_2^2}{R^2 - r^2} & \frac{x_3 x_2}{R^2 - r^2} \\ \frac{x_1 x_3}{R^2 - r^2} & \frac{x_2 x_3}{R^2 - r^2} & 1 + \frac{x_3^2}{R^2 - r^2} \end{vmatrix}$$

256 B modification : "supposer".

257 B suppression : parenthèses supprimées.

Ce déterminant est égal à

$$\gamma = \frac{R^2}{R^2 - r^2}$$

On peut écrire d'après (6)

$$dx_1\, dx_2\, dx_3 = 4\pi r^2\, dr$$

Le volume total de l'espace est donc

$$\int_0^\infty \frac{4\pi R\, r^2\, dr}{\sqrt{R^2 - r^2}} = 2\pi R^3$$

La masse totale d'un univers sans borne de densité constante ρ est donc

(9) $$M = 2\pi\rho R^3$$

ou d'après (6)

(10) $$M = 4\pi \frac{R}{\kappa} = \sqrt{\frac{32\pi^2}{\kappa^3 \rho}}$$

Chapitre V. Les masses électriques.

[p. 123] La définition du champ électromagnétique par le mouvement qu'il communique à des particules électrisées, a été étudiée au § 6. Nous avons ainsi donné aux lois qui impriment[258] l'action des champs sur les masses une forme indépendante du mode de repérage utilisé pour étudier les phénomènes.

Il est facile de se rendre compte que les quantités que nous avons introduites alors sont des tenseurs.

Nous avons considéré en plus que [*sic*] l'invariant (ds)[259], un autre invariant

$$\varphi_1\, dx_1 + \varphi_2\, dx_2 + \varphi_3\, dx_3 + \varphi_4\, dx_4 = \sum_\alpha \varphi_\alpha\, dx_\alpha$$

Les potentiels électromagnétiques φ_σ se transforment, lorsqu'on fait un changement de coordonnées, par des équations de la forme

$$(1) \qquad \varphi'_\sigma = \sum_\alpha \frac{\partial x_\alpha}{\partial x'_\sigma}\, \varphi_\alpha$$

C'est en cela que consiste le caractère tensoriel d'un covariant φ_σ.
Le champ électro-magnétique

$$(2) \qquad F_{\mu\nu} = \frac{\partial \varphi_\mu}{\partial x_\nu} - \frac{\partial \varphi_\nu}{\partial x_\mu}$$

est aussi un tenseur. On a en effet

$$F'_{\mu\nu} = \frac{\partial \varphi'_\mu}{\partial x'_\nu} - \frac{\partial \varphi'_\nu}{\partial x'_\mu}$$

et d'après (1)

$$\frac{\partial \varphi'_\mu}{\partial x'_\nu} = \sum_\beta \frac{\partial x_\beta}{\partial x'_\nu}\, \frac{\partial}{\partial x_\beta} \sum_\alpha \frac{\partial x_\alpha}{\partial x'_\mu}\, \varphi_\alpha$$

[p. 124] et en effectuant la dérivation

$$\frac{\partial \varphi'_\mu}{\partial x'_\nu} = \sum_\alpha \sum_\beta \frac{\partial x_\alpha}{\partial x'_\mu}\, \frac{\partial x_\beta}{\partial x'_\nu}\, \frac{\partial \varphi_\alpha}{\partial x_\beta} + \sum_\alpha \sum_\beta \frac{\partial x_\beta}{\partial x'_\nu}\, \frac{\partial}{\partial x_\beta} \left(\frac{\partial x_\alpha}{\partial x'_\mu} \right) \cdot \varphi_\alpha$$

Le dernier terme peut s'écrire

$$\sum_\alpha \frac{\partial^2 x_\alpha}{\partial x'_\mu \partial x'_\nu}\, \varphi_\alpha$$

il ne change pas lorsqu'on permute μ et ν.

---

[258] B modification : "expriment".

[259] B suppression : parenthèses supprimées.

On obtient donc bien

$$F'_{\mu\nu} = \sum_{\alpha}\sum_{\beta} \frac{\partial x_\alpha}{\partial x'_\mu} \frac{\partial x_\beta}{\partial x'_\nu} F_{\alpha\beta}$$

Le courant électrique et la masse d'électricité ont été représentés par

(3) $$\mathfrak{J}^\alpha = e\frac{dx_\alpha}{ds}$$

où [260] la densité d'électricité.

Cette quantité n'est pas un tenseur. Il faudra pour obtenir un tenseur procéder comme nous l'avons fait dans le cas des masses matérielles.

$$J^\alpha = \frac{e}{\sqrt{-g}} \frac{dx_\alpha}{ds}$$

sera un tenseur. $\mathfrak{J}^\alpha$ est la densité tensorielle correspondante.
Enfin la force électro-magnétique

(4) $$f_\mu = -\sum_\sigma F_{\mu\sigma}\mathfrak{J}^\sigma$$

est aussi une densité tensorielle.

f_μ (§ 6-9) est le [261] premier membre de l'équation (§ 5-7) obtenue dans l'étude du mouvement d'un point libre. Nous avons désigné par f_μ le [262] premier membre de l'équation (§ 5-10) du même paragraphe[263]. La manière dont on passe d'une forme à l'autre montre que [p. 125] f_μ et f^μ sont deux densités torentielles[264] associées.

On aura ainsi

$$f_\mu = \sum_\alpha g_{\mu\alpha} f^\alpha$$

Nous avons exprimé f^α en fonction du tenseur d'énergie matérielle. Nous avons trouvé (§ 11-11)[265]

$$f^\alpha = \sum_\sigma \frac{\partial \mathfrak{T}^{\alpha\sigma}}{\partial x_\sigma} + \sum_\sigma\sum_\tau \left\{ \begin{matrix} \sigma\tau \\ \alpha \end{matrix} \right\} \mathfrak{T}^{\sigma\tau}$$

[260] B ajout : "*e* est".

[261] B ajout : "produit par *m* du".

[262] B ajout : "produit par *m* du".

[263] C correction (voir Annexe 2, p. 249) : "au lieu de : 'f_μ le produit par *m*' lire : 'f_μ le produit par $-m$'. Il faut, en conséquence, changer le signe des expressions de f_μ en fonction de $\mathfrak{T}_\mu^\nu$ ([p. 125] – voir p. 235 à p. 236, [p 129] formule (11) – voir p. 239); le second membre de la formule (13) ([p. 130] – voir p. 240) change ainsi de signe. Le texte qui suit cette formule est alors exact."

[264] B modification : "tensorielles".

[265] Il faut lire bien sûr : (§ 11-11'), en utilisant la numérotation attribuée dans ce texte-ci à l'équation considérée.

f_μ s'exprime donc en fonction du tenseur d'énergie matérielle par l'équation

$$f_\mu = \sum_\alpha \sum_\sigma g_{\mu\alpha} \frac{\partial \mathfrak{T}^{\alpha\sigma}}{\partial x_\sigma} + \sum_\alpha \sum_\sigma \sum_\tau g_{\alpha\mu} \left\{ \begin{matrix} \sigma\tau \\ \alpha \end{matrix} \right\} \mathfrak{T}^{\sigma\tau}$$

Cette expression peut se transformer en introduisant la densité tensorielle

$$\mathfrak{T}_\mu^\sigma = \sum_\alpha g_{\mu\alpha} \mathfrak{T}^{\alpha\sigma}$$

associée à $\mathfrak{T}^{\alpha\sigma}$.
[266]

$$\sum_\alpha g_{\alpha\mu} \left\{ \begin{matrix} \sigma\tau \\ \alpha \end{matrix} \right\} = \left[\begin{matrix} \sigma\tau \\ \mu \end{matrix} \right]$$

et à cause de la symétrie de $\mathfrak{T}^{\sigma\tau}$, on peut écrire

$$f_\mu = \sum_\sigma \frac{\partial \mathfrak{T}_\mu^\sigma}{\partial x_\sigma} - \sum_\alpha \sum_\sigma \frac{\partial g_{\mu\alpha}}{\partial x_\sigma} \mathfrak{T}^{\alpha\sigma} + \sum_\tau \sum_\sigma \frac{\partial g_{\tau\mu}}{\partial x_\sigma} \mathfrak{T}^{\sigma\tau} - \frac{1}{2} \sum_\sigma \sum_\tau \frac{\partial g_{\sigma\tau}}{\partial x_\mu} \mathfrak{T}^{\sigma\tau}$$

Les deuxième et troisième termes se détruisent. On le voit immédiatement en écrivant l'indice sommatoire τ au lieu de α.
Il reste donc

$$f_\mu = \sum_\sigma \frac{\partial \mathfrak{T}_\mu^\sigma}{\partial x_\sigma} - \frac{1}{2} \sum_\sigma \sum_\tau \frac{\partial g_{\sigma\tau}}{\partial x_\mu} \mathfrak{T}^{\sigma\tau} \tag{5}$$

Les équations (4) et (5) sont les équations du mouvement des [p. 126] masses électriques en fonction du tenseur d'énergie matérielle.

Les lois que nous venons de rappeler interprètent complètement l'action des champs sur les masses. Il nous faut maintenant aborder l'étude de l'action des masses électriques sur les champs.

Rappelons tout d'abord ce que nous apprend l'expérience à ce sujet.

La loi par laquelle le potentiel scalaire φ dépend de la répartition des masses est analogue à la loi de Newton. On peut l'écrire en employant une unité convenable pour la masse électrique (unités Heaviside).

$$\varphi = \sum \frac{1}{4\pi} \frac{e}{r}$$

ou inversement, en utilisant l'équation de Poisson :

$$\Box \varphi = -e \tag{6}$$

Ces lois sont équivalentes à la loi de Coulomb.

---

[266] B ajout : "En remarquant (§ 5 – 17′) que".

L'action des courants ou des particules matérielles sur le champ est exprimée par la loi de Laplace. En employant le potentiel électromagnétique

$$-\varphi_x, \quad -\varphi_y, \quad -\varphi_z$$

on peut l'écrire

$$-\varphi_x = \frac{1}{4\pi}\sum \frac{e\,v_x}{r}$$

et les deux équations analogues, ou en employant l'équation de Poisson.

$$(7) \qquad \begin{cases} \square\,\varphi_x = e\,v_x \\ \square\,\varphi_y = e\,v_y \\ \square\,\varphi_z = e\,v_z \end{cases}$$

Le potentiel électro-magnétique ne serait pas déterminé par les [p. 127] équations que nous venons d'écrire. Pour lever cette indétermination il faut ajouter une équation. On peut lui donner avec Maxwell la forme

$$(8) \qquad \frac{\partial \varphi_x}{\partial x} + \frac{\partial \varphi_y}{\partial y} + \frac{\partial \varphi_z}{\partial z} - \frac{\partial \varphi}{\partial t} = 0$$

Le terme $\frac{\partial \varphi}{\partial t}$ est nécessaire, lorsqu'on emploie des potentiels retardés.

Il nous faut rechercher une équation tensorielle qui se réduise aux équations précédentes, lorsqu'on utilise les coordonnées qui y sont employées. Ces coordonnées sont celles pour lesquelles les dérivées des $g_{\mu\nu}$ sont nulles (au moins approximativement) et où ds^2 se réduit à

$$ds^2 = -dx^2 - dy^2 - dz^2 + dt^2$$

Nous avons trouvé (§ 11-13) que

$$\sum_\sigma \frac{\partial \mathfrak{T}^{\mu\sigma}}{\partial x_\sigma}$$

est une expression tensorielle si

$$\mathfrak{T}^{\mu\sigma} = -\mathfrak{T}^{\sigma\mu}$$

À $F_{\alpha\beta}$ correspond la densité tensorielle $\mathfrak{F}_{\alpha\beta}$. La densité contrevariante associée est

$$\mathfrak{F}^{\mu\nu} = \sum_\alpha \sum_\beta g^{\mu\alpha}\, g^{\nu\beta}\, \mathfrak{F}_{\alpha\beta}$$

L'expression

$$\sum_\sigma \frac{\partial \mathfrak{F}^{\mu\sigma}}{\partial x_\sigma}$$

est donc un tenseur.

Lorsqu'on emploie les coordonnées utilisées dans les équations (6, 7, 8) cette expression se simplifie. En effet,

$$g^{11} = g^{22} = g^{33} = -1, \qquad g^{44} = 1, \qquad \sqrt{-g} = 1$$

les autres $g_{\mu\nu}$ et toutes les dérivées [267] sont nuls. On obtient [p. 128]

$$\sum_\sigma \frac{\partial \mathfrak{F}^{\mu\sigma}}{\partial x_\sigma} = \sum_\sigma \pm \frac{\partial \mathfrak{F}_{\mu\sigma}}{\partial x_\sigma}$$

ou[268] l'on doit prendre le signe + si l'indice 4 ne figure pas parmi les deux indices μ et (σ)[269] et le signe − dans le cas contraire. En remplaçant $F_{\mu\sigma}$ par sa valeur en fonction des φ (2), cette expression s'écrit

$$\sum_\sigma \pm \frac{\partial^2 \varphi_\mu}{\partial x_\sigma^2} - (\pm)\frac{\partial}{\partial x_\mu} \sum_\sigma \frac{\partial \varphi_\sigma}{\partial x_\sigma}$$

ou pour $\mu = 1$

$$\frac{\partial^2 \varphi_x}{\partial x^2} + \frac{\partial^2 \varphi_x}{\partial y^2} + \frac{\partial^2 \varphi_x}{\partial z^2} - \frac{\partial^2 \varphi_x}{\partial t^2} + \frac{\partial}{\partial x}\left(\frac{\partial \varphi_x}{\partial x} + \frac{\partial \varphi_y}{\partial y} + \frac{\partial \varphi_z}{\partial z} - \frac{\partial \varphi}{\partial t}\right) = \Box \varphi_x$$

à cause de l'équation supplémentaire de Maxwell (8).

On aurait pour $(\mu = 2, 3)$ des expressions analogues en (y)[270] et en (z)[271]. Pour $(\mu = 4)$ on trouve

$$-\Box \varphi$$

D'autre part lorsque les vitesses sont de l'ordre de celles qui se rencontrent dans les expériences

$$e \quad , \quad ev_x \quad , \quad ev_y \quad , \quad ev_z$$

peuvent être représentés par la densité tensorielle

$$\mathfrak{J}^\mu = e\frac{dx_\mu}{ds}$$

Les formules (6) et (7) peuvent donc être mises sous la forme tensorielle

$$\sum_\sigma \frac{\partial \mathfrak{F}^{\mu\sigma}}{\partial x_\sigma} = \mathfrak{J}^\mu \tag{9}$$

C'est la loi générale de l'action des masses sur les champs électriques.

---

[267] B ajout : "premières".

[268] B modification : "où".

[269] B suppression : parenthèses supprimées.

[270] B suppression : parenthèses supprimées.

[271] B suppression : parenthèses supprimées.

Pesanteur des champs électriques.

En tenant compte de la loi que nous venons de [p. 129] trouver, l'équation du mouvement des corps électrisés (4) s'écrit en remplaçant $\mathfrak{J}^{\mu}$ par sa valeur

$$(10) \qquad f_{\mu} + \sum_{\sigma}\sum_{\tau} F_{\mu\sigma} \frac{\partial \mathfrak{F}^{\sigma\tau}}{\partial x_{\tau}} = 0$$

f_{μ} peut s'écrire[272] au moyen du tenseur d'énergie matérielle (5). Nous avons vu que des équations de la gravitation on peut déduire que f_{μ} s'annule identiquement. Il s'en suit que f_{μ} s'annule aussi. Les équations que nous venons de trouver signifient donc que les équations de la gravitation ne sont plus appliquables [*sic*]. Elles doivent être modifiées, lorsque l'action électromagnétique subie par une particule chargée n'est pas nulle.

Dans les équations de la gravitation, nous devrons ajouter au tenseur d'énergie matérielle un terme supplémentaire, le tenseur d'énergie électrique.

L'expression

$$(11) \qquad f_{\mu} = \sum_{\sigma} \frac{\partial \mathfrak{T}_{\mu}^{\sigma}}{\partial x_{\sigma}} - \frac{1}{2} \sum_{\sigma}\sum_{\tau} \frac{\partial g_{\sigma\tau}}{\partial x_{\mu}} \mathfrak{T}^{\sigma\tau}$$

s'annule lorsque $\mathfrak{T}_{\mu}{}^{\sigma}$ représente le tenseur d'énergie matérielle seul. Lorsqu'on remplace $\mathfrak{T}_{\mu}{}^{\sigma}$ par le tenseur d'énergie électrique, que nous cherchons f_{μ} doit se réduire à (10). On obtient ce résultat en posant l'équation tensorielle

$$(12) \qquad \mathfrak{T}_{\mu}^{\sigma} = -\sum_{\alpha} F_{\mu\alpha} \mathfrak{F}^{\sigma\alpha} + g_{\mu}^{\sigma} \frac{1}{4} \sum_{\alpha}\sum_{\beta} F_{\alpha\beta} \mathfrak{F}^{\alpha\beta}$$

Il suffit de le démontrer en un point (M)[273] en employant des coordonnées particulières. Nous supposerons donc que les dérivées des $g_{\mu\nu}$ s'annulent en ce point et que les $g_{\mu\nu}$ se réduisent à

$$g_{\mu\nu} = g_{\mu}^{\nu}$$

Dans ce cas les équations (10) (11) et (12) se réduisent [p. 130] respectivement à

$$(10') \qquad f_{\mu} + \sum_{\sigma}\sum_{\tau} F_{\mu\sigma} \frac{\partial F_{\sigma\tau}}{\partial x_{\tau}} = 0$$

et[274]

$$(11') \qquad f_{\mu} = \sum_{\sigma} \frac{\partial T_{\mu}^{\sigma}}{\partial x_{\sigma}}$$

---

[272] B modification : "s'exprimer".

[273] B suppression : parenthèses supprimées.

[274] B suppression : "et".

d'autre part[275]

$$T_\mu^\sigma = -\sum_\alpha F_{\mu\alpha} F_{\sigma\alpha} + \frac{1}{4} g_\mu^\sigma \sum_\alpha \sum_\beta F_{\alpha\beta}^2 \tag{12'}$$

Remplaçons T_μ^σ par sa valeur et effectuons la dérivation

$$f_\mu = -\sum_\sigma \sum_\alpha \frac{\partial F_{\mu\alpha}}{\partial x_\sigma} F_{\sigma\alpha} - \sum_\sigma \sum_\alpha F_{\mu\alpha} \frac{\partial F_{\sigma\alpha}}{\partial x_\sigma} + \frac{1}{4} \times 2 \sum_\alpha \sum_\beta F_{\alpha\beta} \frac{\partial F_{\alpha\beta}}{\partial x_\mu} \tag{13}$$

Le second terme est identique à celui que nous devons trouver. Il faut donc montrer que les deux autres se déterminent[276].

Remplaçons les $F_{\alpha\beta}$ par leurs valeurs la[277] fontion [*sic*] des φ (2), le premier terme devient

$$-\sum_\sigma \sum_\alpha \left(\frac{\partial^2 \varphi_\mu}{\partial x_\alpha \partial x_\sigma} - \frac{\partial^2 \varphi_\alpha}{\partial x_\mu \partial x_\sigma} \right) \left(\frac{\partial \varphi_\sigma}{\partial x_\alpha} - \frac{\partial \varphi_\alpha}{\partial x_\sigma} \right)$$

$$= -\sum_\sigma \sum_\alpha \left[\begin{array}{c} \frac{\partial^2 \varphi_\mu}{\partial x_\alpha \partial x_\sigma} \frac{\partial \varphi_\sigma}{\partial x_\alpha} - \frac{\partial^2 \varphi_\mu}{\partial x_\alpha \partial x_\sigma} \frac{\partial \varphi_\alpha}{\partial x_\sigma} \\ - \frac{\partial^2 \varphi_\alpha}{\partial x_\mu \partial x_\sigma} \frac{\partial \varphi_\sigma}{\partial x_\alpha} + \frac{\partial^2 \varphi_\alpha}{\partial x_\mu \partial x_\sigma} \frac{\partial \varphi_\alpha}{\partial x_\sigma} \end{array} \right]$$

Les termes de la première ligne se détruisent car ils ont une même valeur absolue si on échange les indices sommatoires.

Calculons le dernier terme de (13)

$$\frac{1}{2} \sum_\alpha \sum_\beta \left(\frac{\partial \varphi_\alpha}{\partial x_\beta} - \frac{\partial \varphi_\beta}{\partial x_\alpha} \right) \left(\frac{\partial^2 \varphi_\alpha}{\partial x_\beta \partial x_\mu} - \frac{\partial^2 \varphi_\beta}{\partial x_\alpha \partial x_\mu} \right)$$

[p. 131] ou en réduisant les termes qui ne diffèrent que par la notation des indices sommatoires

$$\sum_\alpha \sum_\beta \left[\frac{\partial \varphi_\alpha}{\partial x_\beta} \frac{\partial^2 \varphi_\alpha}{\partial x_\beta \partial x_\mu} - \frac{\partial \varphi_\alpha}{\partial x_\beta} \frac{\partial^2 \varphi_\beta}{\partial x_\mu \partial x_\alpha} \right]$$

En écrivant l'indice sommatoire σ au lieu de β, on voit que tous les termes se détruisent.

---

[275] B modification : "d'autre part" remplacé par "et".

[276] B modification : "détruisent".

[277] B modification : "en". Notons que cette correction a été faite au crayon et ne semble pas être de la main de G. Lemaître.

Nous avons utilisé des tenseurs covariants pour écrire les équations de la gravitation. Nous devons donc mettre le tenseur d'énergie électrique sous forme covariante. Il suffit d'utiliser le tenseur associé à $\mathfrak{T}_\mu^\sigma$

$$T_{\mu\nu} = \sum_\sigma \frac{g_{\sigma\nu}}{\sqrt{-g}} \mathfrak{T}_\mu^\sigma$$

Dans les équations générales de la gravitation $R_{\mu\nu} = -\kappa(T_{\mu\nu} - \frac{1}{2}g_{\mu\nu}T)$ nous devons considérer $T_{\mu\nu}$ comme la somme du tenseur d'énergie matérielle et du tenseur d'énergie électrique. Le tenseur contracté d'énergie électrique est nul. On s'en rend compte immédiatement en le calculant au moyen de l'équation (12).

$$T = \frac{1}{\sqrt{-g}} \sum_\sigma \mathfrak{T}_\sigma^\sigma = 0$$

Louvain, le 31 Mai 1922,
G. Lemaître

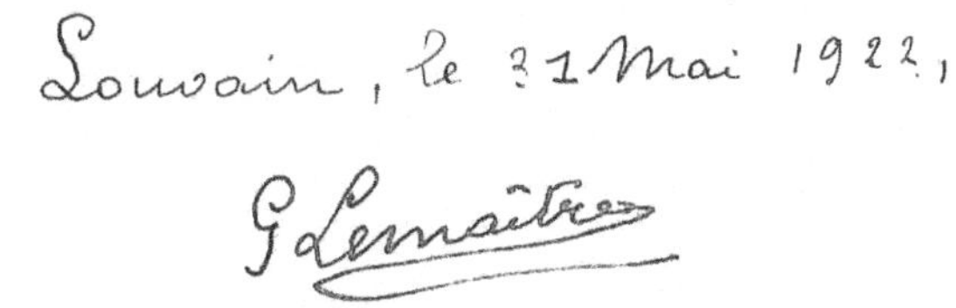

La correspondance entre Georges Lemaître et Maurice Alliaume conservée aux Archives Georges Lemaître

Éléments contextuels au concours des bourses de voyage

En octobre 1920, Georges Lemaître entre au séminaire Saint-Rombaut, annexe du grand séminaire de l'archidiocèse de Malines ayant pour finalité d'accueillir des séminaristes dont le parcours fut perturbé par la guerre afin de les conduire rapidement à l'ordination sacerdotale. La discipline y est plus souple que dans un autre séminaire. C'est ainsi que Georges Lemaître reçut la permission d'y étudier de la physique. De plus, il est toujours en contact avec le milieu scientifique, et particulièrement avec ses professeurs : Maurice Alliaume,[1] mathématicien, lequel dispose de la plupart des livres consacrés à la relativité dont les travaux d'Albert Einstein, de Théophile de Donder et d'Arthur Eddington ; Edouard Goedseels, mathématicien lui aussi, qui rejoindra d'ailleurs le séminaire en 1923 ; et le chanoine René De Muynck, docteur en sciences physiques et mathématiques, duquel viendra la proposition de participer au concours des bourses de voyage.[2]

---

[1] Maurice Alliaume (1882-1931), chevalier de l'Ordre de Léopold, officier de l'Ordre de la Couronne, est ingénieur civil des mines et docteur en sciences physiques et mathématiques belge. Professeur à l'Université catholique de Louvain, il y enseigna l'astronomie, la géodésie, la topographie, la géographie mathématique, la théorie des probabilités et la géométrie descriptive appliquée, succédant à Edouard Goedseels (1857-1928, mathématicien, professeur à l'Université catholique de Louvain, se concentrant sur la théorie des erreurs d'observation). Membre de plusieurs sociétés scientifiques (dont l'Union astronomique internationale, la Société scientifique de Bruxelles, la Fondation), il traita en 1902, *e.a.*, des diverses hypothèses concernant l'état de mouvement ou de repos du Soleil et développa celle d'un centre d'attraction autour duquel graviterait le système solaire tout entier. Cf. Collard A., *Le professeur Maurice Alliaume (1882-1931)*, in *Ciel & Terre*, vol. 48, 1932, pp. 1-2. Son cours d'astronomie axé sur ses aspects mathématiques fera dire à Georges Lemaître que "sa mort inopinée a été une catastrophe pour l'enseignement des mathématiques appliquées et de l'astronomie à Louvain". Cf. Mawhin J., *Une brève histoire des mathématiques à l'Université catholique de Louvain*, in *Revue des questions scientifiques*, 163, 1992, p. 373. Ce sont d'ailleurs Alliaume et Goedseels qui, le 26 octobre 1922, font nommer Georges Lemaître en tant que membre de la Société scientifique de Bruxelles à laquelle ils appartiennent déjà tous les deux. Cf. Lambert D., *Un Atome d'Univers. La vie et l'œuvre de Georges Lemaître*, Bruxelles, Lessius & Racine, 2000, p. 56.

[2] Cf. Lambert D., *Un Atome d'Univers...*, p. 55.

G. Lemaître, *Learning the Physics of Einstein with Georges Lemaître*,
https://doi.org/10.1007/978-3-030-22030-3

Le concours, proposé chaque année par le Ministre des Sciences et des Arts à destination des étudiants ayant obtenu leur doctorat en deux ans, comporte deux épreuves : le dépôt d'un mémoire original dans la discipline de son choix ainsi que trois thèses annexes, non développées, portant sur la même matière mais indépendantes de l'objet du mémoire, et une défense orale publique. La bourse d'un an voit son montant réparti sur deux années et un rapport d'activités doit être déposé après le voyage.

Georges Lemaître, dès 1920-1921, commence la rédaction d'une synthèse personnelle de la relativité restreinte et générale qu'il termine le 31 mai 1922 sous le titre *La Physique d'Einstein* et soumet au jury composé de Maurice Alliaume, Théophile de Donder (1872-1957) et Henry Janne d'Othée (1884-1966).

Maurice Alliaume annonce, dans un courrier du 4 août 1922 à Georges Lemaître,[3] qu'il a été désigné, par l'Université de Louvain, membre de son jury, et qu'il est chargé également d'examiner un second candidat, qui, travaillant sur les probabilités, ne lui paraît pas "extérieurement bien terrible". Cette lettre, portant principalement sur un hommage d'auteur et traitant de l'intensité de l'atome terrestre et de l'atome solaire, témoigne de l'existence d'échanges scientifiques entre Alliaume et Lemaître et ce indépendamment de cette candidature. Ils s'estiment, se questionnent, discutent, réfléchissent, Lemaître s'adressant toujours à son "professeur".

Le 5 février 1923, Alliaume lui demande de lui adresser une "note dans laquelle vous signalerez les points de votre travail qui sont originaux en quelque manière".[4] Et le 13 mars 1923, il le recontacte en lui précisant que les résultats des deux candidats sont très proches.[5] Il lui précise les cotes obtenues sur base de la lecture des travaux déposés et lui communique les remarques de M. de Donder et "des autres membres du Jury". Il lui conseille *e.a.* la lecture de l'ouvrage d'Eddington, *The Mathematical Theory of Relativity*. Sans attendre, dès le jour suivant, le 14 mars, Théophile de Donder se manifeste en transmettant des précisions quant au mémoire d'Adriaan Fokker (1887-1972), physicien hollandais, intitulé *De geodetische Precessie*, 1920, et mettant à disposition du candidat un nouvel exemplaire de l'ouvrage d'Eddington, information que Alliaume s'empresse encore le même jour de communiquer à Lemaître.[6]

Fidèle à lui-même, Georges Lemaître répond "consciencieusement"[7] en faisant parvenir à Maurice Alliaume une première note (voir Annexe 1) le 16 février et en lui promettant une seconde note signalant les erreurs de calcul laissées dans le manuscrit soumis. La seconde note (voir Annexe 2) arrive dix jours plus tard,

---

[3] Archives de l'Université catholique de Louvain | Archives Georges Lemaître, BE A4006 FG LEM-425.

[4] Archives de l'Université catholique de Louvain | Archives Georges Lemaître, BE A4006 FG LEM-426.

[5] Archives de l'Université catholique de Louvain | Archives Georges Lemaître, BE A4006 FG LEM-428.

[6] Archives de l'Université catholique de Louvain | Archives Georges Lemaître, BE A4006 FG LEM-430.

[7] Archives de l'Université catholique de Louvain | Archives Georges Lemaître, BE A4006 FG LEM-427.

le 26 février. Elle relève les inexactitudes comme annoncé et traite des points originaux de son travail. Ici aussi, Georges Lemaître reste fidèle à lui-même et précise qu'il a "été plus occupé d'obtenir une solution vraiment satisfaisante des problèmes que de faire la part exacte de ce que je trouvais dans les auteurs et de ce que j'y ajoutais de personnel" ayant essayé "de vous satisfaire en étant aussi prudent que possible".

Finalement le 9 mars Lemaître adresse une dernière lettre à Alliaume afin d'encore préciser des conventions de signe portant sur les tenseurs de Riemann et de Ricci.[8]

Et c'est ainsi préparé et ayant effectué les lectures demandées qu'il ira défendre son mémoire, gagnant ainsi cette bourse de voyage qui lui ouvrira, conjuguée avec le *Fellowship* de la *Commission for the relief in Belgium* de l'*American Educational Foundation*, les portes de son aventure scientifique.

---

[8] Archives de l'Université catholique de Louvain | Archives Georges Lemaître, BE A4006 FG LEM-432.

Annexe 1

Note signalant les points du mémoire *"La Physique d'Einstein"* qui sont originaux en quelque manière[1]

1. Einstein a établi sa théorie par généralisations successives de la relativité restreinte à la relativité générale. Il était désirable de reprendre la théorie sous forme synthétique. C'est ce qu'a fait Mr de Donder en partant du principe d'Hamilton généralisé. L'exposé actuel réalise cette même synthèse, en prenant pour point de départ le principe de relativité générale.

2. Les mesures effectuées sur un corps sont définies directement en fonction des potentiels d'univers et des coordonnées arbitraires. On donne sous forme explicite, non seulement la définition du temps (temps propre), mais celle de la simultanéité et de l'élément de longueur (p. 21[2]). Il en résulte, 1) une nouvelle solution des difficultés soulevées par l'expérience de Mr Sagnac (cf. Langevin c.r. 1921- 831). Cette expérience montre d'une façon intuitive que la simultanéité dépend du chemin suivant lequel on la définit (p. 33[3]) ; 2) une solution immédiate des problèmes géométriques. Les résultats sont interprétés en faisant la carte de l'espace dans un espace euclidien (p. 10[4]) ; l'ellipsoïde des échelles met en évidence les déformations que subit dans la carte l'image d'un mètre rigide déplacé dans l'espace. Ces cartes peuvent être construites d'une infinité de manières. Chacune d'elle corespond [*sic*] à une déformation apparente spéciale. Les déformations signalées par Einstein, sur un corps tournant (Das Relativitäts-prinzip 3° éd. p. 75) et au voisinage d'une masse matérielle (l.c.p. 122) ne sont qu'une solution particulière. On peut ainsi bien considérer des déformations indépendantes de la direction des règles : (p. 114[5]) dans un champ de gravitation quelconque (en 1° approximation) et (p. 70[6]) sur un disque tournant. On calcule aussi (p. 71[7]) les équations d'une surface de révolution sur laquelle la géométrie est euclidienne lorsque la surface est animée d'une rotation donnée autour de son axe.

[1] Archives de l'Université catholique de Louvain | Archives Georges Lemaître, BE A4006 FG LEM-446.

[2] Pagination du tapuscrit. Voir p. 144 du présent ouvrage.

[3] Pagination du tapuscrit. Voir p. 154 du présent ouvrage.

[4] Pagination du tapuscrit. Voir p. 136 du présent ouvrage.

[5] Pagination du tapuscrit. Voir p. 225 du présent ouvrage.

[6] Pagination du tapuscrit. Voir p. 185 du présent ouvrage.

[7] Pagination du tapuscrit. Voir p. 186 du présent ouvrage.

3. La contraction de Lorentz et le retard apparant [*sic*] des chronomètres sont considérés comme des erreurs systématiques inhérentes aux procédés indirects de mesures (p. 26[8]). Les mesures effectuées en portant les instruments sur les corps à mesurer ont seules une signification absolue. Ces résultats sont établis non seulement pour mouvement uniforme [*sic*] dans un champ de Galilée, mais pour un mouvement quelconque dans un champ quelconque. Cette démonstration est basée sur l'emploi des coordonnées propres d'un corps ; leur existence est établie en chaque point (p. 23[9]).

4. L'étude du mouvement uniformément accéléré d'un solide dans du [*sic*] champ d'inertie uniforme qui résulte de ce mouvement a joué un role [*sic*] décisif dans le développement de la théorie de la relativité. Cependant personne, à ma connaissance, n'a donné une solution rigoureuse de ce problème. On se contente de l'approximation "provisoire" donnée par Einstein (l.c.p. 73) qui néglige la contraction et le retard de Lorentz. (Le mouvement uniformément accéléré d'un point a été étudié au point de vue de la relativité restreinte : Pauli, encycl. math. p. 627). Le problème du mouvement d'un solide dans un champ de Galilée est posé dans le cas général (p. 60[10]), et résolu dans le cas d'un mouvement rectiligne quelconque (p. 62[11]). On en déduit l'équation du champ d'inertie. Lorsque on suppose que le champ doit être stationnaire, on obtient les équations du mouvement uniformément accéléré d'un solide (p. 65[12]). et la transformation de coordonnées correspondante (p. 66[13]). L'équation du champ d'inertie est écrite sous une forme qui met en évidence l'analogie avec le champ de Scharzschild [*sic*] (p. 64[14] et 115[15]).

Les équations du mouvement d'un électron sont écrites sous la forme classique de la mécanique, sans faire intervenir de masses variables, seule l'expression de la force électrique est modifiée (p. 57[16]).

6. On démontre que la trajectoire d'un point libre est une ligne de temps maximum, mais de maximum relatif seulement. On donne un exemple où ce maximum n'est pas absolu (p. 18[17]).

7. Einstein a utilisé le calcul tensoriel pour développer le principe de relativité. Il est plus naturel d'établir le caractère tensoriel par simple vérification algébrique. C'est ce qu'a fait Mr de Donder, sous le nom de "covariences [*sic*] de la Gravifique". Cette méthode donne malheureusement lieu à des calculs assez longs. Elle

---

[8] Pagination du tapuscrit. Voir p. 149 du présent ouvrage.

[9] Pagination du tapuscrit. Voir p. 146 du présent ouvrage.

[10] Pagination du tapuscrit. Voir p. 177 du présent ouvrage.

[11] Pagination du tapuscrit. Voir p. 178 du présent ouvrage.

[12] Pagination du tapuscrit. Voir p. 181 du présent ouvrage.

[13] Pagination du tapuscrit. Voir p. 182 du présent ouvrage.

[14] Pagination du tapuscrit. Voir p. 180 du présent ouvrage.

[15] Pagination du tapuscrit. Voir p. 226 du présent ouvrage.

[16] Pagination du tapuscrit. Voir p. 174 du présent ouvrage.

[17] Pagination du tapuscrit. Voir p. 142 du présent ouvrage.

se simplifie beaucoup grâce à l'emploi systématique de coordonnées particulières dont l'existence en chaque point est démontrée (p. 46[18]). La méthode est résumée sous forme de théorème (p. 81[19]). Elle est établie ou utilisée, en particulier pour le tenseur de Riemann (p. 73[20]), pour la divergence d'un tenseur (p. 100[21]), pour montrer que la divergence du tenseur de Riemann contracté est identiquement nulle (p. 104[22]), pour calculer le tenseur d'énergie électrique (p. 129[23]). Cette méthode fournit immédiatement l'expression simplifiée à laquelle se réduit chaque tenseur, en un point et pour un système de coordonnées particulier. Ces résultats connus sont démontrés d'une manière nouvelle (spécialement pour $R_{\mu\nu}$, p. 86[24]).

8. La méthode d'approximations successives utilisée par Einstein (l.c.p. 119 et Pauli l.c.p. 736) est poussée plus loin, en introduisant un potentiel newtonnien [*sic*] du [*sic*] à des masses fictives (p. 112[25]). On obtient ainsi, sous la forme de la mécanique classique, et dans le cas de masses attirantes quelconques, des équations qui rendent compte du mouvement du périhélie des planètes (p. 108[26] et 112[27]). Enfin, on signale l'influence du mouvement des masses attirantes et on indique la méthode suivant laquelle cette influence peut être calculée (p. 112[28]).

---

[18] Pagination du tapuscrit. Voir p. 167 du présent ouvrage.

[19] Pagination du tapuscrit. Voir p. 194 du présent ouvrage.

[20] Pagination du tapuscrit. Voir p. 188 du présent ouvrage.

[21] Pagination du tapuscrit. Voir p. 211 du présent ouvrage.

[22] Pagination du tapuscrit. Voir p. 215 du présent ouvrage.

[23] Pagination du tapuscrit. Voir p. 239 du présent ouvrage.

[24] Pagination du tapuscrit. Voir p. 199 du présent ouvrage.

[25] Pagination du tapuscrit. Voir p. 223 du présent ouvrage.

[26] Pagination du tapuscrit. Voir p. 219 du présent ouvrage.

[27] Pagination du tapuscrit. Voir p. 223 du présent ouvrage.

[28] Pagination du tapuscrit. Voir p. 223 du présent ouvrage.

Annexe 2

Errata de "*La Physique d'Einstein*"[1]

p. 53[2], formule (7), au lieu de $F_{\alpha\sigma}$, lire $F_{\sigma\alpha}$.
La démonstration qui suit doit être modifiée, en changeant le signe dans la formule (8).

p. 74[3] Le tenseur de Riemann est défini par une expression de signe contraire à l'expression habituelle.
Cette définition est conservée dans tout le mémoire, sauf : § 12 p 102[4] et p 109[5]. Dans ce dernier cas il en résulte une erreur dans le calcul de ρ !
En rectifiant, on obtient (p. 112[6], 1ere formule)

$$\rho' = \left(1 + \tfrac{2V}{c^2}\right)\rho + \tfrac{4}{\kappa c^2}\, g^2$$

p 108[7] formule (11), lire

$$\frac{d^2x}{dt^2} + \frac{1}{2}(1 - 3v_x^2 + v_y^2 + v_z^2)\frac{\partial\varpi}{\partial x} - 2v_x v_y \frac{\partial\varpi}{\partial y} - 2v_x v_z \frac{\partial\varpi}{\partial z} - \frac{1}{2}v_x(3 - v^2)\frac{\partial\varpi}{\partial t} = 0$$

p 112[8] dernière formule, lire

$$T_{\mu\nu} = \begin{matrix} \rho v_x^2 & , \cdots & , \cdots & , -\rho v_x \\ \cdots & , \cdots & , \cdots & , -\rho v_y \\ \cdots & , \cdots & , \cdots & , -\rho v_z \\ -\rho v_x & , -\rho v_y & , -\rho v_z & , \rho(1 + v^2) \end{matrix}$$

---

[1] Archives de l'Université catholique de Louvain | Archives Georges Lemaître, BE A4006 FG LEM-432.

[2] Pagination du tapuscrit. Voir p. 171 du présent ouvrage.

[3] Pagination du tapuscrit. Voir p. 188 du présent ouvrage.

[4] Pagination du tapuscrit. Voir p. 212 du présent ouvrage.

[5] Pagination du tapuscrit. Voir p. 220 du présent ouvrage.

[6] Pagination du tapuscrit. Voir p. 223 du présent ouvrage.

[7] Pagination du tapuscrit. Voir p. 219 du présent ouvrage.

[8] Pagination du tapuscrit. Voir p. 223 du présent ouvrage.

p 124[9] 2e ligne en remontant au lieu de : "f_μ le produit par m"
lire : "f_μ le produit par $-m$". Il faut, en conséquence, changer le signe des expressions de f_μ en fonction de $\mathcal{T}_\mu^\nu$ (p. 125[10], p 129[11] formule (11)) le second membre de la formule (13) (p. 130[12]) change ainsi de signe. Le texte qui suit cette formule est alors exact.

Le 26 Février 1923

[9] Pagination du tapuscrit. Voir p. 234 du présent ouvrage.
[10] Pagination du tapuscrit. Voir p. 235 du présent ouvrage.
[11] Pagination du tapuscrit. Voir p. 239 du présent ouvrage.
[12] Pagination du tapuscrit. Voir p. 239 du présent ouvrage.

Annexe 3 : Correspondance de Maurice Alliaume[1]

Louvain, le 4 août 1922

Cher Monsieur,

La liste des dissertations adressées au Ministère pour le prochain concours Bourses de voyage a paru récemment au Moniteur. Sans doute l'aurez vous vue : il y a deux concurrents pour les mathématiques : vous et l'auteur d'une dissertation intitulée "Quelques questions de Probabilités". Cela ne paraît pas, extérieurement, bien terrible. L'Université de Louvain m'a désigné comme membre du Jury pour l'un et l'autre de ces deux travaux.

Encore que tardivement, je vous remercie bien vivement de votre lettre. Dans l'hommage d'auteur que je vous enverrai dans une huitaine de jours, vous verrez que j'ai largement utilisé vos indications, spécialement en ce qui concerne l'intensité de l'atome terrestre et de l'atome solaire considérés. – Quant à l'ordre de grandeur du terme auquel est dû le déplacement du périhélie de Mercure, j'ai préféré ne pas soulever cette objection pour le moment : d'une part je la crois très sérieuse, et d'autre part je pense qu'il est possible d'y répondre sans modifier ma conclusion, mais je n'ai pas eu le loisir de mettre cette réponse au point ; l'idée est à peu près ceci ; le terme en question est, il est vrai, dans l'équation différentielle, petit du deuxième ordre, mais, dans l'intégration, il remonte au premier ordre par résonance. Nous verrons cela ensemble à l'occasion.

Où en est l'impression de votre ouvrage ?

Votre bien dévoué

Alliaume

Pendant les vacances mon adresse sera :

Rochefort

---

[1] Successivement les documents 425, 426, 428, 430 et 429 des Archives de l'Université catholique de Louvain | Archives Georges Lemaître, BE A4006 FG LEM.

Louvain, le 5 février 1923

Cher Monsieur,

Je viens de terminer une deuxième lecture de votre mémoire, et je vais l'expédier à M. Janne de Liège, qui sera le dernier à l'examiner.

En vue de la réunion du Jury, je vous prie de m'adresser une note dans laquelle vous signalerez les points de votre travail qui sont originaux en quelque manière.

Votre bien dévoué

Alliaume

Louvain, le 13 mars 1923

Cher Monsieur,

Dans la séance d'hier du Jury des Bourses de voyage les mémoires présentés ont été l'un et l'autre classés et admis à la défense publique. Sur 100 points, le vôtre a obtenu 76, et l'autre 72. Vous êtes donc peu distants l'un de l'autre et la défense publique sera de haute importance.

M. De Donder et les autres membres du Jury m'autorisent à attirer votre attention sur les points suivants :

1. A propos des coordonnées propres, voir un mémoire de Fokker (de Delft) paru en 1921 dans les Verslag d'Amsterdam.
2. Il y aura lieu d'éclaircir votre notion de simultanéité.

A propos de l'ensemble, voir l'ouvrage récent d'Eddington : Mathematical Theory of Relativity.

C'est à propos du 2° ci-dessus, que je vous envoie la note ci-jointe de M. De Donder.

Au cours de la défense publique, vous serez probablement admis à parler des recherches que avez poursuivies depuis le dépôt du mémoire.

Votre bien dévoué

Alliaume

Envoi par carte postale adressée à

Monsieur G. Lemaître
Maison S. Rombaut
Rue de Mérode, 84
Malines

Louvain, le 14 mars 1923

Cher Monsieur,

M. De Donder m'envoie à votre intention des précisions sur le mémoire de Fokker.
Titre : De geodetische Precessie, dans les Verslag d'Amsterdam, Deel XXIX, 30 oct. 1920.

M. De Donder met à ma disposition un exemplaire du nouvel ouvrage d'Eddington. Je l'aurai probablement jeudi. Je compte vous voir mardi[(1)] à l'occasion de la réunion du Cercle mathématique. Je vous prie de faire connaître le titre de votre conférence au plus tôt à M. Wibaut, Secrétaire du Cercle, Rue Notre Dame, 18.

[(1)] mardi prochain

Votre tout dévoué

Alliaume

Blankenberghe, le 7 avril 1923

Cher Monsieur,

J'ai laissé passer bien longtemps sans plus penser à ma promesse de vous donner l'adresse de M. Desmedt en Hollande pour vous faciliter l'acquisition du mémoire de Fokker. Je m'en excuse. Comme je ne connais pas encore cette adresse, j'espère que vous aurez, dans l'entretemps, pris vous-même vos mesures.

Votre bien dévoué

Alliaume

Annexe 4 : Correspondance de Georges Lemaître[1]

Commentaire

Les Archives Georges Lemaître conservent deux brouillons, recto-verso sur une même feuille, d'une ou deux lettres préparées par Lemaître en réponse à la demande de M. Alliaume exprimée dans la lettre de ce dernier datée du 5 février 1923 (voir p. 252). Un premier brouillon (le premier texte reproduit ci-dessous) ne débute pas par la formule d'adresse personnelle usuelle, mais se clôture par une expression de considération comme il convient. Le deuxième brouillon (le second texte reproduit ci-dessous) débute avec la formule usuelle de "Monsieur le professeur", mais n'est pas clôturé par une formule de politesse de convention. Ces deux brouillons, tous deux de la main de Lemaître, ont clairement été rédigés à des moments différents, à l'évidence le premier avec plus de précipitation que le second.

Sur le premier brouillon ont été ajoutés, clairement encore de la main de Lemaître, en haut à gauche le nom de "Alliaume" et en haut à droite la date du 16 Février 1923. Ce premier brouillon qui fait référence à la lettre du 5 février 1923 de Maurice Alliaume mais sans en préciser la date, indique que la note demandée (voir l'Annexe 1, p. 246) est jointe à cette correspondance, et annonce une seconde note (voir l'Annexe 2, p. 249) reprenant des errata.

Le second brouillon, qui ne porte pas de date, mentionne explicitement la date du 5 février 1923 pour la lettre de Maurice Alliaume qui demande une note faisant état des points originaux de la dissertation de Lemaître. Ce brouillon indique non seulement que cette note (l'Annexe 1 dans le présent ouvrage) accompagne cette lettre, mais qu'y est jointe également une seconde note (l'Annexe 2 dans le présent ouvrage) signalant "quelques inadvertances [...] relevées [...] en vérifiant les calculs".

Précisons par ailleurs que l'Annexe 1 ne porte pas de date, tandis que l'Annexe 2 est datée du 26 février 1923. Ajoutons également que l'original de l'Annexe 1 conservé dans les Archives Georges Lemaître est une copie d'un texte dactylographié à la machine à écrire, tandis que l'Annexe 2, telle que conservée, est une note manuscrite de la main de Lemaître.

Aucun autre document conservé par les Archives Georges Lemaître ne permet de déterminer lequel de ces deux brouillons a été envoyé à Maurice Alliaume, pour accompagner ces deux notes, ni si les deux l'ont été. Cependant, puisque dix jours séparent ces deux dates, et vu le contenu du second brouillon, tout porte à croire qu'ayant préparé le premier brouillon le 16 février, le temps que l'Annexe 1 soit dactylographiée, Lemaître a encore eu la possibilité de préparer l'Annexe 2, de telle sorte que finalement une seule lettre ait été adressée à Maurice Alliaume afin d'accompagner les deux notes en un envoi unique en date du 26 février 1923.

C'est donc pour toutes ces raisons que les contenus de ces deux brouillons sont reproduits ici.

---

[1] Successivement les documents 427 et 432 des Archives de l'Université catholique de Louvain | Archives Georges Lemaître, BE A4006 FG LEM.

Par ailleurs, notons que les Archives Georges Lemaître conservent tout un ensemble de notes personnelles de Lemaître, soit réellement des brouillons avec de nombreuses ébauches de calculs et de vérifications, soit des textes faisant état de réflexions et d'équations sous des formes déjà plus structurées, manifestement en préparation de la rédaction des Annexes 1 et 2, ainsi que de la dernière lettre datée du 9 mars 1923 que Lemaître adressait à M. Alliaume à la veille de la réunion du Jury (voir la lettre en p. 257). Ces notes personnelles, non reproduites ici, sont fort fournies (tout en reprenant des éléments des unes et des autres de ces notes au fur et à mesure de la préparation des documents finaux reproduits dans cet ouvrage) et témoignent d'une intense activité de la part de Lemaître visant à rencontrer au mieux la demande de Maurice Alliaume, et certainement aussi celle du Jury, datée du 5 février 1923. Ces documents originaux additionnels sont disponibles aux Archives Georges Lemaître, ainsi que sous forme numérisée *via* le portail du Service des Archives de l'UClouvain, voir `https://archives.uclouvain.be/`.

---

16 Février 1923

[Alliaume]

Je vous envoie ci-joint la note que vous avez bien voulu me demander. Je l'ai rédigée consciencieusement, mais je ne puis naturellement en garantir l'exactitude absolue; je n'ai pas lu tout ce qui a paru sur la question et il m'est difficile de préciser les influences que j'ai subies.

Si vous le jugez utile je pourrais venir un de ces jours à Louvain, compléter verbalement les indications forcément laconiques que je vous envoie. Je suis libre toutes les après-midi, sauf le mercredi.

Je compte vous envoyer une seconde note où je vous signalerai les principales erreurs de calcul que j'ai relevées en révisant mon manuscrit.

Je possède encore un exemplaire identique à celui que j'ai déposé. Je puis vous l'apporter si vous le désirez.

Veuillez agréer monsieur le professeur l'assurance de mon entier dévouement

---

Monsieur le professeur,

Dans votre lettre du 5 courant vous me demandez d'attirer votre attention sur les points de mon travail qui me paraissent originaux.

Je ne puis naturellement vous garantir l'exactitude de mes impressions à cet égard. Je n'ai pas lu tout ce qui a paru sur la relativité et j'ai été plus occupé d'obtenir une solution vraiment satisfaisante des problèmes que de faire la part exacte de ce que je trouvais dans les auteurs et de ce que j'y ajoutais de personnel.

Ces réserves faites je vais essayer de vous satisfaire en étant aussi prudent que possible.

J'ai cru utile de joindre une note où je vous signale quelques inadvertances que j'ai relevées dans mon mémoire en en vérifiant les calculs.

[G. Lemaître]

Malines, le 9 Mars 1923

Monsieur le professeur,

Dans l'errata que je vous ai envoyé, je vous signale une erreur au sujet du signe des tenseurs $R^{\delta}_{\alpha\beta\gamma}$et $R_{\alpha\beta}$. Je tiens à vous faire remarquer que les auteurs ne sont pas d'accord pour le choix de ces signes, non plus que pour la relation qui lie les deux tenseurs.

Einstein, de Donder, Eddington et Becquerel sont d'accord.

Pauli et Weyl écrivent

$$R_{\alpha\beta} = \sum_{\sigma} R^{\sigma}_{\alpha\sigma\beta} \tag{1}$$

au lieu de

$$R_{\alpha\beta} = \sum_{\sigma} R^{\sigma}_{\alpha\beta\sigma} \tag{2}$$

Il en résulte le changement du signe d'un des deux tenseurs. Pauli change $R^{\delta}_{\alpha\beta\gamma}$, Weyl change $R_{\alpha\beta}$.

On peut résumer ceci dans le tableau suivant :

	$R^{\delta}_{\alpha\beta\gamma}$	$R_{\alpha\beta}$
Einstein etc.	+	+
Pauli	−	+
Weyl	+	−

J'ai pris pour $R^{\delta}_{\alpha\beta\gamma}$ la définition de Pauli ; mais j'ai conservé la relation (2). Cela me conduit à la valeur de $R_{\alpha\beta}$ de Weyl, sauf aux endroits signalés dans l'errata où j'ai introduit directement la valeur ordinaire de $R_{\alpha\beta}$.

Veuillez agréer, monsieur le professeur, l'expression de mes sentiments distingués.

The manufacturer's authorised representative in the EU is Springer Nature Customer Service Centre GmbH, Europaplatz 3, 69115 Heidelberg, Germany. If you have any concerns regarding our products, please contact ProductSafety@springernature.com

Printed and bound by CPI Group (UK) Ltd, Croydon, CR0 4YY

30/09/2026

02225450-0016